S0-DJP-906

Electrical Properties of the Earth's Mantle

Edited by
Wallace H. Campbell

1987

Birkhäuser Verlag
Basel · Boston

Reprint from Pure and Applied Geophysics
(PAGEOPH), Volume 125 (1987), No. 2/3

Editor's adress:

Wallace H. Campbell
United States Departement of the Interior
Geological Survey
Mail Stop 964
Federal Center Box 25046
Denver, CO 80225
USA

Library of Congress Cataloging in Publication Data

Electrical properties of the Earth's mantle.
Bibliography: p.
1. Earth-Mantle. 2. Electric conductivity.
I. Campbell, Wallace H. (Wallace Hall), 1926–
QE509.E44 1987 551.1'3 87-15771
ISBN 0-8176-1901-1 (U.S.)

CIP-Kurztitelaufnahme der Deutschen Bibliothek

Electrical properties of the Earth's mantle
ed. by Wallace H. Campbell. – Basel; Boston:
Birkhäuser, 1987
Aus: Pure and applied geophysics (PAGEOPH);
Vol. 125, 1987
NE: Campbell, Wallace H. [Hrsg.]

ISBN 3-7643-1901-1
ISBN 0-8176-1901-1

Contents

PAGEOPH, Vol. 125, Nos. 2/3 (1987)

0033-4553/87/030193-12$1.50+0.20/0

Introduction to *Electrical Properties of the Earth's Mantle*

WALLACE H. CAMPBELL [1]

The mantle is the largest of the concentric regions that constitute the Earth's interior. Lying beneath an oceanic or continental crust which may be 6 to 75 (or more) km thick, the mantle region extends to the boundary with the Earth's liquid core. The most detailed characterization of the mantle is provided by the velocity-depth profiles obtained from the seismic waves (DZIEWONSKI and ANDERSON, 1981). A generally abrupt rise in velocity, the Mohorovicic discontinuity, bounds the crust and the 'upper mantle'. Between about 100 and 220 km deep, a low velocity zone is encountered beyond which the velocity increases continuously and in steps near 400, 670, and perhaps 770 km. The name 'lower mantle' is assigned to the region from about 1000 km to the start of the core at 2890 km deep.

Motions of crustal plates indicate that sections of an outer, relatively rigid shell of the Earth (the lithosphere) drift over a weak, yielding layer (the asthenosphere). This plastic layer corresponds to the low velocity zone and is presumably a region of partial melting. Only the seismic velocity, density, and pressure of the Earth's interior are rather well established. Much mantle research effort is now focused on such topics as composition, phase changes, partial melting, temperature profiles, and thermal convection. Each of these subjects could benefit from a better understanding of the electrical properties of this region.

At least five characteristics of mantle material can affect the observed electrical properties. (1) Laboratory studies demonstrated that for a particular composition and phase the conductivity rises almost exponentially with the negative reciprocal of the temperature (TOZER, 1970). Because the Earth's temperature, in general, increases with depth, the conductivities increase with depth accordingly. (2) At increased depths within the Earth, the enhanced temperatures and pressures cause a re-adjustment of the mineral structure; such major changes are recorded as phase transition steps in the seismic velocity (ANDERSON, 1967). The observed conductivity is modified by such phase steps. (3) The major composition and phase changes within the Earth have been identified seismically and used to delineate some internal

[1] U.S. Geological Survey, Denver Federal Center, Mail Stop 964, P.O. Box 25046, Denver, Colorado 80225.

boundaries. The mantle is considered to be mostly silicates in which magnesium-rich olivine dominates down to the 400 km level where a transition to spinel may occur; perovskite and oxides of magnesium and iron may be important in the lower mantle. The electrical properties must exhibit analagous changes (SHANKLAND, 1975). (4) There seems to be increasing evidence that a partial melt exists in the upper mantle region with corresponding seismic velocity decreases and conductivity increases (SHANKLAND, O'CONNELL and WAFF, 1981). (5) Known lateral inhomogeneities in the Earth should show covariant electrical behavior. For example, the cold, poorly conducting oceanic lithosphere can extend several hundred kilometers into the upper mantle in a subduction zone (TOKSÖZ, 1975). Also, surface heat flow observations indicate lateral inhomogeneities (JARVIS and PELTIER, 1986) implying convection involving asthenospheric materials (POLLACK and CHAPMAN, 1977; ANDERSON and DZIEWONSKI, 1984).

The Earth's silicates that are insulators at surface temperatures are semiconducting at mantle temperatures; three conditions contribute to this property. One is the presence of impurities with a misfitting valency in the crystal lattice. It is generally assumed that this 'impurity semiconduction' is important only at the very top of the upper mantle. The second is the existence of ions that are free to move. This 'ionic semiconduction' increases with depth rapidly because of the increasing temperature, but the corresponding pressure increase probably quenches the 'ionic' contribution beneath about 400 km and almost surely by 670 km. The third is the presence of free electrons because of thermal agitation. This 'electronic semiconduction' increases with temperature thus becoming the dominant process deeper than about 400 km.

Almost a century ago, SCHUSTER (1890) concluded from his spherical harmonic analysis of the daily variations of the quiet magnetic field that '... the Earth does not behave as a uniformly conducting sphere, but the upper layers must conduct less than the inner layers'. The first quantitative estimates of the deep-Earth conductivity had to await the mathematical capability and geomagnetic curiosity of Chapman about 30 years later (CHAPMAN, 1919; CHAPMAN and WHITEHEAD, 1922; CHAPMAN and PRICE, 1930; CHAPMAN and BARTELS, 1940). Since that time, refined mathematical techniques, improved data bases, and the application of computers have constantly increased our understanding of the electrical properties of the inaccessible Earth structure. A review of the early studies of deep conductivity may be found in Price's 1970 article in which he emphasized some of the special difficulties such as the effects of highly conducting surface layers upon the computations.

To determine the conductivity of the Earth's mantle, two analysis methods are currently in use. In 'forward' modeling, a trial Earth conductivity profile is investigated for its electromagnetic response to an external field; then the profile is adjusted until the response is in agreement with the observations over the study region. This method can be quite demanding of computer time for two- or three-dimensional models. Ambiguities arise occasionally with more than one conductivity profile

Table 1

Earth conductivity profiles

Figure	File Name	Apx. Depth	Remarks	References
1.	Global Models 1939–69			
	LAPR39	0–1250	global Sq, Dst	LAHIRI and PRICE, 1939; PRICE, 1973
	RIKI50	0–1400	misc. data sources	RIKITAKE. 1950; 1966
	MCDO57	0–2900	LAPR39 + secular change	MCDONALD, 1957
	CANT60	100–600	see ECKHARDT *et al.*, 1963	CANTWELL, 1960
	YUKU65	380–1900	ring current	YUKUTAKE, 1965
	BANK69	0–1700	ring current	BANKS, 1969; 1972
2.	Global Models 1970–74			
	BFRS70	100–700	Sq, Dst 27-d variations	BERDICHEVSKY *et al.*, 1970; 1973
	PRKR70	0–3200	rework BANKS, 1969, data	PARKER, 1970
	SCJA72	0–1000	pulsations, bays, Sq, Dst	SCHMUCKER and JANKOWSKI, 1972
	BANK72	230–1250	model summary	BANKS, 1972
	JADY74	0–2951	Sq, 27-d, annual variations	JADY, 1974
	FARO74	300–1500	with BFRS70	FAINBERG and ROTANOVA, 1974
	SCHM74	0–1000	see HAAK, 1980	SCHMUCKER, 1974
	DMRB77	0–1450	all available data	DMITRIEV *et al.*, 1977
3.	Global Models 1974–1983			
	PRKN74	60–430	Sq	PARKINSON, 1974
	DUCM80	0–2900	annual means	DUCRUIX *et al.*, 1980
	ISIK80	320–2020	Sq, Dst, annual, solar cycle	ISIKARA, 1980
	ACMC81	0–2875	secular impulse	ACACHE *et al.*, 1980
	ROKI82	350–1200	various methods	ROKITYANSKY. 1982
	JAPA83	0–1200	Dst	JADY and PATERSON, 1983
4.	Pacific Models			
	LAUN74	0–500	near Calif.; see DRURY, 1978	LAUNAY, 1975
	LAHA75	0–800	Hawaii	LARSEN, 1975
	FILL80	7–1350	NE Pacific	FILLOUX, 1980
	LWGR81	0–200	Juan de Fuca	LAW and GREENHOUSE, 1981
	OLJA84	0–250	Juan de Fuca	OLDENBURG *et al.*, 1984
	OLCA84	0–250	near Calif.	OLDENBURG *et al.*, 1984
	OLNC84	0–250	N. cent. Pacific	OLDENBURG *et al.*, 1984
5.	Atlantic Models			
	POVH76	0–215	NW Atlantic	POEHLS and VON HERZEN, 1976
	BEBJ78	7–125	Iceland	BEBLO and BJÖRNSSON, 1978
	CFGL80	0–1000	Bermuda	COX *et al.*, 1980
6.	North American Models			
	SWIF67	70–300	SW USA; see PARKINSON & JONES, 1979	SWIFT, 1967
	COHY70	0–450	W Canada	COCHRANE and HYNDMAN, 1970
	PORA71	0–560	W USA; see GOUGH, 1974	PORATH, 1971
	GOUG73	0–475	W North America	GOUGH, 1973
	BEGK74	0–350	E Canada	BAILEY *et al.*, 1974

Table 1—*continued*

Figure	*File Name*	*Apx. Depth*	*Remarks*	*References*
	LATU75	60–1000	Tucson, USA	LARSEN, 1975
	CSNA86	10–610	N America region	CAMPBELL and SCHIFFMACHER, 1986
7.	European Models			
	VBFF77	40–450	E Europe	VANYAN *et al.*, 1977
	KOPO80	40–2000	Russian platform	KOVTUN and POROKHOVA, 1980
	JOKE82	0–150	N Finland and Norway	JONES, 1982
	ADKM83	3–280	Karelia	ADAM *et al.*, 1983
	ADPB83	10–560	Pannonia	ADAM *et al.*, 1983
	JOSA83	0–250	Scandinavia	JONES *et al.*, 1983
	CSEU86	10–590	Europe region	CAMPBELL and SCHIFFMACHER, 1986
8.	Other Continental Models			
	HAAK77	0–700	Ethiopia	HAAK, 1977
	LSEA81	0–700	SE Australia	LILLEY *et al.*, 1981
	LCEA81	0–700	Central Australia	LILLEY *et al.*, 1981
	CSAU86	250–550	Australia region	CAMPBELL and SCHIFFMACHER, 1987
	CSEA86	190–560	E Asia region	CAMPBELL and SCHIFFMACHER, 1986
	CSCA86	90–640	Central Asia region	CAMPBELL and SCHIFFMACHER, 1986

accommodating the same surface field observations. Magnetometer array and magnetotelluric techniques favor 'forward' modeling to investigate the crust and topmost mantle regions. In the inverse modeling, a conducting substructure profile is obtained from a 'transfer function' that translates, on a theoretical basis, either the orthogonal electric and magnetic fields or the separated internal and external magnetic fields (observed at the Earth's surface) into depth and conductivity of an equivalent substitute layered Earth structure that would provide the observed response. Full mantle determinations are limited only by the available natural field periods (hours to months); but the resolution is generally poor, lateral heterogeneity is not accommodated, and results differ somewhat for the various mathematical approaches in use. Often for both 'forward' and 'inverse' methods, constraints imposed by the mathematical representation itself affect the form (and therefore the validity) of the conductivity profile that is produced. Excellent reviews of the analysis techniques have been published by WAIT (1982) and ROKITYANSKY (1982).

The following eight figures (itemized in Table 1) summarize some of the published conductivity results using a common scale display. At the present time, the values that were determined for depths greater than 1000 km seem so highly speculative that they are not shown in these examples. At the left half of each figure, the conductivity scale is linear, and the depth is displayed to 600 km to compare the small scale features reported at the top of the upper mantle. The right half of each figure is a logarithmic conductivity display to depths of 1000 km. The code identification of each plot represents the name of the author(s), the year of publication and sometimes the location, as referenced in Table 1. To produce this common plot of results, it was necessary to take some liberties with the original publications. For example, continuous line plots were drawn through point values, and occasionally a centerline of an author's distribution of values was estimated. The interested reader should refer to the original publication of each result.

In the first three figures, grouped chronologically, are the whole-Earth models of conductivity. Between 1939 and 1969 (Figure 1), we see a great variation in the estimated conductivities, particularly at shallow depths. The 400 km step that appears in most of the models seemed to have been created by the authors to accommodate the seismic-velocity discontinuity that was known to exist at that level. Many recent researchers find values in the range of 0.1 to 0.001 S/m for the topmost mantle to about 400 km; most of these early global models would be acceptable. At depths below this step, recent works seem to find values of about 0.1 to 3.0 S/m; models RIKI50 and BANK69 would be anticipating the present values for that region. In Figure 2, model SCJA72 seems to have quite low values at shallow-mantle depths compared to the other conductivity results. Interestingly, model BFRS70 may be the first global model to indicate an abnormal high-conducting layer near the topmost mantle. A value of approximately 1 S/m near 1000 km is obtained by all these models. In Figure 3, models DMRB77 and JAPA83 appear to be in disagreement with most of the others for the upper mantle. It seems to me that the present global model would be one with a gradually rising conductivity from about 50 km down to about 400 km and with values between 0.01 and 0.1 S/m. A rapid rise in conductivity near 400 to 500 km could cause an order of magnitude

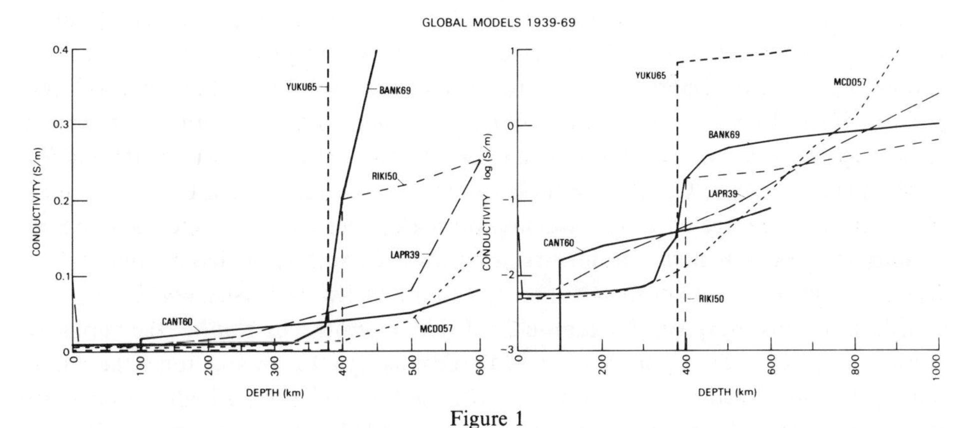

Figure 1
Global conductivity profiles 1939 to 1969. Refer to model code list in Table 1. Left, linear conductivity (S/m) versus depth (km). Right, log conductivity versus depth.

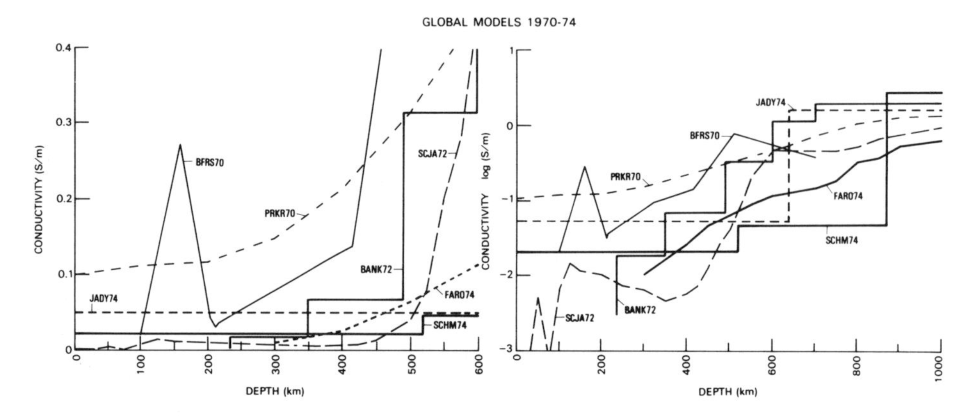

Figure 2
Similar to Figure 1 only 1970 to 1974 profiles.

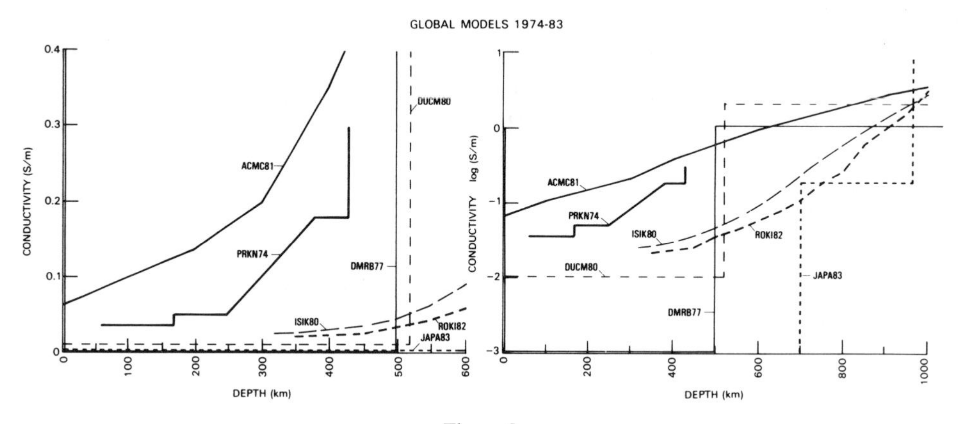

Figure 3
Similar to Figure 1 only 1974 to 1983 profiles.

increase in conductivity over a 50 to 100 km transition zone. Below this transition, it seems that the conductivity should rise gradually to a value near 1.0 to 3.0 S/m.

A group of Pacific Ocean model-conductivity profiles is illustrated in Figure 4. The interesting feature of these is the appearance of narrow layers of relatively high conductivity, usually near the low velocity zone of the upper mantle. The great differences between profiles may be ascribed to the lateral heterogeneity between the variety of regions sampled. For example, LAHA75 was thought to be representative of a mantle plume, and LWGR81 and OLJD84 sampled a subduction region near active volcanoes. Most models favor a high conducting layer about 10 times its surrounding levels, ranging 10 to 50 km thick somewhere in the depth range of about 50 to 150 km.

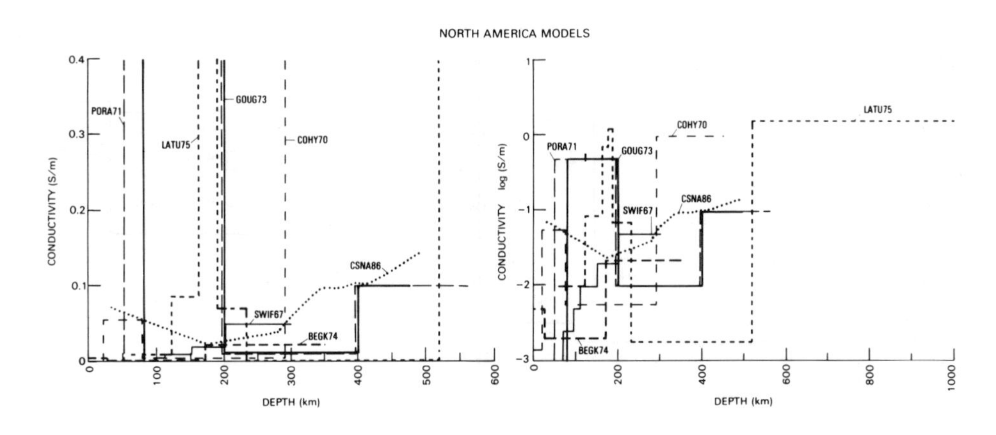

Figure 4
North America continental region conductivity profiles. Refer to model code list in Table 1. Left, linear conductivity (S/m) versus depth (km). Right, log conductivity versus depth.

Only a few conductivity determinations have been made in the Atlantic Ocean region. Figure 5 illustrates the results that are generally similar to those reported for the Pacific region. Note that the BEBJ78 Iceland model has a shallow location of the high conducting layer; this island is near the mid-Atlantic ridge of hot upwelling mantle magma.

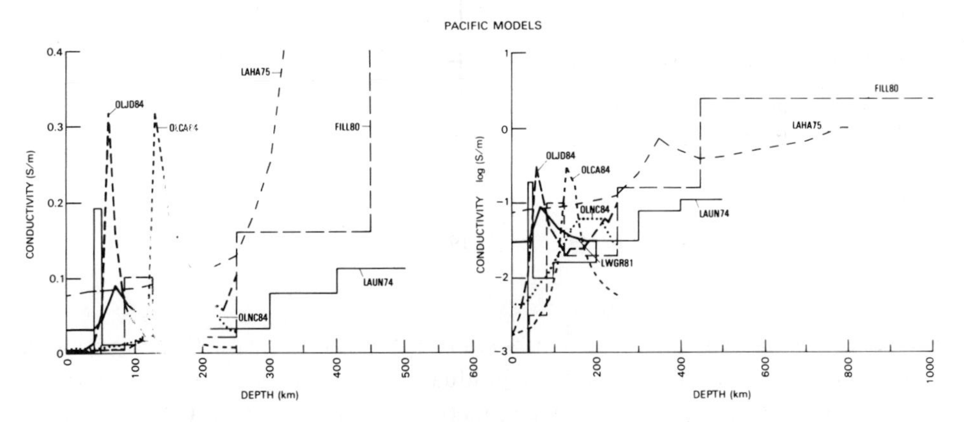

Figure 5
Similar to Figure 4 except that the Pacific Ocean region is shown.

Figure 6 shows the North American data set. The variety of conductivity profiles probably reflects the distribution of study locations. The CSNA86 profile describes a rather broad continental region, but most other profiles are quite local representations of unique geologic conditions. Many of these examples show a high conducting layer near the top of the upper mantle, as did the ocean models.

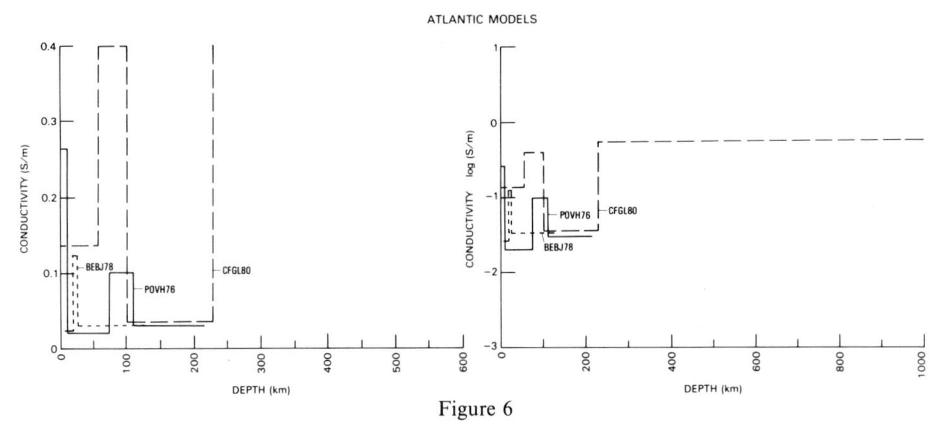

Figure 6
Similar to Figure 4 except that the Atlantic Ocean region is shown.

Figure 7 shows the European profiles. Note that models VBFF77, KOPO80, and ADPB83 indicate extremely low conductivity values at the top of the upper mantle and no special high conducting layer; this situation may be typical of cold shield regions. Except for these three models, the range of reported values is similar to that for North America. CSEU86 is a regional model, whereas the others are more locally determined (see Table 1).

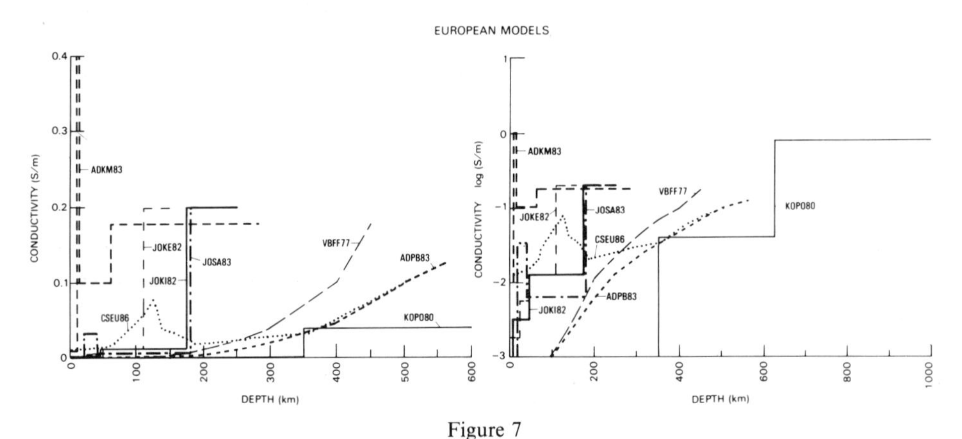

Figure 7
Similar to Figure 4 except that the Europe continental region is shown.

Figure 8 is a collection of conductivity determinations for Asia, Africa, and Australia. It is interesting that no high conducting layers appear in the three models created from data over the Australian shield. The three regional models, CSCA86, CSEA86, and CSAU86, indicate lower conductivity below the 400 depth than the other models. The East African model (HAAK77), with the shallow, high conducting layer, is near a continental plate spreading center where hot, upwelling magma is expected to be near the surface.

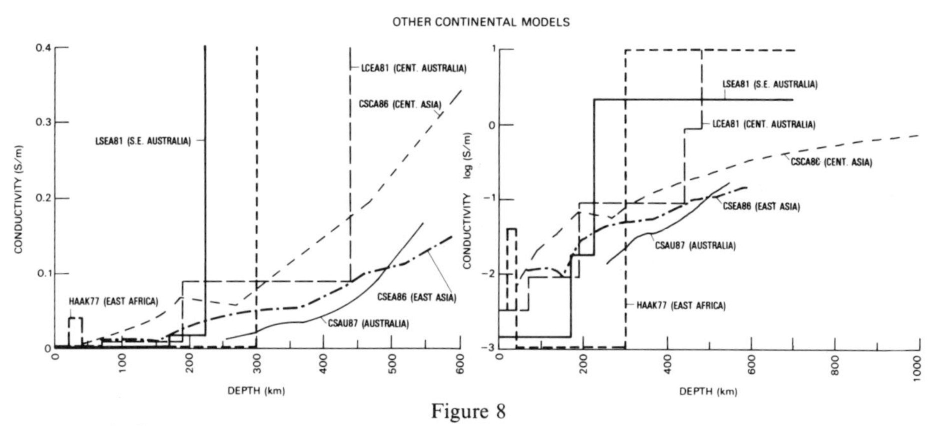

Figure 8
Similar to Figure 4 except that other miscellaneous continental regions are shown.

Although the 50 model profiles described above certainly show a great diversity in appearance, within an order of magnitude there seems to be an agreement in the general level of conductivity to be expected for the mantle region. Many of the profiles may have been selected with guidance from the seismic data on transition locations. However, not all the profile differences can be ascribed to noise, field-source variations, or modeling techniques. There is strong reason to believe that the profiles give valid evidence of a heterogeneous electrical structure of the mantle, especially in the uppermost regions. In the more recent publications, there is a greater interest in exploring the interrelationship of conductivity profiles with the composition, thermal, and seismic properties of the mantle that justify this lateral inhomogeneity.

This special issue of PAGEOPH is a collection of research papers representative of the recent international efforts to understand the electrical properties of the Earth's mantle. The incentive to prepare such a collection grew from a special scientific session on this topic at an assembly of the International Association of Geomagnetism and Aeronomy that convened at Prague, Czechoslovakia, in August 1985. Most of the presentations printed here have been expanded since that meeting, and several new works have been added to complete the topic coverage. The collection will open with a review of the seismic properties important to an understanding of the conductivity profiles. Next, the papers concerning interrelations between mantle conductivity and other basic properties of the Earth will be presented. Discussions of some conductivity modeling methods are in the central part of this issue. The final topic is the review of the special research programs in various countries. The purpose of this volume is to bring together a sample of representative research on an interesting topic at the growing edge of geophysics. We ask the reader's forgiveness for not being able to cover all aspects of the Earth's electrical properties in this limited space.

Bibliography

ACHACHE, J., COURTILLOT, V., DUCRUIX J. and LE MOUËL J. L. (1980), *The late 1960s secular variation impulse: further constraints on deep mantle conductivity.* Phys. Earth Planet. Int. *23*, 72–75.

ADAM, A., VANYAN, L. L., HJELT, S. E., KAIKKONEN, P., SHILOVSKY, P. P. and PALSHIN, N. A. (1983), *The comparison of deep geoelectric soundings in the Pannonian Basin and on the Baltic shield.* J. Geomag. Geoelectr *35*, 829–830.

ANDERSON, D. L. (1967), *Phase changes in the upper mantle.* Science *157*, 1165–1173.

ANDERSON, D. L. and DZIEWONSKI, A. M. (1984), *Seismic Tomography.* Sci. Amer. *251*, 58–66.

BAILEY, R. C., EDWARDS, R. N., GARLAND, G. D., KURTZ, R. and PITCHER D. (1974), *Electrical conductivity studies over a tectonically active area in eastern Canada.* J. Geomag. Geolectr. *26*, 125–146.

BANKS, R. J. (1969), *Geomagnetic variations and electrical conductivity of the upper mantle.* Geophys. J. Roy. Astr. Soc. *17*, 457–487.

BANKS, R. J. (1972), *The overall conductivity distribution of the Earth.* J. Geomag. Geolectr. *24*, 337–351.

BEBLO, M. and BJÖRNSSON, A. (1978), *Magnetotelluric investigation of the crust and upper mantle beneath Iceland.* J. Geophys. *45*, 1–16.

BERDICHEVSKY, M. N., OBUKHOV, G. G. and FAINBERG, E. B. (1973), *Frequency magnetovariational sounding of the Earth, using the ratio of potentials.* Geomag. Aeron. *13*, 117–122 (English edn.).

BERDICHEVSKY, M. N., VANYAN, L. L., LAGUTINSKAYA, L. P., ROTANOVA, N. M. and FAINBERG, E. B. (1970), *Experience in frequency sounding of the Earth from the results of spherical analysis of geomagnetic field variations.* Geomag. Aeron. *10*, 294–297 (English edn.).

CAMPBELL, W. H. and SCHIFFMACHER, E. R. (1986), *A comparison of upper mantle subcontinental electrical conductivity for North America, Europe, and Asia.* J. Geophys. *59*, 56–61, (see also correction p. 204–205).

CAMPBELL, W. H. and SCHIFFMACHER, E. R. (1987), *Quiet ionospheric currents and Earth conductivity profile computed from quiet time geomagnetic field changes in the region of Australia.* Australian J. Phys. 40, (in press).

CANTWELL, T. (1960), *Detection and analysis of low-frequency magnetotelluric signals*, Ph.D. thesis, Mass. Inst. Tech., 170 pp.

CHAPMAN, S. (1919), *The solar and lunar diurnal variation of the Earth's magnetism.* Phil. Trans. Roy. Soc. A218, 1–118.

CHAPMAN, S. and BARTELS, J. (1940), *Geomagnetism.* Oxford University Press, London, 1049 p.

CHAPMAN, S. and PRICE, A. T. (1930), *The electric and magnetic state of the interior of the Earth as inferred from terrestrial magnetic variations.* Phil. Trans. Roy. Soc., London *A229*, 427–460.

CHAPMAN, S. and WHITEHEAD, T. T. (1922), *The influence of electrically conductivity material within the Earth as inferred from terrestrial magnetic variations.* Trans. Cambridge Phil. Soc. *22*, 463–482.

COCHRANE, N. A. and HYNDMAN, R. D. (1970), *A new analysis of geomagnetic depth-sounding data from western Canada.* Canadian J. Phys. 7, 1208–1218.

COX, C. S., FILLOUX, J. H., GOUGH, D. I., LARSEN, J. C., POEHLS, K. A., VON HERZEN, P., and WINTER, R. (1980), *Atlantic lithosphere sounding.* J. Geomag. Geoelectr. *32*, Suppl. 1, 13–32.

DMITRIEV, V. I., ROTANOVA, N. M., ZAKHAROVA, O. K. and BALYKINA, O. N. (1977), *Geoelectric and geothermal interpretation of the results of deep magnetic-variation sounding.* Geomag. Aeron *17*, 210–213 (English edn.).

DRURY, M. J. (1978), *Partial melt in the asthenosphere: evidence from electrical conductivity data.* Phys. Earth Planet. Int. *17*, 16–20.

DUCRUIX, J., COURTILLOT, V. and LE MOUËL, J. L. (1980), *The late 1960s variation impulse, the eleven year magnetic variation and the electrical conductivity of the deep mantle.* Geophys. J. Roy. Astr. Soc. *61*, 73–94.

DZIEWONSKI, A. M. and ANDERSON, D. L. (1981), *Preliminary reference Earth model.* Phys. Earth Planet. Int. *25*, 279–356.

ECKHARDT, D., LARNER, K. and MADDEN, T. (1963), *Long-period magnetic fluctuations and mantle electrical conductivity estimates.* J. Geophys. Res. *68*, 6279–6286.

FAINBERG, E. B. and ROTANOVA, N. M. (1974), *Distribution of electrical conductivity and temperature in the interior of the Earth according to deep electromagnetic soundings.* Geomag. Aeron. *14*, 603–607 (English edn.).

FILLOUX, J. H. (1980), *Magnetotelluric soundings over the northeast Pacific may reveal spatial dependence of depth and conductance of the asthenosphere.* Earth and Planet Sci. Letters *46*, 244–252.

GOUGH, D. I. (1973), *The geophysical significance of geomagnetic variation anomalies.* Phys. Earth Planet. Int. *7*, 379–388.

GOUGH, D. I. (1974), *Electrical conductivity under western North America in relation to heat flow, seismology, and structure.* J. Geomag. Geoelectr. *26*, 105–123.

HAAK, V. (1977), *The electrical resistivity of the upper 300 km of the Afar-depression in Ethiopia derived from magnetotelluric measurements.* Acta Geodaet., Geophys. et Montanist. Acad. Sci. Hung. *12*, 7–10.

HAAK, V. (1980), *Relations between electrical conductivity and petrological parameters of the crust and upper mantle.* Geophysical Surveys *4*, 57–69.

ISIKARA, A. M. (1980), *Long period variations of the geomagnetic field and inferences about the deep electric conductivity.* J. Geomag. Geoelectr. *32*, Suppl. 1, 155–157.

JARVIS, G. T. and PELTIER, W. R. (1986), *Lateral heterogeneity in the convecting mantle.* J. Geophys. Res. *91*, 435–451.

JADY, R. J. (1974), *The conductivity of spherically symmetric layered Earth models determined by Sq and longer period magnetic variations.* Geophys. J. Roy. Astr. Soc. *36*, 399–410.

JADY, R. J. and PATERSON, G. A. (1983), *Inversion methods applied to Dst data.* J. Geomag. Geoelectr. *35*, 733–746.

JONES, A. G. (1982), *Observations of the electrical asthenosphere beneath Scandinavia.* Tectonophys. *90*, 37–55.

JONES, A. G., OLAFSDOTTIR, B. and TIKKAINEN, J. (1983), *Geomagnetic induction studies in Scandinavia.* J. Geophys. *54*, 35–50.

KOVTUN, A. A. and POROKHOVA, L. N. (1980), *Deep conductivity distribution on Russian platform from the results of combined magnetotelluric and global magnetovariational data interpretation.* J. Geomag. Geoelectr. *32*, Suppl. 1, 105–113.

LAHIRI, B. N. and PRICE, A. T. (1939), *Electromagnetic induction in nonuniform conductors, and the determination of the conductivity of the Earth from terrestrial magnetic variations.* Phil. Trans. Roy. Soc., London *A237*, 509–540.

LARSEN, J. C. (1975), *Low frequency* (0.1–6.0 cpd) *electromagnetic study of deep mantle electrical conductivity beneath the Hawaiian Islands.* Geophys. J. Roy. Astr. Soc. *43*, 17–46.

LAUNAY, L. (1974), *Conductivity under the oceans: interpretation of magnetotelluric sounding 630 km off the California coast. Phys. Earth Planet. Int. 8*, 83–86.

LAW, L. K. and GREENHOUSE, J. P. (1981), *Geomagnetic variation sounding of the asthenosphere beneath the Juan de Fuca ridge.* J. Geophys. Res. *86*, 967–978.

LILLEY, F. E. M., WOODS, D. V. and SLOANE, M. N. (1981), *Electrical conductivity profiles and implications for the absence or presence of partial melting beneath central and southeast Australia.* Phys. Earth Planet. Int. *25*, 419–428.

MCDONALD, K. (1957), *Penetration of the geomagnetic secular field through a mantle with variable conductivity.* J. Geophys. Res. *62*, 117–141.

OLDENBERG, D. W., WHITTALL, K. P. and PARKER, R. L. (1984), *Inversion of ocean bottom magnetotelluric data revisited.* J. Geophys. Res. *89*, 1829–1833.

PARKER, R. L. (1970), *The inverse problem of electrical conductivity in the mantle.* Geophys. J. Roy. Astr. Soc. *22*, 121–138.

PARKINSON, W. D. (1974), *The reliability of conductivity derived from diurnal variation.* J. Geomag. Geoelectr. *26*, 281–284.

PARKINSON, W. D. and JONES F. W. (1979), *The geomagnetic coast effect.* Rev. Geophys. Space Phys. *17*, 1999–2015.

POEHLS, K. A. and VON HERZEN, R. P. (1976), *Electrical resistivity structure beneath the north-west Atlantic Ocean.* Geophys. J. Roy. Astr. Soc. *47*, 331–346.

POLLOCK, H. N. and CHAPMAN, D. S. (1977), *The flow of heat from the Earth's interior.* Sci. Amer. *237*, 60–76.

PORATH, H. (1971), *Magnetic variation anomalies and seismic low-velocity zone in western United States.* J. Geophys. Res. *76*, 2643–2648.

PRICE, A. T. (1970), *The electrical conductivity of the Earth.* Quart. J. Roy. Astr. Soc. *11*, 23–42.

PRICE, A. T. (1973), *The theory of geomagnetic induction.* Phys. Earth Planet. Int. *7*, 227–233.

RIKITAKE, T. (1950), *Electromagnetic induction within the Earth and its relation to the electrical state of the Earth's interior. Part II.* Tokyo Univ. Bull. Earthquake Res. Inst. *28*, 263–283.

RIKITAKE, T. (1966), *Electromagnetism and the Earth's Interior* (Elsevier Pub. Co., Amsterdam) Chap. 15, pp. 221–230.

ROKITYANSKY, I. I. (1982), *Geoelectromagnetic Investigation of the Earth's Crust and Mantle* (Springer-Verlag, Berlin) 381 p.

SCHMUCKER, U. (1974), *Erdmagnetische Teifensondierung mit lang periodischen Variationen.* In proceedings of the Colloquium, *Erdmagnetische Tiefensondierung* at Grafth, Bavaria, 313–342.

SCHMUCKER, U., and JANKOWSKI, J. (1974), *Geomagnetic induction studies and electrical state of the upper mantle.* Tectonophysics *13*, 233–256.

SCHUSTER, A. (1890), *The diurnal variations of terrestrial magnetism.* Phil. Trans. Roy. Soc. *A180*, 467–512.

SHANKLAND, T. J. (1975), *Electrical conduction in rocks and minerals: parameters for interpretation.* Phys. Earth Planet. Int. *10*, 209–219.

SWIFT, C. M. (1967), *A magnetotelluric investigation of an electrical conductivity anomaly in southwestern United States.* Mass. Inst. Tech., Ph.D. thesis, Cambridge, Mass.

TOKSÖZ, M. N. (1975), *The subduction of the lithosphere.* Sci. Amer. *233*, 88–89.

TOZER, D. C. (1970), *Temperature, conductivity, composition and heat flow.* J. Geomag. Geoelectr. *22*, 35–51.

VANYAN, L. L., BERDICHEWSKI, M. N., FAINBERG, E. B. and FISKINA, M. V. (1977), *Study of the asthenosphere of the east European platform by electromagnetic sounding*, Phys. Earth Planet. Int. *14*, 1–2.

WAIT, J. R. (1982), *Geo-Electromagnetism* (Academic Press, New York) 268 p.

YUKUTAKE, T. (1965), *The solarcycle contribution to secular change in the geomagnetic fields*, J. Geomag. Geoelectr. *17*, 287–309.

PAGEOPH, Vol. 125, Nos. 2/3 (1987) 0033-4553/87/030205-35$1.50+0.20/0

Crustal and Upper Mantle Structure of Stable Continental Regions in North America and Northern Europe

ROBERT P. MASSÉ[1]

Abstract—From an analysis of many seismic profiles across the stable continental regions of North America and northern Europe, the crustal and upper mantle velocity structure is determined. Analysis procedures include ray theory calculations and synthetic seismograms computed using reflectivity techniques. The P wave velocity structure beneath the Canadian Shield is virtually identical to that beneath the Baltic Shield to a depth of at least 800 km. Two major layers with a total thickness of about 42 km characterize the crust of these shield regions. Features of the upper mantle of these regions include velocity discontinuities at depths of about 74 km, 330 km, 430 km and 700 km. A 13 km thick P wave low velocity channel beginning at a depth of about 94 km is also present.

A number of problems associated with record section interpretation are identified and a generalized approach to seismic profile analysis using many record sections is described. The S wave velocity structure beneath the Canadian Shield is derived from constrained surface wave data. The thickness of the lithosphere beneath the Canadian and Baltic Shields is determined to be 95–100 km. The continental plate thickness may be the same as the lithospheric thickness, although available data do not exclude the possibility of the continental plate being thicker than the lithosphere.

Key words: Seismic velocity, crystal structure, upper mantle structure.

Introduction

The most detailed information available today on the interior of the Earth has been obtained from studies of seismic wave data. Over the last few decades, studies of both body and surface waves have yielded considerable information on the complex velocity structure of the Earth's crust and upper mantle. The purpose of this paper is to review the general features of this structure for the stable regions of North America and northern Europe. The structure of these stable regions is considerably simpler and probably better known than that of geologically young regions of recent tectonism.

A substantial amount of the information which will be presented in this review has been previously published in a series of papers on crustal and upper mantle

[1] U.S. Geological Survey, National Earthquake Information Center, MS 967, Box 25046, Denver Federal Center, Denver, Colorado 80225.

structure. However, some new conclusions will be made relating to the procedures and problems associated with interpreting seismic data to determine Earth structure. The thickness of the lithosphere in North America and Europe will also be determined. Finally, a correlation will be made between structural features of these stable interior regions and conductivity data obtained from the analysis of geomagnetic records (e.g., BANKS, 1969; ANDERSSEN *et al.*, 1979; CAMPBELL and ANDERSSEN, 1983).

The history of our knowledge of Earth structure can conveniently be divided between discoveries made before 1962 and those made after this date. The year 1962 marks the beginning of many improvements in seismic instrumentation and networks around the world, including the deployment of the Worldwide Standard Seismograph Network (WWSSN). After these networks were installed, our knowledge of Earth structure advanced rapidly in conjunction with concurrent developments in electronic data processing.

Early Information on Earth Structure

One of the first important discoveries about Earth structure obtained from seismic data was made by a Yugoslav seismologist A. Mohorovičić. From a study of the seismic records of an earthquake which occurred in 1909, he discovered two arrivals for both the P and S body waves at near-regional distances (150–600 km). He proposed a seismic velocity discontinuity at a depth of around 50 to 60 km to explain these separate arrivals. This discontinuity is today considered to be the boundary between the Earth's crust and mantle and is known as the Mohorovičić discontinuity. For the purposes of most crustal and upper mantle structure studies, the Mohorovičić discontinuity can be modeled as a sharp discontinuity. However, there is some evidence from near vertical deep crustal profiling which suggests a more laminated and somewhat broader discontinuity (HALE and THOMPSON, 1982). The exact depth of this discontinuity varies with different tectonic regions. Another much weaker discontinuity (the Conrad or Riel discontinuity) dividing the continental crust into two layers was proposed later.

The boundary between the Earth's outer core and mantle was determined by Gutenberg in 1914. This boundary, at a depth of about 2,900 km, is distinguished by a sudden decrease in compressional wave velocity from about 13.2 to 8.5 km/s and by the absence of shear wave propagation into the outer core. The Earth's mantle is bounded, therefore, by relatively sharp velocity discontinuities, both above and below.

In the years between 1914 and 1962, studies by Jeffreys, Bullen, Lehmann, Gutenberg, and others began to define some of the details of the mantle structure. Overall, the seismic velocity was thought to increase gradually with depth in the mantle. However, a velocity discontinuity in the depth range 200 to 400 km was suspected, and evidence was found for a low velocity zone in the upper mantle, not too far below the crust. This then was the general state of knowledge concerning the main features of the Earth's velocity distribution at the beginning of the 1960s.

New Data Sources and Analysis Techniques

The last two decades have seen a remarkable expansion in our knowledge of Earth structure. This increase in knowledge is the consequence of several factors: improvements in seismic instruments, deployment of worldwide standardized seismic networks, availability of analog and digital tape recording for seismic data, advent of computers capable of processing large quantities of data, development of more sophisticated data processing techniques, and the use of sources of seismic energy such as underground nuclear explosions, vibroseis, and underwater chemical explosions.

Beginning in the early 1960s, the United States began to install the WWSSN. Stations of the WWSSN are equipped with similar instrumentation (POWELL and FRIES, 1966; OLIVER and MURPHY, 1971) and are distributed over most continental areas of the world. By 1968, the number of stations in the network reached 123.

Another network deployed in the early 1960s which has provided data extremely valuable for crustal and upper mantle studies is the Long Range Seismic Measurements (LRSM) network. The LRSM stations were generally deployed in long profiles

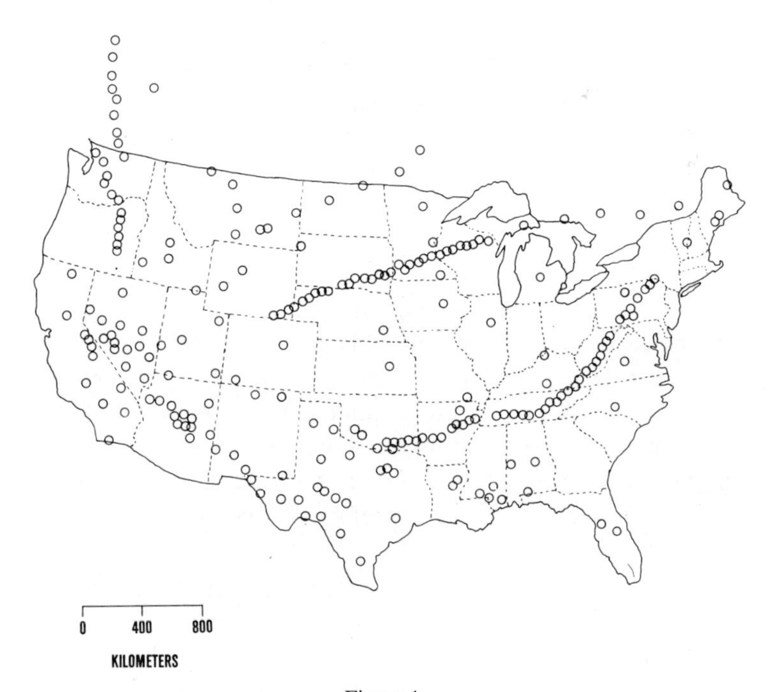

Figure 1
Location of LRSM stations.

(Figure 1) extending from the Nevada Test Site (NTS). An important feature of the LRSM network was the continuous analog recording on magnetic tape of all the seismic data. These tapes could then be digitized and the data studied using time series analysis techniques.

Seismic profile data, such as that obtained from the LRSM network, can be displayed in the form of record sections to provide a very useful analysis tool. Many experiments have been conducted employing one or more seismic profile lines from which record sections have been created. A classic example is the Early Rise Experiment, an investigation of seismic signals generated by a series of underwater chemical explosions in Lake Superior (WARREN *et al.*, 1968; MASSÉ, 1973a). The profile lines for this experiment are shown in Figure 2. Some of the better seismic record sections from this experiment are shown in Figures 3, 4, and 5.

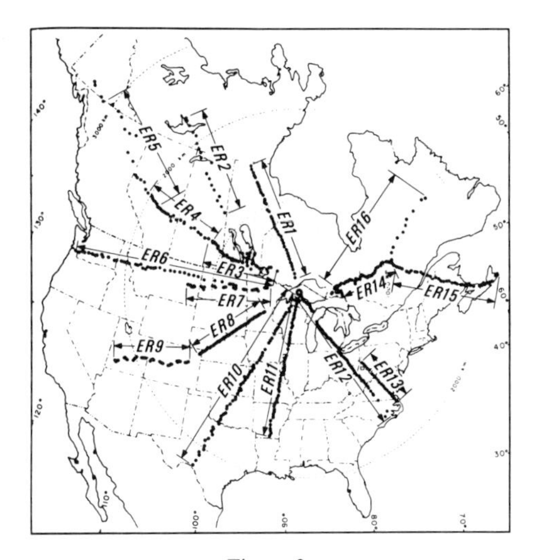

Figure 2
Locations of recording profiles for Project Early Rise (after WARREN *et al.*, 1968).

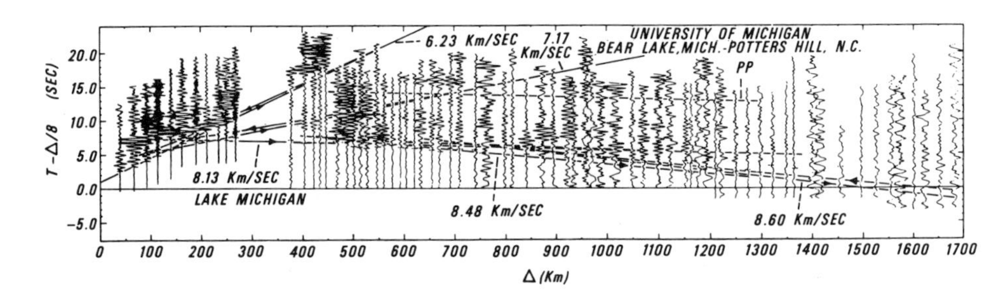

Figure 3
Early Rise profile ER12 record section adapted from WARREN *et al.* (1968). Refraction and reflection lines determined by MASSÉ (1973a).

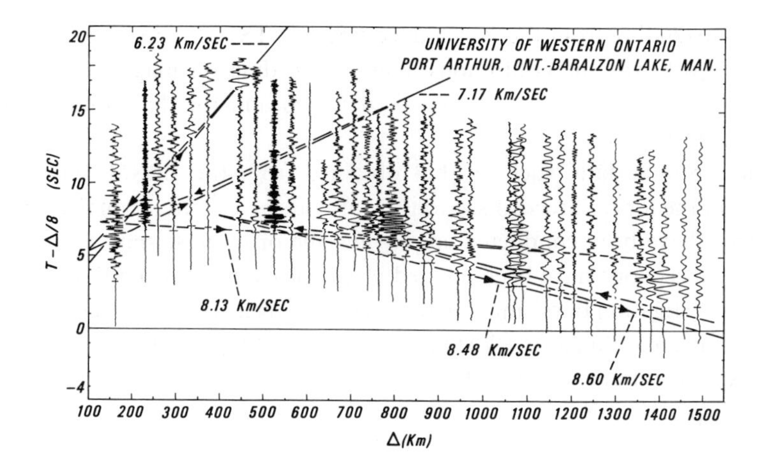

Figure 4
Early Rise profile ER1 record section adapted from WARREN *et al.* (1968). Refraction and reflection lines determined by MASSÉ (1973a).

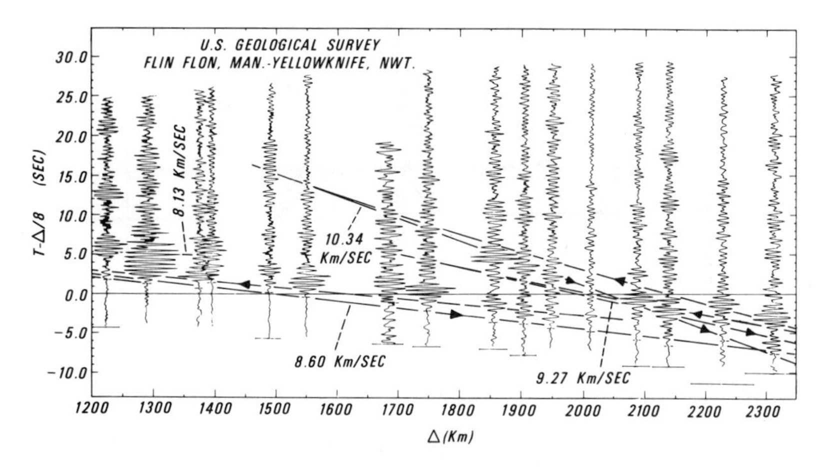

Figure 5
Early Rise profile ER2 record section adapted from WARREN *et al.* (1968). Refraction and reflection lines determined by MASSÉ (1973a).

In the 1960s and until the mid-1970s, the interpretation of such record sections consisted primarily of devising velocity models that would yield the observed travel time branches and would qualitatively fit the observed amplitudes. For a layered Earth, both refraction and reflection time curves contain information about the structure (BRAILE, 1973). Figure 6 shows a suite of refraction and reflection lines which agree with the observed record section data for the Early Rise Experiment

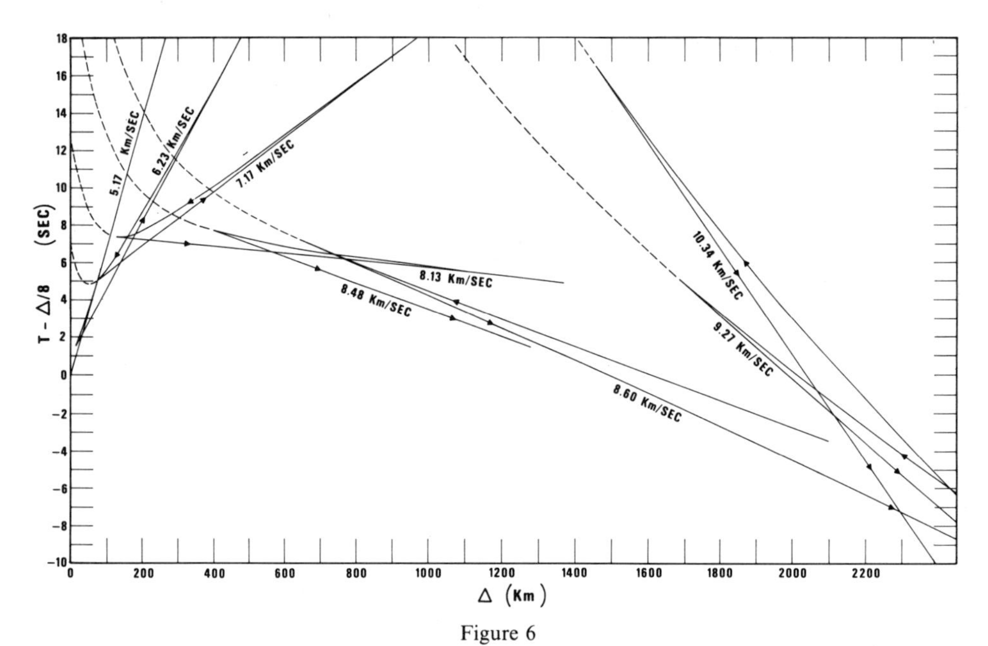

Figure 6

Refraction and reflection lines for P wave energy propagating through the stable platform regions of North America (MASSÉ, 1973a).

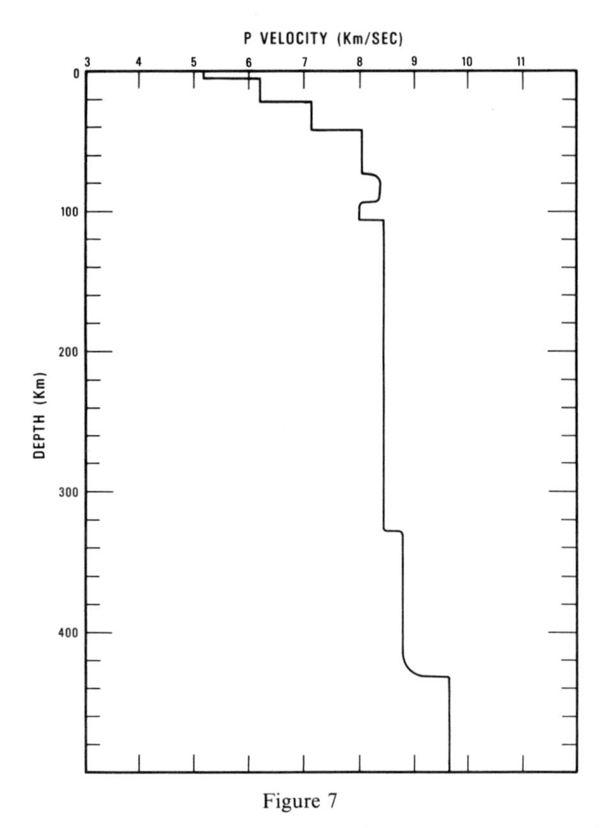

Figure 7

P velocity distribution for the stable platform regions of North America.

(MASSÉ, 1973a). The velocity model corresponding to these travel time curves is shown in Figure 7. From the location of the Early Rise profiles (Figure 2), this model can be seen appropriate for a stable region which includes part of the Canadian Shield and the burial extension of the shield into the central United States.

To extend this model to deeper depths of the upper mantle, it is necessary to analyze record section data recorded at greater distances from the energy source point. To obtain a good signal-to-noise ratio at the greater distances, a large source of energy is valuable. Therefore, recordings from nuclear explosions have often been used. In one such study, MASSÉ (1973b) used data from NTS explosions recorded at

Table 1

P velocity model for Central and Eastern North America

Depth (km)	P velocity (kms^{-1})
0.0	5.170
4.7	5.170
4.7	6.220
21.8	6.220
21.8	7.140
42.2	7.140
42.2	8.060
72.6	8.060
72.7	8.100
73.5	8.200
77.5	8.370
93.5	8.370
94.0	8.050
107.0	7.980
107.0	8.430
325.0	8.435
328.0	8.450
328.5	8.550
328.6	8.770
332.5	8.780
410.0	8.783
420.0	8.810
425.0	8.847
430.0	8.950
432.0	9.625
440.0	9.640
590.0	9.647
610.0	9.720
670.0	10.200
695.0	10.520
707.0	10.800
710.0	11.240
800.0	11.260

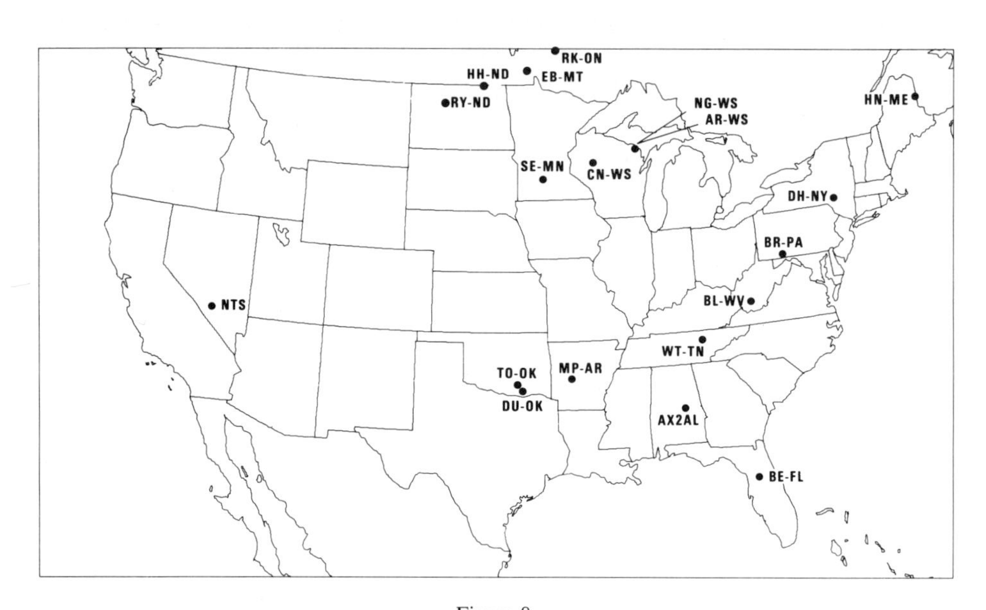

Figure 8
Location of the Nevada Test Site (NTS) and the LRSM stations used to determine P velocity in the 450–800 km depth range.

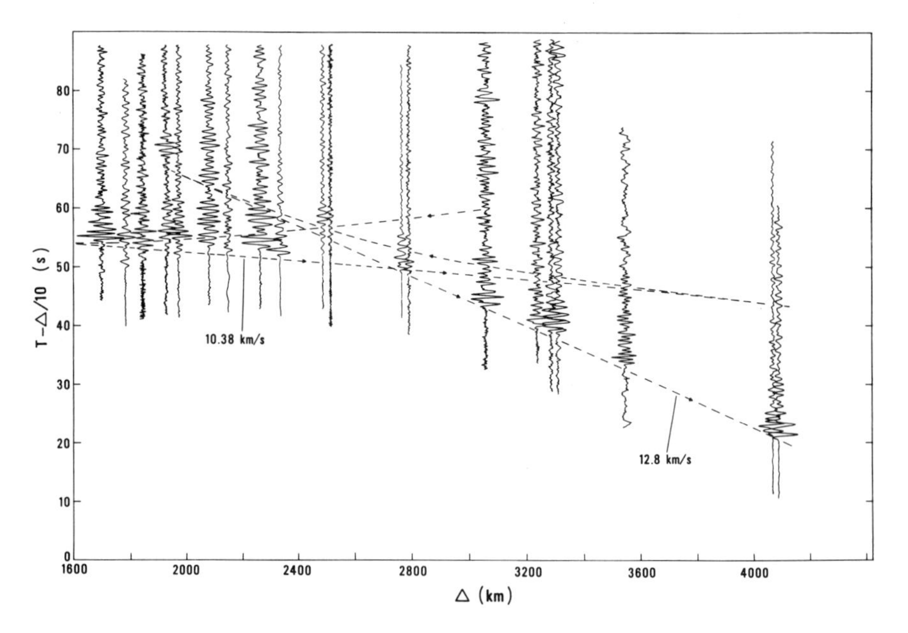

Figure 9
Record section of seismic signals from NTS explosions recorded at LRSM stations.

distant LRSM stations (see Figure 8) to extend the Canadian Shield velocity model (Figure 7) to deeper depths. The resulting record section and corresponding travel time curves are shown in Figure 9. Figure 10 shows the Canadian Shield velocity model extended to a depth of 800 km, using the information obtained from the record section in Figure 9. The corresponding velocities and depths are listed in Table 1.

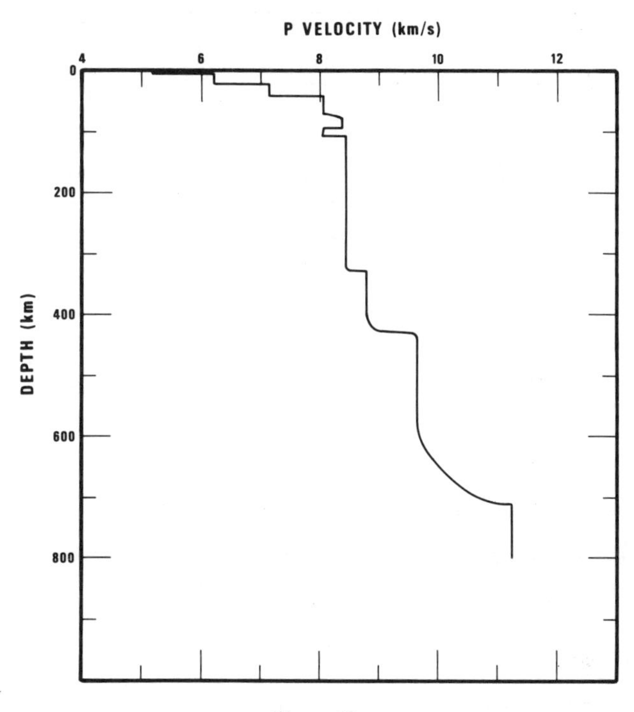

Figure 10
P velocity for the stable platform regions of North America.

The velocity model given in Figures 7 and 10 represents a refinement in our knowledge of Earth structure over that which existed before 1962. This model is, however, only one of many that have been determined for North America in the last two decades. Fortunately, there is additional information that can be examined which pertains to the travel time curves shown in Figures 6 and 9 and the velocity model shown in Figures 7 and 10. It has been found that similar tectonic regions expectantly have similar crustal and upper mantle structures. For example, areas of northern Europe expectedly have a similar structure to that of the Canadian Shield and the central United States since, in the pre-drift configuration, northern Europe and North America were contiguous parts of Pangaea. Therefore, it is interesting to superimpose the travel times derived for the stable regions of North America (Figures

6 and 9) on record sections from Baltic Shield crustal profiles (Figure 11) and from a western Russia upper mantle profile.

Crustal refraction profiles for northern Europe were made as part of a Trans-Scandinavian Deep Seismic Sounding Project (GREGERSEN, 1971; DAHLMAN, 1971a, 1971b; KANESTRØM and HAUGLAND, 1971; PENTTILÄ, 1971b; VOGEL and LUND, 1971).

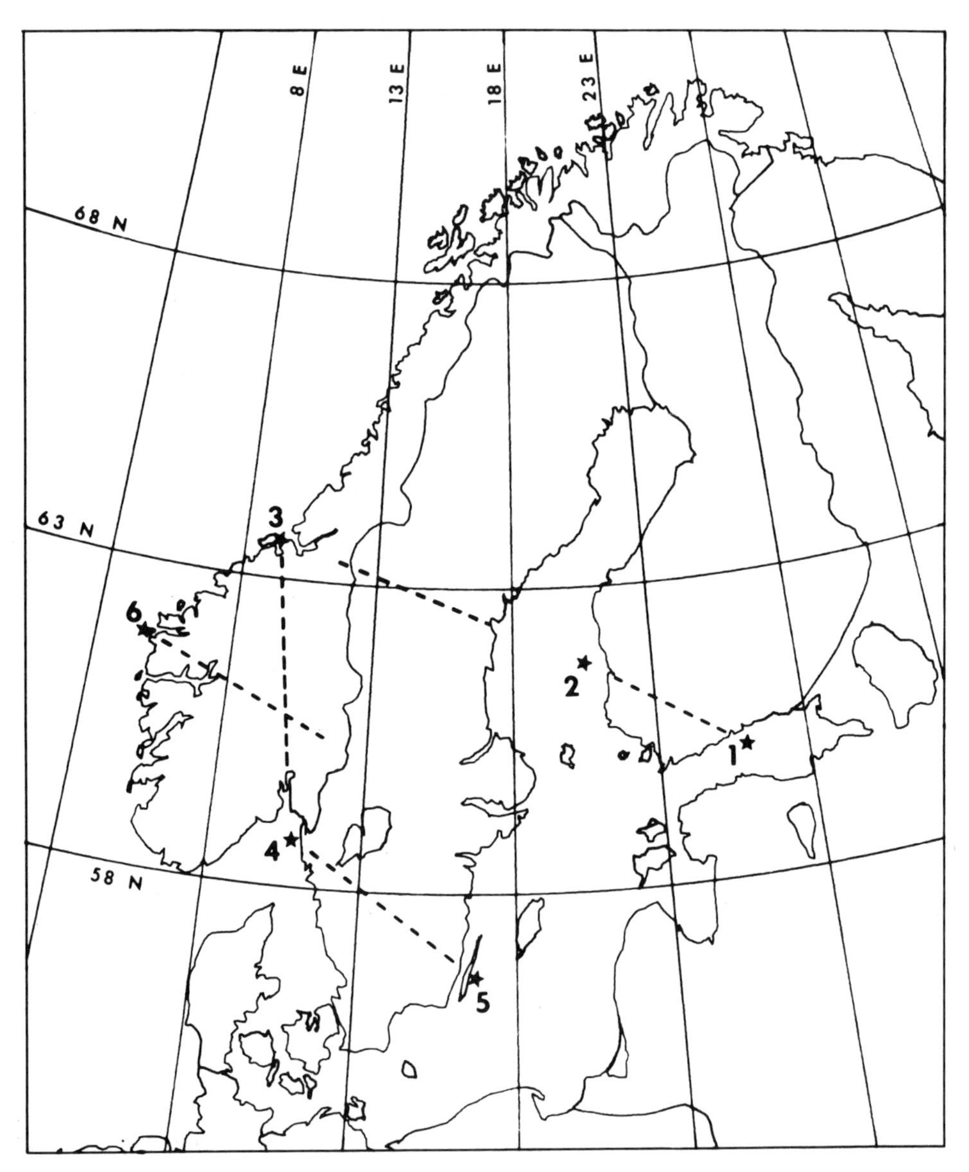

Figure 11
Selected profiles in the Scandinavian region.

A mantle record section was also constructed using recordings made at the seismic array NORSAR in Norway of signals from Russian nuclear explosions (MASSÉ and ALEXANDER, 1974). Some of these record sections with the North American travel time curves superimposed are shown in Figures 12, 13 and 14 (MASSÉ, 1975; MASSÉ and ALEXANDER, 1974). The surprisingly good fit of these travel time curves to the Baltic Shield data obviously increases our confidence in the Earth model initially derived from North American data. Velocity models are always open to different interpretations however. We will return to consider velocity models for these regions again later in this paper.

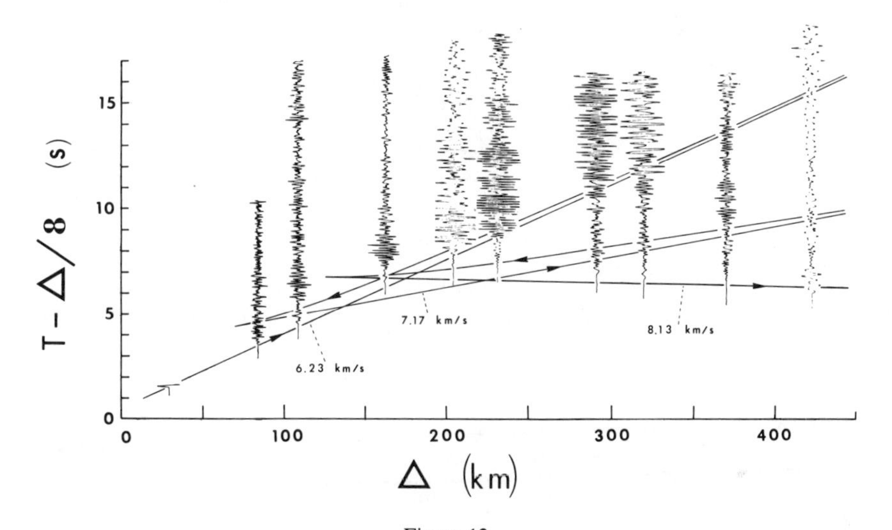

Figure 12
Scandinavian profile 3–4 record section adapted from KANESTRØM and HAUGLAND (1971). Travel time curves of reflected and refracted P waves determined by MASSÉ (1973a, 1975).

Beginning in the 1970s, sophisticated computational techniques were developed which made possible more realistic calculations (FUCHS and MÜLLER, 1971; HELMBERGER, 1972, 1973; HELMBERGER and ENGEN, 1974; CHOY, 1977; CHAPMAN, 1978; KIND, 1978; CORMIER, 1980; KENNETT, 1980). With these techniques, it is possible to calculate synthetic seismograms which can then be compared with recorded seismograms. Many studies have used these techniques over the last few years to develop Earth models and to investigate earthquake source mechanisms. In Earth structure studies, one of the principal benefits of the new techniques is the constraints they provide on the velocity gradients in the crust and upper mantle. Seismic amplitudes are strongly influenced by the velocity gradient.

Synthetic seismograms for the Canadian and Baltic Shield model given in Figure 7 are shown in Figure 15. These seismograms were calculated using the reflectivity techniques of FUCHS and MÜLLER (1971) and KIND (1978). The travel time curves

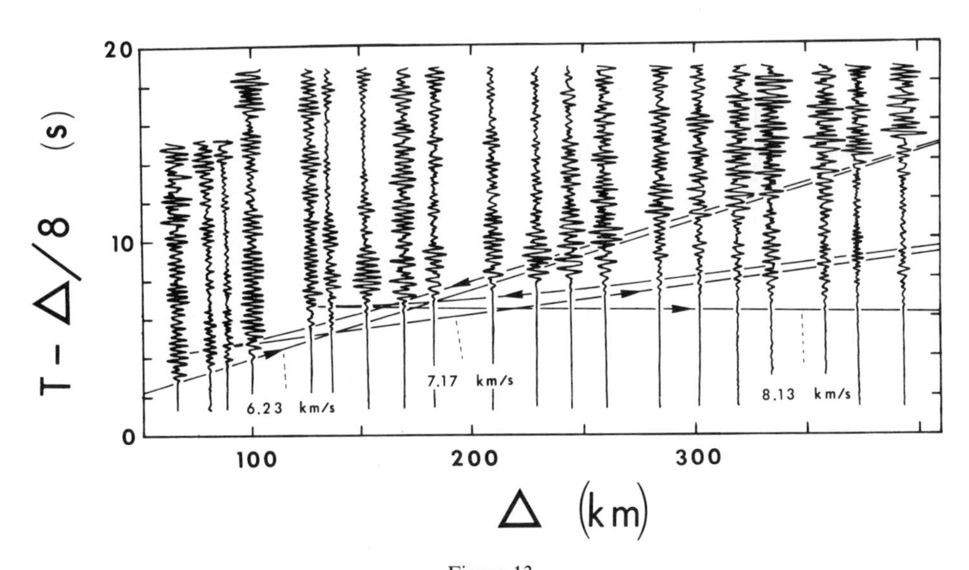

Figure 13
Scandinavian profile 3–2 record section adapted from VOGEL and LUND (1971). Refraction and reflection lines determined by MASSÉ (1973a, 1975).

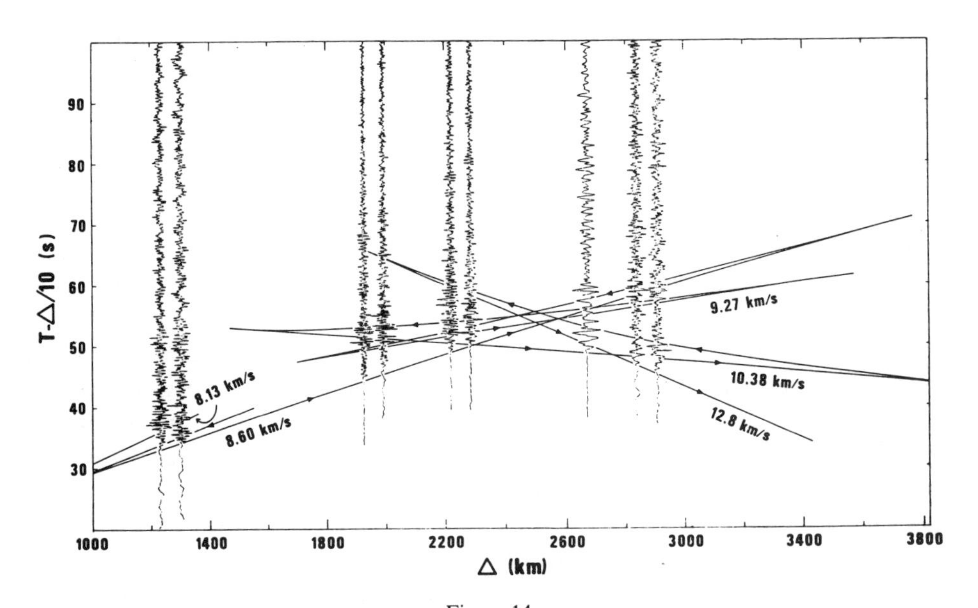

Figure 14
Fennoscandian and western Russia upper mantle record section. Refraction and reflection lines determined by MASSÉ and ALEXANDER (1974).

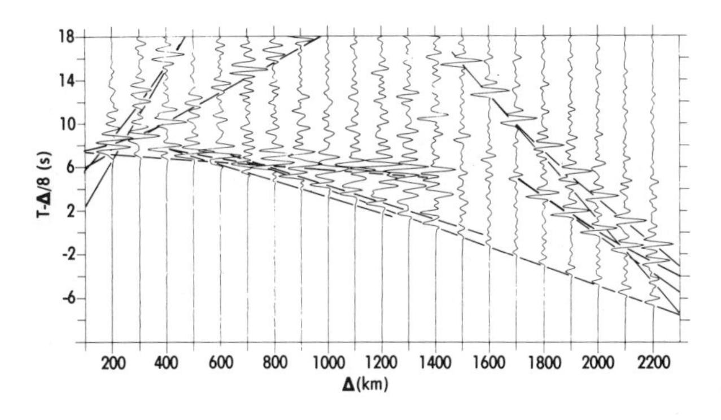

Figure 15
Synthetic seismogram record section for crustal and upper mantle *P* velocity model of MASSÉ (1973a). Reflectivity method of FUCHS and MÜLLER (1971) used to compute seismograms. Travel time lines same as in Figure 6.

from Figure 6 are superimposed upon these synthetic seismograms. While the calculated amplitudes are instructive, actual data will show some variation in amplitude caused by slight variations in velocity gradients and by other factors which will be discussed below.

Although the new processing techniques provide much useful information, it is now realized that these techniques have been unable to remove as much of the uncertainty from the crustal and upper mantle models as had been anticipated. While modeling of long-period waveforms for source mechanism studies has been very successful, observed short-period seismic waveforms are complex, and synthetic seismograms never provide perfect agreement.

Contributing to this problem is the uncertainty in the actual seismic source function. Any errors in the assumed source function will strongly affect the synthetic seismograms. This is particularly true with respect to distance ranges where seismic arrivals are closely spaced in time. The frequency dependence of seismic signal attenuation is another important unknown factor which influences synthetic waveform analysis of Earth structure. Finally, much of the actual uncertainty in structure models derives from the fact that subjective decisions must be made concerning what characteristics of a seismogram will be modeled closely, while other characteristics are not as closely matched. It is this subjectivity, together with the Earth's complexity and the inherent uncertainty of the inversion process that has resulted in a number of different crustal and upper mantle models being proposed for what is often the same region of the Earth.

What can really be said then about the structure of the crust and upper mantle? To seek an answer of this question, we will look again at results of a number of refraction studies for stable continental regions in North America and northern Europe.

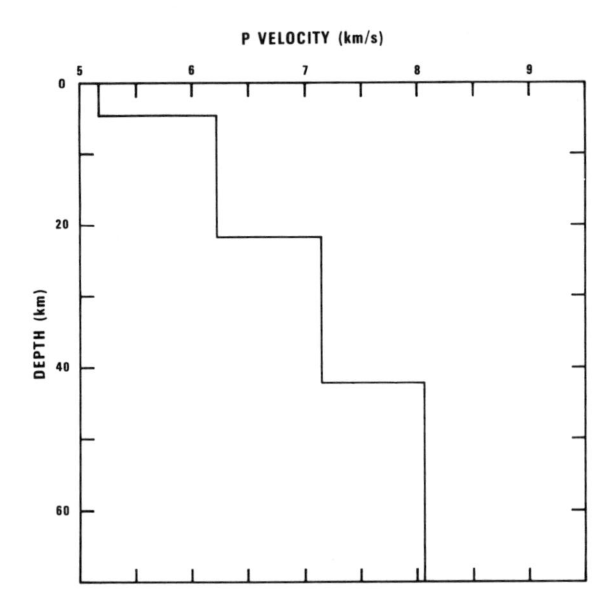

Figure 16
Crustal P velocity model for the Canadian Shield (MASSÉ, 1973a).

Crustal P Wave Structure

The Earth's continental crust generally has a complex structure. Particularly at shallow depths, many complexities have been found by detailed reflection profiling programs such as the Consortium for Continental Reflection Profiling (COCORP). Therefore, any crustal model which attempts to represent the velocity distribution in the crust with a few constant-velocity layers is, by necessity, a simplification. In shield regions, the complexities may be less, but they still exist. As a result, shield crustal models, such as shown in Figure 16, only represent average conditions in the crust. The lower part of the crust is thought to be more uniform than the upper part.

The crustal layers containing P wave velocities of 6.22 and 7.14 km/s in the model shown in Figure 16, are based on apparent refraction velocities of 6.23 and 7.17 km/s respectively, determined for the phases P_g and P^* from the Early Rise data (MASSÉ, 1973a). The possible range of P wave velocities for these crustal layers is suggested by the variation found in different studies of P_g and P^*.

For example, compared to the 6.23 km/s refraction velocity for P_g indicated in Figure 6, refraction studies of the Canadian and Baltic shields have found P_g velocities between generally 6.0 and 6.3 km/s (HODGSON, 1953; HALL and BRISBIN, 1961; SELLEVOLL and PENTTILÄ, 1964; LUOSTO, 1967; SELLEVOLL and POMEROY, 1968; HALL and HAJNAL, 1969; MEREU and HUNTER, 1969; BÅTH, 1971; GREGERSEN, 1971;

KANESTRØM and HAUGLAND, 1971; MEREU and JOBIDON, 1971; PENTTILÄ, 1971a, 1971b; SELLEVOLL and WARRICK, 1971; VOGEL and LUND, 1971; KANESTRØM, 1973; HIRSCHLEBER *et al.*, 1975; WAHLSTRÖM, 1975; GREEN *et al.*, 1980). Compared to the 7.17 km/s refraction velocity for P^* shown in Figure 6, many other studies have determined P^* values ranging from 6.5 to 7.3 km/s for these same shield regions (HODGSON, 1953; HALL and BRISBIN, 1961; SELLEVOLL and PENTTILÄ, 1964; ROLLER *et al.*, 1965; ROLLER and JACKSON, 1966; LUOSTO, 1967; SELLEVOLL and POMEROY, 1968; HALL and HAJNAL, 1969; GREGERSEN, 1971; KANESTRØM and HAUGLAND, 1971; SELLEVOLL and WARRICK, 1971; VOGEL and LUND, 1971; BERRY and FUCHS, 1973; KANESTRØM, 1973; WAHLSTRÖM, 1975; GREEN *et al.*, 1980).

Subcritical reflection seismic surveys have been made in the Canadian Shield by GREEN *et al.* (1978) and GREEN *et al.* (1979). From these near-vertical reflections, they determined P_g values between 6.1 and 6.2 km/s as compared to the 6.23 km reflection velocity in Figure 6. For P^*, they found a value of 7.2 km/s compared to 7.17 km/s in Figure 6. They also proposed that a mid-crustal layer might exist above the lower crustal layer. This mid-crustal layer would be only several kilometers thick and would have a velocity of about 6.9 km/s. The total thickness of the crust above the lower crustal layer would then be 21–22 km as compared to the 21.8 km determined by MASSÉ (1973a).

From the range of crustal P velocities estimated in the studies listed above, we can be sure that there remains some uncertainty in the actual velocities and in the layer thicknesses. We can, however, have considerable confidence that two major crustal layers with P wave refraction velocities at least near 6.23 and 7.17 km/s respectively are present in the shield regions studied. The layers themselves may, of course, represent average values of properties that vary slightly with depth or with horizontal position. There may also be a thin mid-crustal layer beneath some areas of the Canadian Shield.

It is possible to obtain an estimate of total crustal thickness using three-component long-period data and the spectral method of PHINNEY (1964). This information can provide an additional check on the shield crustal model shown in Figure 16, which has a total crustal thickness of 42 km. A number of studies have used Phinney's method to derive the crustal thickness of the Canadian and Fennoscandian Shields. For Coppermine in the northern Canadian Shield, UTSU (1966) determined a crustal thickness of 45 km. LEONG (1975) found a crustal thickness of 42 km for the Baltic Shield beneath Umeå, Sweden. For central and northern Fennoscandia, BUNGUM *et al.* (1980) found a crustal thickness of about 40 to 47 km.

Estimates of crustal thickness have also been obtained for the Canadian and Baltic shields from a number of refraction studies. Compared to the 42 km thickness of the model shown in Figure 16, BERRY and FUCHS (1973) determined an average thickness of about 45 km along a seismic profile in the Superior and Grenville provinces of the northeastern Canadian Shield. GREEN *et al.* (1980) estimated the crustal thickness to be between 41 and 46 km in the region of southwestern Manitoba and southeastern Saskatchewan where the Canadian Shield is covered by 1–2 km of younger rocks. In northern Scandinavia, LUND (1979) found an average crustal thickness of about 44–45 km. For the same area, GOLDFLAM *et al.* (1977) determined a crustal thickness of 42 km. From a study of the crustal and upper mantle structure of Finland, NOPONEN *et al.* (1967) estimated the crustal thickness to be 42 km.

Upper Mantle P Wave Structure

In the Canadian Shield, the upper mantle immediately below the Mohorovičić discontinuity was determined to have an apparent refraction velocity of 8.13 km/s (MASSÉ, 1973a). Comparison with many other studies of the Canadian and Baltic Shield areas indicates that the sub-Moho compressional velocity varies from 8.0 to 8.2 km/s (HODGSON, 1953; HALL and BRISBIN, 1961; SELLEVOLL and PENTTILÄ, 1964; ROLLER *et al.*, 1965; ROLLER and JACKSON, 1966; BARR, 1967; LUOSTO, 1967; GREEN and HALES, 1968; SELLEVOLL and POMEROY, 1968; HALL and HAJNAL, 1969; MEREU and JOBIDON, 1971; PENTTILÄ, 1971a, 1971b; SELLEVOLL and WARRICK, 1971; VOGEL and LUND, 1971; BERRY and FUCHS, 1973; KANESTRØM, 1973).

There is now general agreement that relatively abrupt increases in velocity occur in the upper mantle at depths of about 430 and 700 km (Figure 10). Other features of upper mantle structure remain more uncertain. There may be a velocity discontinuity near 330 km, although there is no general agreement concerning this feature. NOPONEN *et al.* (1967) found a velocity discontinuity at a depth of 70 km where the *P* velocity increases from 8.03 to 8.30 km/s. This compares with the discontinuity near 74 km in Figure 7 where the velocity increases from approximately 8.06 to 8.37 km/s. While it is clear that a *P* wave low velocity channel exists in the upper mantle beneath tectonic regions such as the Basin and Range Province (NIAZI and ANDERSON, 1965; ARCHAMBEAU *et al.*, 1969; MASSÉ *et al.*, 1972), there is not yet a consensus on whether such an upper mantle low velocity channel exists beneath most shield regions. Where it exists, it is apparently far more subdued than that under the western United States. The Early Rise profiles provide good evidence for a small low velocity channel 10–15 km thick at a depth of about 100 km.

Interpreting Seismic Record Sections

The two major features of the upper mantle *P* velocity distribution beneath stable platform regions of North America and Europe are discontinuities at depths near 430 km and 700 km (Figure 10). Most studies of seismic record sections have been able to identify these major upper mantle features. In addition to these major features, a fine structure exists in the velocity distribution between the Mohorovičić discontinuity and the discontinuity at 430 km depth (Figure 7). To date, there has been little agreement on the nature of this fine structure, although most studies acknowledge the existence of some velocity discontinuities in this depth range. We will now examine the causes of the failure of the inversion process to arrive at some consistent interpretation of any but the major features of upper mantle P velocity structure.

To focus the discussion, we will use the velocity model shown in Figure 10, which has prominent velocity discontinuities at depths of 430 km and 700 km. These features are found whether only ray theory travel time and amplitude calculations are made or synthetic seismograms are also computed. Differences in the analysis techniques influence only the travel time gradients, but have little effect on the depth determined for any discontinuity. Basically, the analysis technique chosen can be considered to have only a second order effect on the model determined. Therefore, whatever problem exists in record section interpretation, exists regardless of whether ray theory or synthetic seismogram techniques are used to analyze the record section.

As we shall now see, the problem in record section interpretation is very fundamental and very simple. It is nothing more complex than deciding what features of a record section correspond to distinct branches of the travel time curve that need to be modeled (using whatever technique we choose). This problem is obvious, and yet when uncertainties in model velocity distributions are computed, it is generally ignored. Instead, the interpretation of which refraction and reflection lines exist is assumed to have been made without error.

To make this point clear and, at the same time, to suggest a means of minimizing this problem, two structural features of the upper mantle model shown in Figure 10 will now be discussed. These features are the velocity discontinuity at 330 km depth and the thin low velocity channel at approximately 100 km depth.

The velocity discontinuity at 330 km depth in the model shown in Figure 10 is based on an interpretation that concludes that the 9.27 km/s refraction line (Figure 6) exists in record sections from stable interior regions of North America. This refraction line is not equally obvious in all record sections for these regions. For example, only the record sections for Early Rise profiles ER2 (Figure 5) and ER4 (Massé, 1973a) show any evidence for this refraction line, and that evidence is not conclusive. Using only these two record sections, it is unlikely that the 9.27 km/s refraction line would have been identified. For example, McMechan (1975) was not able to identify this refraction line from Early Rise data. One of the factors adding to the difficulty in identifying the 9.27 km/s refraction line from record sections is that this line never corresponds to first arrivals.

Previously, however, a record section constructed from three long profiles stretching from NTS across the Eastern United States and part of the Canadian

Shield had been analyzed (MASSÉ, 1972; MASSÉ *et al.*, 1972). This record section, presented in Figure 17, shows very strong evidence for the 9.27 km/s refraction line. Moreover, it is obvious from this record section that a separate subsequent refraction line corresponding to the 430 km discontinuity exists. Therefore, there must be a velocity discontinuity somewhat shallower than 430 km depth. Based on this evidence, it was possible to include the 9.27 km/s refraction line in the Early Rise data interpretation.

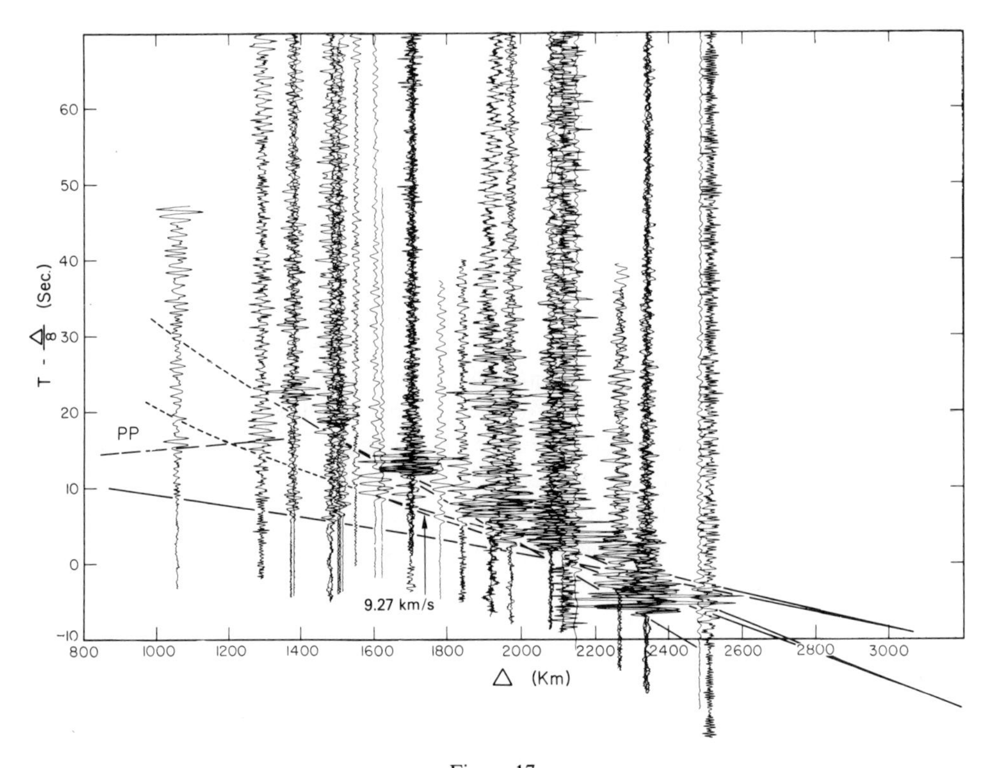

Figure 17
Record section from NTS explosions recorded at LRSM stations located at sites across the eastern United States and Canadian Shield.

The above analysis suggests that any given refraction line may not be equally obvious in all record sections. The degree to which a refraction line is evident in a record section may depend on many factors. One of the more important of these factors is probably the spectral energy characteristics of the seismic source producing the waves which illuminate the Earth structure. As a further example of this, the travel time lines from the Early Rise data were superimposed on a Fennoscandian record section (Figure 14). In this record section, there is good evidence for the 9.27 km/s line. The Soviet nuclear explosions used to create the record section in Figure

14 have somewhat similar source spectral characteristics to the NTS explosions used as seismic wave sources for the record section in Figure 17.

As regards the thin low velocity channel at a depth of about 100 km, this structure is defined by two refraction lines: the 8.48 km/s line which terminates at a distance near 1,350 km, and the 8.60 km/s line. Both lines are clearly evident in profile ER1 (Figure 4) as well as in a number of the other Early Rise profiles. The 8.60 km/s refraction line is evident in the Fennoscandian record section in Figure 14, but data were not available for distances less than 1,200 km and so the 8.48 km/s could not be identified. However, the entire set of North American travel time lines (Figure 6), when superimposed on the Baltic Shield crustal and upper mantle record sections, are in excellent agreement with all major features of the seismograms. CASSELL and FUCHS (1979) also argue that some type of low velocity structure is required beneath the Baltic Shield area.

We can conclude, therefore, that one of the most important steps in the construction of Earth models from seismic record sections is making the correct initial decisions as to which refraction and retrograde reflections exist in the data. The result of omitting an entire refraction line is generally to simplify the Earth model. Once a refraction line is omitted (the data are excluded from the inversion process), neither ray theory nor synthetic seismogram calculations will recover the correct model. Moreover, both ray theory and synthetic seismogram calculations will generally produce results that appear to fit the remaining set of data chosen for inversion.

We can also conclude that often a single record section, and sometimes even a group of record sections from the same type of seismic source, will not be sufficient to identify important refraction lines. If a number of record sections are available from different types of seismic sources having different spectral characteristics (e.g., chemical explosions and nuclear explosions), it is more likely that the interpretation will be complete and all refraction lines will be identified.

Over the last decade, a number of studies using synthetic seismograms have concentrated on fitting computed amplitudes to observed amplitudes with the maximum precision possible to derive an upper mantle model from an individual record section. Two major problems are often encountered in such an approach, however. First, as discussed above, all refraction and reflection lines are usually not obvious in any one record section, so the whole modeling process can begin with the wrong premise which cannot be remedied by amplitude calculations (for example, MCMECHAN, 1979). Second, the amplitude data are strongly affected by a number of factors including: local structure at the receiver, path attenuation, near-source structure and upper mantle structure. Even amplitudes of different arrivals in a recording from a single station can be affected differently by these factors, due to the P wave energy traveling slightly different paths. Therefore, careful judgment must be exercised in deciding how the amplitude data are used.

All of the above considerations have important implications when trying to determine how similar the upper mantle structure beneath North America is to that beneath northern Europe. If record sections from each of the two regions are analyzed independently, there is an excellent chance that two different velocity models will be determined (at least with respect to the fine details). In interpreting the record section data from one region, it is important to recognize the record section data from the other region. This was done in the studies by MASSÉ and ALEXANDER (1974) and MASSÉ (1975) and so it was possible to show that a common P velocity distribution will fit data from both regions. Therefore, the crustal and upper mantle structures of these regions indeed could be shown in similitude.

With the above procedures in mind, we will review a few other studies of crustal and upper mantle structure. The study of the upper mantle beneath Fennoscandian and western Russia by KING and CALCAGNILE (1976) is particularly unusual in that it failed to discern any of the upper mantle fine structure. In their analysis, King and Calcagnile used many dense segments of record sections created by assembling recordings of Soviet nuclear tests at a seismic array (NORSAR) located in Norway. Yet the only upper mantle features beneath Fennoscandia which they found were the two major discontinuities at depths of about 430 km and 700 km.

King and Calcagnile failed to identify certain refraction and reflection lines which led them to misinterpret other data in the record section. As a result, they were left with the situation that the retrograde reflections for the refraction lines they proposed were 'diffuse'. If the refraction lines were correct and none was missing, it would be difficult to explain why the retrograde reflections were not visible in any record section. Essentially, their model is based only on first arrivals at distances greater than 1,700 km, which helps explain why only the major discontinuities were found. This situation could have been avoided if they had used additional information from other available record sections.

CASSEL and FUCHS (1979) and GIVEN and HELMBERGER (1980) also investigated upper mantle structure beneath northern Europe. They noted that the model of King and Calcagnile appears inconsistent with waveform observations and is too simple above a depth of 420 km. Given and Helmberger concentrated on modeling accurately the long-period waveform data while trying to fit only the general features of the short-period observations. With this approach, the major discontinuities at 430 and 700 km depth were identified. In addition, a low velocity channel between 150 and 200 km depth was proposed by Given and Helmberger. The limited analysis of short-period data in their study may have precluded good resolution of all the fine structure above a depth of 430 km.

A study similar to that of Given and Helmberger was made by BURDICK (1981). In this study, Burdick also used long-period waveforms with synthetics. His results were somewhat similar to those of GIVEN and HELMBERGER (1980) and probably suffered from the same lack of resolution inherent in the use of long-period data.

Data from a seismic profile which stretched across France from the northwest to the southeast has been interpreted to obtain crustal and upper mantle structure

(HIRN *et al.*, 1973; KIND, 1974; HIRN *et al.*, 1975). The energy sources for this record section were chemical explosions detonated in the Atlantic Ocean off northern France. In some distance ranges, refracted arrivals are clearly evident in the record section, while in other distance ranges, they are not (Figure 18). Interpretations of the French profile have generally fit discontinuous refraction lines to the obvious arrivals. The corresponding Earth structure has many narrow low velocity channels (FUCHS, 1986).

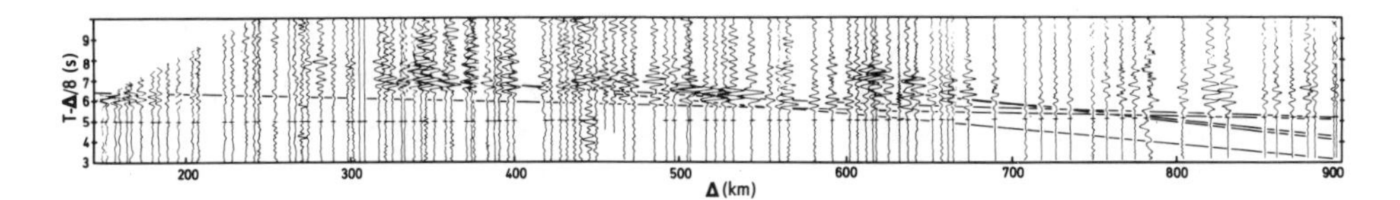

Figure 18
Record section from a seismic profile across France (HIRN *et al.*, 1973). Refraction and reflection lines are those determined for the Canadian Shield by MASSÉ (1973a).

For the French profile, the important question then is whether continuous refraction and reflection lines actually exist or not. They may exist, but the amplitudes in certain distance ranges may be low because of small velocity gradients or local near-surface structure. These low amplitudes may then be masked by the seismic noise.

To address this question, the shield travel time curves in Figure 6 were superimposed on the French profile record section as shown in Figure 18. Other than a common time shift of all curves to allow for near source structure differences, the travel time curves can be seen to fit very well and to explain the major energy arrivals in the seismic records. There is an indication that signals in the first 400 km of the profile may arrive slightly earlier, relative to the shield travel time curves, than do signals for distances greater than 400 km, suggesting minor variations in near-source crustal structure from that of the shield regions. Distinct P_n arrivals are continuously visible to distances of 900 km, with only amplitudes in the distance range 225 to 450 km too small to be clearly observed.

The excellent agreement of the shield travel time curves with the French profile data demonstrates the importance of deriving a velocity model from many record sections involving a wide range of distances and different sources. It also suggests that the P velocity structure beneath France is very similar to the structure beneath the Baltic and Canadian Shields. If this French record section is interpreted by itself without considering data from other seismic profiles, then other interpretations are not only possible, but inevitable.

Previously it was noted that the effect of omitting an entire refraction line was to simplify the model structure derived. Regarding where only segments of actual refraction lines are identified, and these segments are treated as if the rest of the

Table 2

Crustal and Upper Mantle Structure beneath the Canadian Shield

Depth (km)	P Velocity (km/s^{-1})	S Velocity (km/s^{-1})	Density g/cm^3
0	5.170	3.320	2.464
4.7	5.170	3.320	2.464
4.7	6.220	3.650	2.808
21.8	6.220	3.650	2.808
21.8	7.140	4.110	3.110
42.2	7.140	4.110	3.110
42.2	8.060	4.650	3.412
72.6	8.060	4.650	3.412
72.7	8.100	4.670	3.425
73.5	8.200	4.740	3.458
77.5	8.370	4.850	3.513
93.5	8.370	4.850	3.513
94.0	8.050	4.450	3.408
107.0	7.980	4.400	3.385
107.0	8.430	4.800	3.533
130.0	8.430	4.750	3.533
160.0	8.432	4.600	3.534
257.0	8.433	4.500	3.534
300.0	8.434	4.530	3.534
320.0	8.435	4.600	3.535
325.0	8.435	4.680	3.535
328.0	8.450	4.800	3.540
328.5	8.550	4.936	3.572
328.6	8.770	5.063	3.645
332.5	8.780	5.069	3.648
410.0	8.783	5.071	3.649
420.0	8.810	5.086	3.658
425.0	8.847	5.108	3.670
430.0	8.950	5.167	3.704
432.0	9.625	5.557	3.925
440.0	9.640	5.566	3.930
470.0	9.650	5.571	3.933

refraction lines did not exist, the result is to greatly complicate the model structure. A number of low velocity channels must then be added to the model.

Obviously, there exist a number of different philosophies of data interpretation for seismic refraction profiles. One philosophy (adopted in this paper) interprets the disappearance of a refraction line as a simple data gap (not caused by a low velocity channel) whenever evidence of the continuation of the line can be found in *any* of the seismic profiles available for the region. This interpretation yields a relatively simple Earth structure. Another philosophy interprets almost any disappearance of a refraction line in a profile as evidence for a low velocity channel. This interpretation emphasizes the 'shingling' appearance of refractions in many record sections to construct short discontinuous refraction lines. The resulting Earth structure is

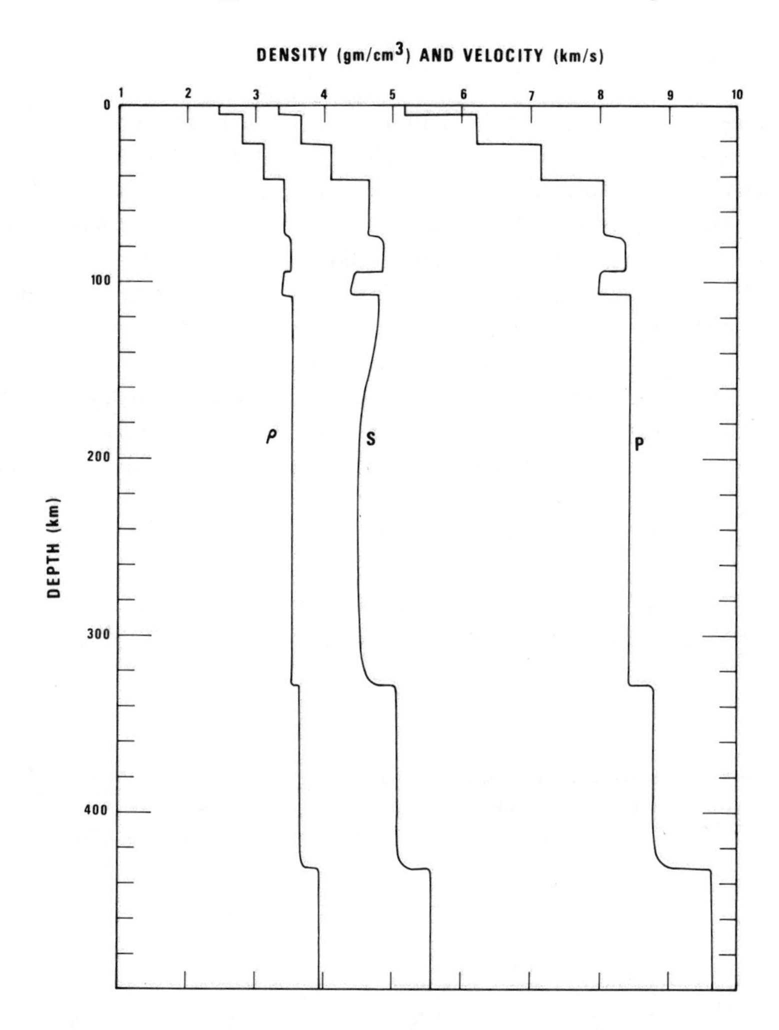

Figure 19
Velocity and density distribution for the Canadian Shield (MASSÉ, 1973c).

relatively complex. This interpretation philosophy is followed by a number of European seismologists (e.g., MUELLER and ANSORGE, 1986).

To choose between these philosophies, examination of a number of different profiles for a given region is important. The generalized approach to seismic interpretation using many profiles over a region may have the deficiency that not all local variations in amplitude are explained. However, it has the great advantage that a more accurate representation of the crustal and upper mantle structure for the region as a whole is obtained.

S-Wave Structure

To complete our review of the seismic velocity structure beneath stable continental regions, the *S* wave velocities must be considered. Although this distribution can be determined by studies of *S* wave refraction, this is more difficult than for *P* waves. The difficulty arises both because explosive sources are not very effective in generating shear wave energy, and because the *S* waves always arrive during the coda for *P* waves. Studies of surface wave dispersion have provided a very important alternative method for defining the Earth's *S* wave velocities (BRUNE and DORMAN, 1963; MASSÉ, 1973c; CARA, 1979; CARA *et al.*, 1980).

The resolving power of surface waves is limited by the relatively long wavelengths involved. However, the depth of first order discontinuities in the crust and upper mantle can be constrained by *P* wave information to make the *S* wave determinations from surface wave data more reliable. This procedure was followed in deriving the *S* wave velocity model for the Canadian Shield shown in Figure 19 (MASSÉ, 1973c). The *S* wave travel times corresponding to this model are shown in Figure 20 and the *P* and *S* wave velocities and the density values are listed in Table 2. The density (ρ) values were obtained from the compressional velocity (α) values using the relationship given by BIRCH (1964):

$$\rho = 0.768 + 0.328\,\alpha.$$

Fundamental Love and Rayleigh wave phase velocity data from the Canadian Shield (BRUNE and DORMAN, 1963), together with the *P* wave depth constraints for

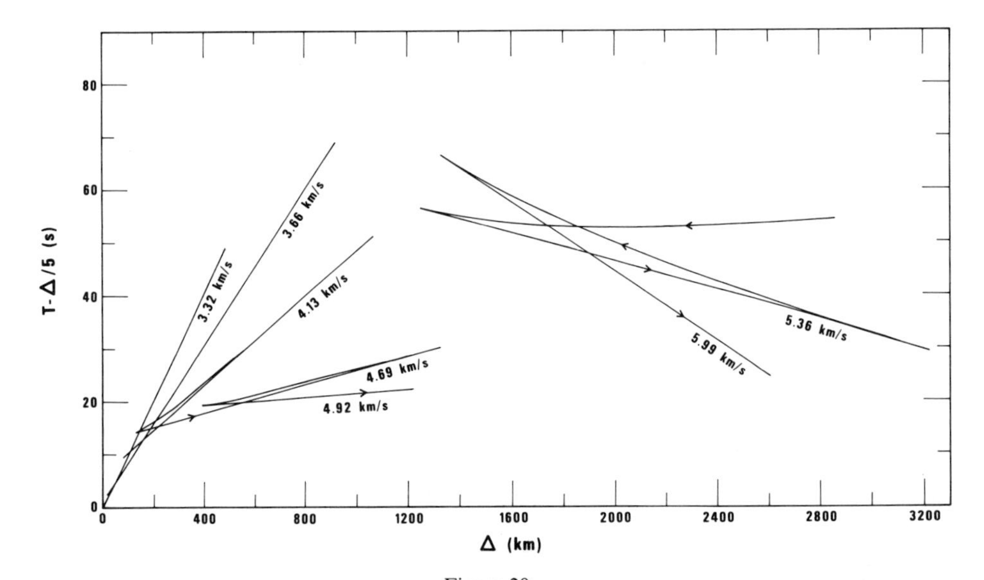

Figure 20
Refraction and reflection lines for *S* wave energy propagating through the Canadian Shield (MASSÉ, 1973c).

the velocity discontinuities require a wide upper mantle low velocity zone for *S* waves such as shown in Figure 19. If the first order velocity discontinuities are not fixed by the *P* wave data, then it is possible to find other solutions which are in agreement with the fundamental mode surface wave data and which do not require a wide low velocity zone. However, for these other solutions, the *S* wave velocity discontinuities would no longer be at the same depths as the *P* wave velocity discontinuities. A comparison of observed Love and Rayleigh wave phase velocity dispersion with that calculated from the Canadian Shield structure model (Figure 19) is shown in Figure 21. In deriving the velocity model shown in Figure 19, the *S* wave velocity had to be restrained for the thin low velocity channel defined by *P* wave data because the calculated phase velocity dispersion was relatively insensitive to the value of the *S* velocity within this channel.

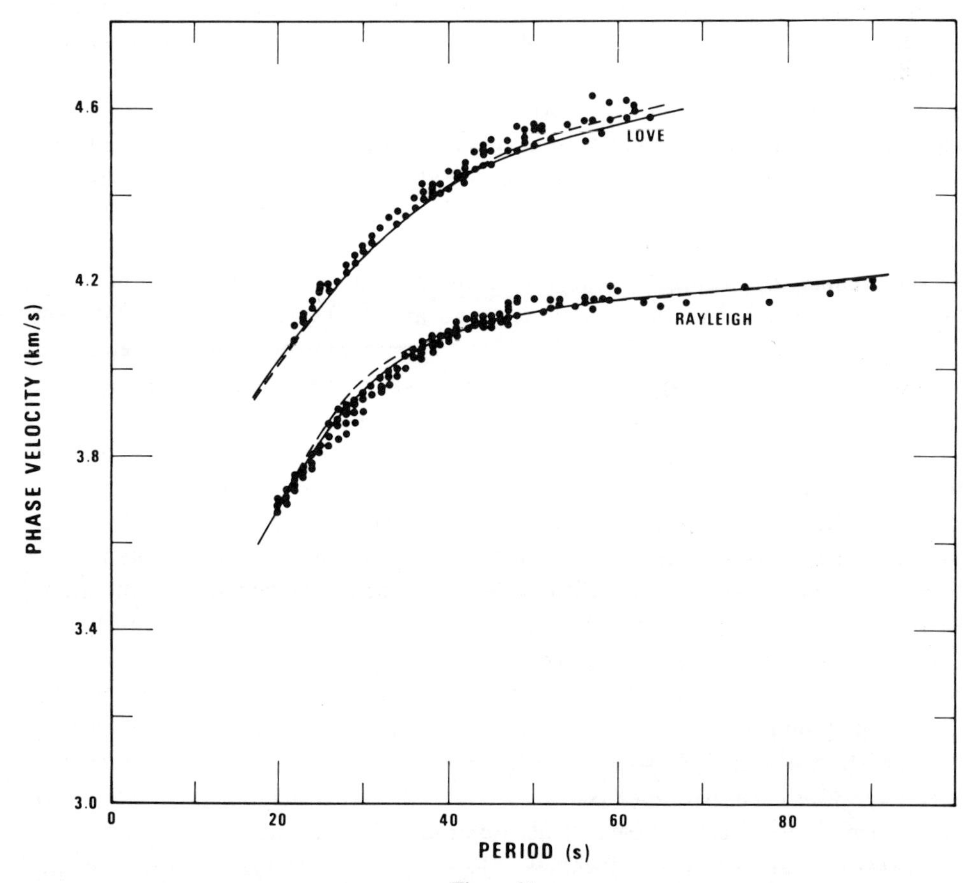

Figure 21
Observed Love and Rayleigh wave phase velocity dispersion (solid circles) for the Canadian Shield (BRUNE and DORMAN, 1963). Solid lines are calculated dispersion from the shield model of MASSÉ (1973c); dashed lines are calculated dispersion from the CANSD model of BRUNE and DORMAN (1963).

GODLEWSKI and WEST (1977) determined fundamental Rayleigh wave group velocity dispersion for a path in the Canadian Shield. From their data, which agree with the BRUNE and DORMAN (1963) data, they found the average crustal thickness of 40 ± 3 km, which is consistent with the thickness of 42 km determined by MASSÉ (1973a).

CARA (1979) noted that if both fundamental and higher mode surface wave data are inverted, an upper mantle low velocity zone is required (even without the *P* velocity constraints). However, a subsequent study by CARA *et al.* (1980) of higher mode surface waves propagating across western Europe and northern Eurasia found that a low velocity zone is not necessary to explain the data observed over this fairly large area (not all shield area). By including constraints from *S* wave refraction profiles, DER and LANDISMAN (1972) found that a low velocity channel for *S* is required in the upper mantle below Fennoscandia.

Although it is difficult to pick *S* wave arrivals precisely, a few studies of *S* wave refraction and reflection across the stable continental regions of North America and Fennoscandia have been made. The results of these studies can be compared with the calculated travel times for *S* shown in Figure 20. The apparent velocity of 3.66 km/s in Figure 20 is consistent with observed refraction velocities of 3.54 to 3.69 km/s associated with the S_g phase (HODGSON, 1953; HALL and BRISBIN, 1961; BRUNE and DORMAN, 1963; GREEN and HALES, 1968; HALL and HAJNAL, 1969; WAHLSTRÖM, 1975; GREEN *et al.*, 1978; GREEN *et al.*, 1979). The apparent velocity of 4.13 km/s in Figure 20 compares favorably with observed refraction velocities of 4.00 and 4.10 km/s for S^* (HALL and BRISBIN, 1961; HALL and HAJNAL, 1969). The apparent velocity of 4.69 km/s, which represents S_n, is comparable to the observed refraction velocities of 4.60 to 4.73 km/s (LEHMANN, 1955; HALL and BRISBIN, 1961; BRUNE and DORMAN, 1963; HALL and HAJNAL, 1969; LUND, 1979).

JORDAN and FRAZER (1975) attempted to use the conversion of *S* to *P* waves at the Mohorovičić discontinuity to define the shear wave velocity contrast across this boundary for eastern Canada. To do this, they analyzed long-period body waves from deep focus earthquakes. Because of the long wavelengths involved, the resolution of this procedure is limited (BULIN, 1980). Modeling of synthetic seismograms to agree with observed seismograms was done to seek improvement of the resolution. From this analysis, Jordan and Frazer calculated that a low velocity channel for both *P* and *S* waves exists at the base of the crust for certain sites in eastern Canada. If this conclusion is valid, however, it holds only for very limited areas of eastern Canada (BURDICK and LANGSTON, 1977). Such a crustal low velocity channel is generally inconsistent with the observed P^* arrivals across the Canadian Shield and the eastern United States. These arrivals are observed at distances which correspond to a lower crustal layer with a thickness at least as large as shown in Figures 16 and 19. Propagation of P^* to these observed distances would not be possible if the thickness of the lower crustal layer is decreased by the presence of a low velocity channel at the base of the crust.

VINNIK (1977) and VINNIK *et al.* (1983) studied the conversion of *P* waves to *SV* waves in the long-period *P* wave coda. From these data, it is possible to estimate the travel time to deep upper mantle discontinuities. Using data from the NORSAR array in the Baltic Shield, Vinnik found discontinuities at depths of roughly 410–440 and 640–690 km. The exact depths estimated for these discontinuities depends on the ratio of *P* to *S* wave velocity assumed.

Thickness of the Lithosphere

As originally defined by BARRELL (1914), the lithosphere is the outer shell of the Earth which can sustain long-term nonhydrostatic stresses. Flows associated with isostatic adjustment occur below the lithosphere. The level of isostatic compensation is generally thought to be at a depth of roundly 100 km, and should be coincident then with the lithosphere-asthenosphere boundary. Therefore, we expect the continental lithosphere to have a thickness of close to 100 km.

From the above definition, the base of the continental lithosphere assumedly is located at a depth coincident with a zone of weakness or low viscosity in the Earth's mantle. Low velocity channels are likely zones of low viscosity. Other possible zones of low viscosity include velocity discontinuities where the velocity increases with depth across the discontinuity. The upper mantle structure beneath shield regions is less complicated than that beneath tectonic regions such as the Basin and Range Province of the western United States. Therefore careful examination of upper mantle structure beneath shield regions may provide the key to determining the average thickness of the continental lithosphere.

In the shield model in Figure 19, several features are evident which may mark the base of the lithosphere. The most probable, considering both its depth and the likely existence of an underlying zone of low viscosity, is the decrease in velocity for both *P* and *S* waves beginning at a depth of 94 km. The low velocity channel in the depth range from 94 km to 107 km may be a particularly weak zone since only in this depth range is there a decrease in both *P* and *S* wave velocities. From these considerations, the thickness of the continental lithosphere is determined to be approximately 95 km. This is in excellent agreement with the expected depth of the level of isostatic compensation. Other much less likely possibilities for the base of the lithosphere are the velocity discontinuities at depths of 330 and 430 km (Figure 19).

While the thickness of the continental lithosphere appears to be well determined, the thickness of the continental plate is very uncertain. Some studies have concluded that the continental plate may be considerably thicker than the continental lithosphere (JORDAN, 1975, 1978). Data obtained from the analysis of body and surface waves provide important information related to this problem.

From the analysis of *P* wave refraction and reflection data presented in this paper, it seems clear that the structure of the Canadian and Baltic Shields are similar down to depths of at least 800 km. A number of surface wave and *S* wave studies have been made comparing ocean and shield region *S* wave structure. From such studies, OKAL and ANDERSON (1975), BUTLER (1979), CARA (1979), LÉVÊQUE (1980), and NAKANISHI (1981) have found that ocean-shield differences exist to a depth of about 250 km. Below this depth, they found no resolvable differences. OKAL (1978) determined that, for very long-period (200–300 s) surface waves, similar *S* wave velocities are observed for ocean and shield regions. However, a few studies have concluded that differences between ocean and shield regions, with regard to *S* wave velocities, may persist to depths of about 400 km or more (SIPKIN and JORDAN, 1975, 1976, 1980).

From these data, a number of hypotheses concerning the thickness of the continental plate can be derived. JORDAN (1975, 1978) proposed a very thick continental plate of the order of 400 km or more. The primary reason for proposing such a thick plate (tectosphere) was to account for differences in *S* wave velocities between old ocean basins and shield regions. From an analysis of long-period *P* precursors to *S* waves, SACKS *et al.* (1979) proposed that the continental lithosphere has a base coincident with a velocity discontinuity at a depth of 250 km. If this is true, presumably the continental plate and continental lithosphere are identical and have a thickness of 250 km. ANDERSON (1979) also proposed a discontinuity at 220 km, but a thinner lithosphere (less than 150–200 km). However, no evidence for a seismic velocity discontinuity near 220–250 km was found in any of the seismic profiles examined in this study. KERR (1986) had reviewed some of the arguments for thick continental plates.

Another possibility is that the continental plate is equivalent to the continental lithosphere and the lithosphere thickness is as determined above: about 95–100 km. Obviously, the fact that structural similarities beneath different shields extend to great depths does not require that continental plates be thick. Only the differences in the *S* wave velocities beneath old ocean basins and shield regions must be explained. Perhaps these differences reflect the different physical regimes imposed on the upper mantle by the presence of a continental versus an oceanic plate. The different pressure and temperature conditions may be sufficient to cause a larger *S* wave low velocity channel beneath the oceans than beneath the shields. If this is true, then differences in *S* wave velocities between old ocean basins and shield regions would not require that continental plates be thick. From an analysis of the flexural rigidity of the lithosphere, WALCOTT (1970) concluded that the lithosphere in Canada, presumably representative of the lithosphere in stable continental areas, is about 110 km thick. He found the lithosphere at Hawaii to be somewhat greater than 75 km thick.

Table 3

Conductivity Data for North America

Depth (km)	Conductivity (S/m)
141	0.008
148	0.026
157	0.027
167	0.015
192	0.013
255	0.045
256	0.051
280	0.083
303	0.053
304	0.058
315	0.082
323	0.059
332	0.059
348	0.075
356	0.062
363	0.099
370	0.073
379	0.091
383	0.097
388	0.081
395	0.120
399	0.156
401	0.122
401	0.126
407	0.099
426	0.186
429	0.166
431	0.134
434	0.137
438	0.190
483	0.161
486	0.198
491	0.192
492	0.183
514	0.178
517	0.177
594	0.503
611	0.430

Conductivity Data

In recent years, a number of studies have presented detailed conductivity data relating to the upper mantle. Figure 22 shows conductivity data (listed in Table 3) from CAMPBELL and ANDERSSEN (1983) obtained for a sector of the Earth which includes North America. Superimposed on the conductivity data in Figure 22 is the

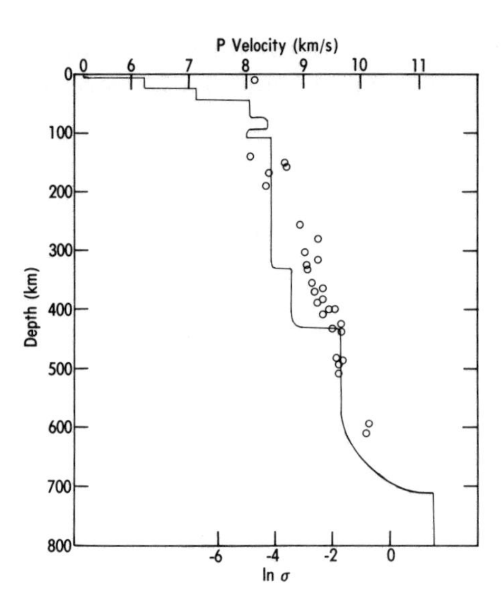

Figure 22
Conductivity data (open circles) for North America and the P velocity distribution for the stable platform regions of North America.

upper mantle model (Figure 10) derived by MASSÉ (1973b) for the Canadian Shield region. The conductivity data suggests the presence of a discontinuity at a depth of 430 km. However, a straight line would also provide a good fit to the conductivity data.

Future Developments

The crustal and upper mantle structure of stable continental areas in North America and northern Europe is much better known today than two decades ago. Nevertheless, much is yet to be learned. For many other areas of the world, the crustal and upper mantle structure is only poorly defined. To address this lack of knowledge, many countries are expanding their permanent networks of seismic stations and are deploying arrays of portable seismograph systems to record special events. As in the past, future advances in our knowledge of the Earth will be tied to the development of new instrumentation and data processing techniques. In the United States, the U.S. Geological Survey is conducting an aggressive program of seismic reflection profiling (HAMILTON, 1986), the COCORP program of vibroseis profiling is continuing, and the PASSCAL program (ANONYMOUS, 1984) has been initiated to provide instrumentation and data for crustal and upper mantle studies.

Within the next few years, additional instrumentation available in many countries will permit high resolution tomography of the Earth (ANDERSON, 1984). This should provide significant new information on lateral heterogeneities and on mantle convection patterns.

Acknowledgments

We wish to thank Dr Rainer Kind for encouraging us to include an interpretation of the seismic profile in France and for providing a copy of his reflectivity program. We also wish to thank Dr Wallace H. Campbell for providing conductivity measurements for North America. Dr David Gordon and Dr William Spence provided valuable reviews of the paper.

REFERENCES

ANDERSON, D. L. (1979), *The deep structure of continents.* J. Geophys. Res. *84*, 7555–7560.

ANDERSON, D. L. (1984), *Surface wave tomography.* EOS *65*, 147–148.

ANDERSSEN, R. S., DEVANE, J. F., GUSTAFSON, S. A. and WINCH, D. E. (1979), *The qualitative character of the global electrical conductivity of the Earth.* Phys. Earth Planet. Inter. *20*, 15–21.

ANONYMOUS, (1984), *PASSCAL, Program for array seismic studies of the continental lithosphere.* Incorporated Research Institutions for Seismology.

ARCHAMBEAU, C. B., FLINN, E. A. and LAMBERT, D. G. (1969), *Fine structure of the upper mantle.* J. Geophys. Res. *74*, 5825–5865.

BANKS, R. J. (1969), *Geomagnetic variations and the electrical conductivity of the upper mantle.* Geophys. J. R. Astr. Soc. *17*, 457–487.

BARR, K. G. (1967), *Upper mantle structure in Canada from seismic observations using chemical explosions.* Can. J. Earth Sci. *4*, 961–975.

BARRELL, J. (1914), *The strength of the Earth's crust.* J. Geology *22*, 655–683.

BÅTH, M. (1971), *Average crustal structure of Sweden.* Pure Appl. Geophys. *88*, 75–91.

BERRY, M. J. and FUCHS, K. (1973), *Crustal structure of the Superior and Grenville Provinces of the northeastern Canadian Shield.* Bull. Seism. Soc. Am. *63*, 1393–1432.

BIRCH, F. (1964), *Density and composition of the mantle and core.* J. Geophys. Res. *69*, 4377–4387.

BRAILE, L. W. (1973), *Inversion of crustal seismic refraction and reflection data.* J. Geophys. Res. *78*, 7738–7744.

BRUNE, J. and DORMAN, J. (1963), *Seismic waves and Earth structure in the Canadian Shield.* Bull. Seism. Soc. Am. *53*, 167–210.

BULIN, N. K. (1980), *Comment on 'Crustal and upper mantle structure from S_p Phases' by Thomas H. Jordan and L. Neil Frazer.* J. Geophys. Res. *85*, 377–380.

BUNGUM, H., PIRHONEN, S. E. and HUSEBYE, E. S. (1980), *Crustal thickness in Fennoscandia.* Geophys. J. R. Astr. Soc. *63*, 759–774.

BURDICK, L. J. and LANGSTON, C. A. (1977), *Modeling crustal structure through the use of converted phases in teleseismic bodywave forms.* Bull. Seism. Soc. Am. 67, 677–691.

BURDICK, L. J. (1981), *A comparison of the upper mantle structure beneath North America and Europe.* J. Geophys. Res. *86*, 5926–5936.

BUTLER, R. (1979), *Shear-wave travel times from SS.* Bull. Seism. Soc. Am. *69*, 1715–1732.

CAMPBELL, W. H. and ANDERSSEN, R. S. (1983), *Conductivity of the subcontinental upper mantle: an analysis using quiet-day geomagnetic records of North America.* J. Geomag. Geoelectr. *35*, 367–382.

CARA, M. (1979), *Lateral variations of S velocity in the upper mantle from higher Rayleigh modes.* Geophys. J. R. Astr. Soc. *57*, 649–670.

CARA, M., NERCESSIAN, A. and NOLET, G. (1980), *New inferences from higher mode data in western Europe and northern Eurasia.* Geophys. J. R. Astr. Soc. *61*, 459–478.

CASSELL, B. R. and FUCHS, K. (1979), *Seismic investigations of the subcrustal lithosphere beneath Fennoscandia.* J. Geophys. *46*, 369–384.

CHAPMAN, C. H. (1978), *A new method for computing synthetic seismograms.* Geophys. J. R. Astr. Soc. *54*, 481–518.

CHOY, G. L. (1977), *Theoretical seismograms of core phases calculated by frequency-dependent full wave theory, and their interpretation.* Geophys. J. R. Astr. Soc. *51*, 275–312.

CORMIER, V. F. (1980), *The synthesis of complete seismograms in an Earth model specified by radially inhomogeneous layers.* Bull. Seism. Soc. Am. *70*, 691–716.

DAHLMAN, O. (1971a), *Organization of the Trans-Scandinavian deep seismic sounding project.* In Proc. Colloquium on Deep Seismic Sounding in Northern Europe, Uppsala, December 1 and 2, 1969, (ed. A. Vogel), (University of Uppsala) pp. 55–57.

DAHLMAN, O. (1971b), *Profile sections 2–4 and 3–5.* In Proc. Colloquium on Deep Seismic Sounding in Northern Europe, Uppsala, December 1 and 2, 1969, (ed. A. Vogel), (University of Uppsala) pp. 96–98.

DER, Z. A. and LANDISMAN, M. (1972), *Theory for errors, resolution, and separation of unknown variables in inverse problems, with application to the mantle and crust in Southern Africa and Scandinavia.* Geophys. J. R. Astr. Soc. *27*, 137–178.

FUCHS, K. and MÜLLER, G. (1971), *Computation of synthetic seismograms with the reflectivity method and comparison with observations.* Geophys. J. R. Astr. Soc. *23*, 417–433.

FUCHS, K. (1986), *Reflections from the subcrustal lithosphere.* In *Reflection Seismology: The Continental Crust*, (eds. M. Barazangi and L. Brown), (Am. Geophys. Un., Washington, D.C.).

GIVEN, J. W. and HELMBERGER, D. V. (1980), *Upper mantle structure of Northwestern Eurasia.* J. Geophys. Res. *85*, 7183–7194.

GOLDFLAM, St., HIRSCHLEBER, H. B. and JANLE, P. (1977), *A refined crustal model and the isostatic state of the Scandinavian Blue Road Area.* J. Geophys. *42*, 419–428.

GODLEWSKI, M. J. C. and WEST, G. F. (1977), *Rayleigh-wave dispersion over the Canadian Shield.* Bull. Seism. Soc. Am. *67*, 771–779.

GREEN, A. G., HALL, D. H. and STEPHENSON, O. G. (1978), *A subcritical seismic crustal reflection survey over the Aulneau batholith, Kenora region, Ontario.* Can. J. Earth Sci. *15*, 301–315.

GREEN, A. G., ANDERSON, N. L. and STEPHENSON, O. G. (1979), *An expanding spread seismic reflection survey across the Snake Bay—Kakagi Lake greenstone belt, northwestern Ontario.* Can. J. Earth Sci. *16*, 1599–1612.

GREEN, A. G., STEPHENSON, O. G., MANN, G. D., KANASEWICH, E. R., CUMMING, G. L., HAJNAL, Z., MAIR, J. A. and WEST, G. F. (1980), *Cooperative seismic surveys across the Superior-Churchill boundary zone in southern Canada.* Can. J. Earth Sci. *17*, 617–632.

GREEN, R. W. E. and HALES, A. L. (1968), *The travel times of P waves to 30° in the central United States and upper mantle structure.* Bull. Seism. Soc. Am. *58*, 267–289.

GREGERSEN, S. (1971), *Profile section 4–5.* In Proc. Colloquium on Deep Seismic Sounding in Northern Europe, Uppsala, December 1 and 2, 1969, (ed. A. Vogel), (University of Uppsala) pp. 92–95.

HALE, L. D. and THOMPSON, G. A. (1982), *The seismic reflection character of the continental Mohorovičić discontinuity.* J. Geophys. Res. *87*, 4625–4635.

HALL, D. H. and BRISBIN, W. C. (1961), *A study of the Mohorovičić discontinuity near Flin Flon, Manitoba.* Final Report for Geophys. Res. Div., Air Force Cambridge Res. Labs., Office Aerospace Res., U.S.A.F., Bedford, Massachusetts.

HALL, D. H. and HAJNAL, Z. (1969), *Crustal structure of Northwestern Ontario: Refraction Seismology.* Can. J. Earth Sci. *6*, 81–99.

HAMILTON, R. M. (1986), *Seismic reflection studies by the U.S. Geological Survey.* In *Reflection Seismology: a Global Perspective*, (eds. M. Barazangi and L. Brown), (Am. Geophys. Un., Washington, D.C.).

HELMBERGER, D. V. (1972), *Long-period body-wave propagation from 4° to 13°.* Bull. Seism. Soc. Am. *62*, 325–342.

HELMBERGER, D. V. (1973), *On the structure of the low-velocity zone.* Geophys. J. R. Astr. Soc. *34*, 251–263.

HELMBERGER, D. V. and ENGEN, G. R. (1974), *Upper mantle shear structure.* J. Geophys. Res. *79*, 4017–4028.

HIRN, A., STEINMETZ, L., KIND, R. and FUCHS, K. (1973), *Long range profiles in western Europe: II. Fine structure of the lower lithosphere in France (southern Bretagne)*. J. Geophys. *39*, 363–384.

HIRN, A., PRODEHL, C. and STEINMETZ, L. (1975), *An experimental test of models of the lower lithosphere in Bretagne (France) (1) (2)*. Ann. Geophys. *31*, 517–530.

HIRSCHLEBER, H. B., LUND, C. E., MEISSNER, R., VOGEL, A. and WEINREBE, W. (1975), *Seismic investigations along the Scandinavian 'Blue Road' Traverse*. J. Geophys. *41*, 135–148.

HODGSON, J. H. (1953), *A seismic survey in the Canadian Shield. II: Refraction studies based on timed blasts*. Publ. Dominion Obs., Ottawa *16*, 169–181.

JORDAN, T. H. (1975), *The continental tectosphere*. Rev. Geophys. Space Phys. *13*, 1–12.

JORDAN, T, H, and FRAZER, L. N. (1975), *Crustal and upper mantle structure from Sp phases*. J. Geophys. Res. *80*, 1504–1518.

JORDAN, T. H. (1978), *Composition and development of the continental tectosphere*. Nature, *274*, 544–548.

KANESTRØM, R. and HAUGLAND, K. (1971), *Profile section 3–4*. In Proc. Colloquium on Deep Seismic Sounding in Northern Europe, Uppsala, December 1 and 2, 1969, (ed. A. Vogel), (University of Uppsala) pp. 76–91.

KANESTRØM, R. (1973), *A crust-mantle model for the NORSAR area*. Pure Appl Geophys. *105*, 729–740.

KENNETT, B. L. N. (1980), *Seismic waves in a stratified halfspace. II Theoretical seismograms*. Geophys. J. R. Astr. Soc. *61*, 1–10.

KERR, R. A. (1986), *The continental plates are getting thicker*. Science *232*, 933–934.

KIND, R. (1974), *Long range propagation of seismic energy in the lower lithosphere*. J. Geophys. *40*, 189–202.

KIND, R. (1978), *The reflectivity method for a buried source*. Geophys. J. R. Astr. Soc. *44*, 603–612.

KING, D. W. and CALCAGNILE, G. (1976), *P-wave velocities in the upper mantle beneath Fennoscandia and Western Russia*. Geophys. J. R. Astr. Soc. *46*, 407–432.

LEHMANN, I. (1955), *The times of P and S in northeastern America*. Ann. Geofis. *8*, 351–370.

LEONG, L. S. (1975), *Crustal structure of the Baltic Shield beneath Umea, Sweden, from the spectral behavior of long-period P waves*. Bull. Seism. Soc. Am. *65*, 113–126.

LÉVÊQUE, J. J. (1980), *Regional upper mantle S-velocity models from phase velocities of great-circle Rayleigh waves*. Geophys. J. R. Astr. Soc. *63*, 23–43.

LUND, C. E. (1979), *The fine structure of the lower lithosphere underneath the Blue Road profile in northern Scandinavia*. Tectonophysics *56*, 111–122.

LUOSTO, U. (1967), *Preliminary results of a seismic refraction study of the Earth's crust in S.W. Finland*. Geophysica *9*, 301–306.

MASSÉ, R. P. (1972), *Upper mantle P velocity beneath the central United States*. Trans. Am. Geophys. Un. *53*, 452 (abstract).

MASSÉ, R. P., LANDISMAN, M. and JENKINS, J. B. (1972), *An investigation of the upper mantle compressional velocity distribution beneath the Basin and Range province*. Geophys. J. R. Astr. Soc. *30*, 19–36.

MASSÉ, R. P. (1973a), *Compressional velocity distribution beneath central and eastern North America*. Bull. Seism. Soc. Am. *63*, 911–935.

MASSÉ, R. P. (1973b), *Compressional velocity distribution beneath central and eastern North America in the depth range 450–800 km*. Geophys. J. R. Astr. Soc. *36*, 705–716.

MASSÉ, R. P. (1973c), *Shear velocity distribution beneath the Canadian Shield*. J. Geophys. Res. *78*, 6943–6950.

MASSÉ, R. P. and ALEXANDER, S. S. (1974), *Compressional velocity distribution beneath Scandinavia and western Russia*. Geophys. J. R. Astr. Soc. *39*, 587–602.

MASSÉ, R. P. (1975), *Baltic shield crustal velocity distribution*. Bull. Seism. Soc. Am. *65*, 885–897.

MCMECHAN, G. A. (1975), *Amplitude constraints and the inversion of Canadian Shield Early Rise explosion data*. Bull. Seism. Soc. Am. *65*, 1419–1433.

MCMECHAN, G. A. (1979), *An amplitude constrained P-wave velocity profile for the upper mantle beneath the eastern United States*. Bull. Seism. Soc. Am. *68*, 1733–1744.

MEREU, R. F. and HUNTER, J. A. (1969), *Crustal and upper mantle structure under the Canadian Shield from Project Early Rise data*. Bull. Seism. Soc. Am. *59*, 147–165.

MEREU, R. F. and JOBIDON, G. (1971), *A seismic investigation of the crust and Moho on a line perpendicular to the Grenville Front*. Canadian J. Earth Sci. *8*, 1553–1583.

MUELLER, St. and ANSORGE, J. (1986), *Long-range seismic refraction profiles in Europe*. In *Reflection*

Seismology: a Global Perspective, (eds. M. Barazangi and L. Brown), (Am. Geophys. Un. Washington D.C.).

NAKANISHI, I. (1981), *Shear velocity and shear attenuation models inverted from the world-wide and pure-path average data of mantle Rayleigh waves* (${}_0S_{25}$ *to* ${}_0S_{80}$) *and fundamental spheroidal modes* (${}_0S_2$ *to* ${}_0S_{24}$). Geophys. J. R. Astr. Soc. *66*, 83–130.

NIAZI, M. and ANDERSON, D. L. (1965), *Upper mantle structure of western North American from apparent velocities of P waves.* J. Geophys. Res. *70*, 4633–4640.

NOPONEN, I., PORKKA, M. T., PIRHONEN, S. and LUOSTO, U. (1967), *The crust and mantle in Finland.* J. Phys. Earth *15*, 19–24.

OKAL, E. A., (1978), *Observed very long period Rayleigh-wave phase velocities across the Canadian Shield.* Geophys. J. R. Astr. Soc. *53*, 663–668.

OKAL, E. A. and ANDERSON, D. L. (1975), *A study of lateral inhomogeneities in the upper mantle by multiple ScS travel time residuals.* Geophys. Res. Lett. *2*, 313–316.

OLIVER, J. and MURPHY, L. (1971), *WWNSS: Seismology's global network of observing stations.* Science *174*, 254–261.

PENTTILÄ, E. (1971a), *Seismic investigations on the Earth's crust in Finland.* In Proc. Colloquium on Deep Seismic Sounding in Northern Europe, Uppsala, December 1 and 2, 1969, (ed. A. Vogel) (University of Uppsala) pp. 9–13.

PENTTILÄ, E. (1971b), *Profile section 1–2.* In Proc. Colloquium on Deep Seismic Sounding in northern Europe, Uppsala, December 1 and 2, 1969, (ed. A. Vogel), (University of Uppsala) pp. 58–61.

PHINNEY, R. A. (1964), *Structure of the Earth's crust from spectral behaviour of long-period body waves.* J. Geophys. Res. *69*, 2997–3017.

POWELL, P. and FRIES, D. (1966), *Handbook: World-Wide Standard Seismograph Network.* Inst. Sci. and Tech., Univ. of Michigan, Ann Arbor.

ROLLER, J. C., HEALY, J. H. and JACKSON, W. H. (1965), *A seismic-refraction profile extending from Lake Superior, Wisconsin, to Agate, Colorado* (abstract). Trans. Am. Geophys. Union *46*, 155.

ROLLER, J. C. and JACKSON, W. H. (1966), *Seismic wave propagation in the upper mantle: Lake Superior, Wisconsin to central Arizona.* J. Geophys. Res. *71*, 5933–5941.

SACKS, I. S., SNOKE, J. A. and HUSEBYE, E. S. (1979), *Lithosphere thickness beneath the Baltic Shield.* Tectonophysics *56*, 101–110.

SELLEVOLL, M. A. and PENTTILÄ, E. (1964), *Seismic refraction measurements of crustal structure in Northern Scandinavia.* Arbok Univ. Bergen, Mat.-Nat. Ser. *9*, 10 pp.

SELLEVOLL, M. A. and POMEROY, P. A. (1968), *A travel-time study for Fennoscandia.* Arbok Univ. Bergen, Mat.-Nat. Ser. *9*, 29 pp.

SELLEVOLL, M. A. and WARRICK, R. (1971), *A refraction study of the crustal structure in southern Norway.* Bull. Seism. Soc. Am. *61*, 457–471.

SIPKIN, S. A. and JORDAN, T. H. (1975), *Lateral heterogeneity of the upper mantle determined from the travel times of ScS.* J. Geophys. Res. *80*, 1474–1484.

SIPKIN, S. A. and JORDAN, T. H. (1976), *Lateral heterogeneity of the upper mantle determined from the travel times of multiple ScS.* J. Geophys. Res. *81*, 6307–6320.

SIPKIN, S. A. and JORDAN, T. H. (1980), *Multiple ScS travel times in the western Pacific: implications for mantle heterogeneity.* J. Geophys. Res. *85*, 853–861.

UTSU, T. (1966), *Variations in spectra of P waves recorded at Canadian Arctic seismograph stations.* Canadian J. Earth Sci. *3*, 597–621.

VINNIK, L. P. (1977), *Detection of waves converted from P to SV in the mantle.* Phys. Earth Planet. Int. *15*, 39–45.

VINNIK, L. P., AVETISJAN, R. A. and MIKHAILOVA, N. G. (1983), *Heterogeneities in the mantle transition zone from observations of P-to-SV converted waves.* Phys. Earth Planet. Int. *33*, 149–163.

VOGEL, A. and LUND, C. E. (1971), *Profile section 2–3.* In Proc. Colloquium on Deep Seismic Sounding in Northern Europe, Uppsala, December 1 and 2, 1969, (ed. A. Vogel), (University of Uppsala) pp. 62–75.

WAHLSTRÖM, R. (1975), *Seismic wave velocities in the Swedish crust.* Pure and Appl. Geophys. *113*, 673–682.

WALCOTT, R. I. (1970), *Flexural rigidity, thickness, and viscosity of the lithosphere.* J. Geophys. Res. *75*, 3941–3954.

WARREN, D. H., HEALY, J. H., HOFFMANN, J. C., KEMPE, R., RAUULA, S. and STUART, D. J. (1968), *Project Early Rise, travel times and amplitudes.* U.S. Geological Survey Open-file Report, Menlo Park, California, 150 pp.

(Received 10th June, 1986, revised 30th June, 1986, accepted 1st July, 1986)

PAGEOPH, Vol. 125, Nos. 2/3 (1987) 0033-4553/87/030241-14$1.50+0.20/0

Properties of the Lithosphere-Asthenosphere System in Europe with a View Toward Earth Conductivity

G. CALCAGNILE[1,2] and G. F. PANZA[3,4]

Abstract—Over the past 20 years the study of *P*- and *S*-wave velocities in the upper mantle of the Mediterranean area and continental Europe has been the subject of intensive research work. We present a summary of results based on the inversion of available surface-wave dispersion data and *P*-wave travel time observations. For areas characterized by different tectonic settings and very large lateral variations, a discussion is made about structural models based on seismological, geothermal and electrical conductivity data.

Key words: Europe, upper Mantle, *P*- *S*-wave Velocities, heat flow, electrical conductivity.

1. *Introduction*

The knowledge of the Earth's composition and the physical properties of the Earth's interior can be improved by piecing together the evidence from many different sources. Though the most important part of information undoubtedly comes from seismological investigations, electromagnetic studies are becoming increasingly a valuable independent method for probing the Earth's upper mantle. Currently methods for measuring the electrical conductivity, in the relevant range of depths, are beginning to reach a level of resolution that puts interesting limits on speculation.

The quality and the amount of data available in the European area make timely a review aiming at the delineation of possible directions for future investigations.

2. *Seismological Models*

The results of the systematic inversion of the available collection of surface wave dispersion data pertinent to studies of the upper mantle are shown in Figure 1, a

[1] Dipartimento di Geologia e Geofisica, Università di Bari, Bari, Italy.
[2] Osservatorio di Geofisica e Fisica Cosmica, Università di Bari, Bari, Italy
[3] Istituto di Geodesia e Geofisica, Università di Trieste, Trieste, Italy.
[4] Scuola Internazionale Superiore di Studi Avanzati, Trieste, Italy 1.

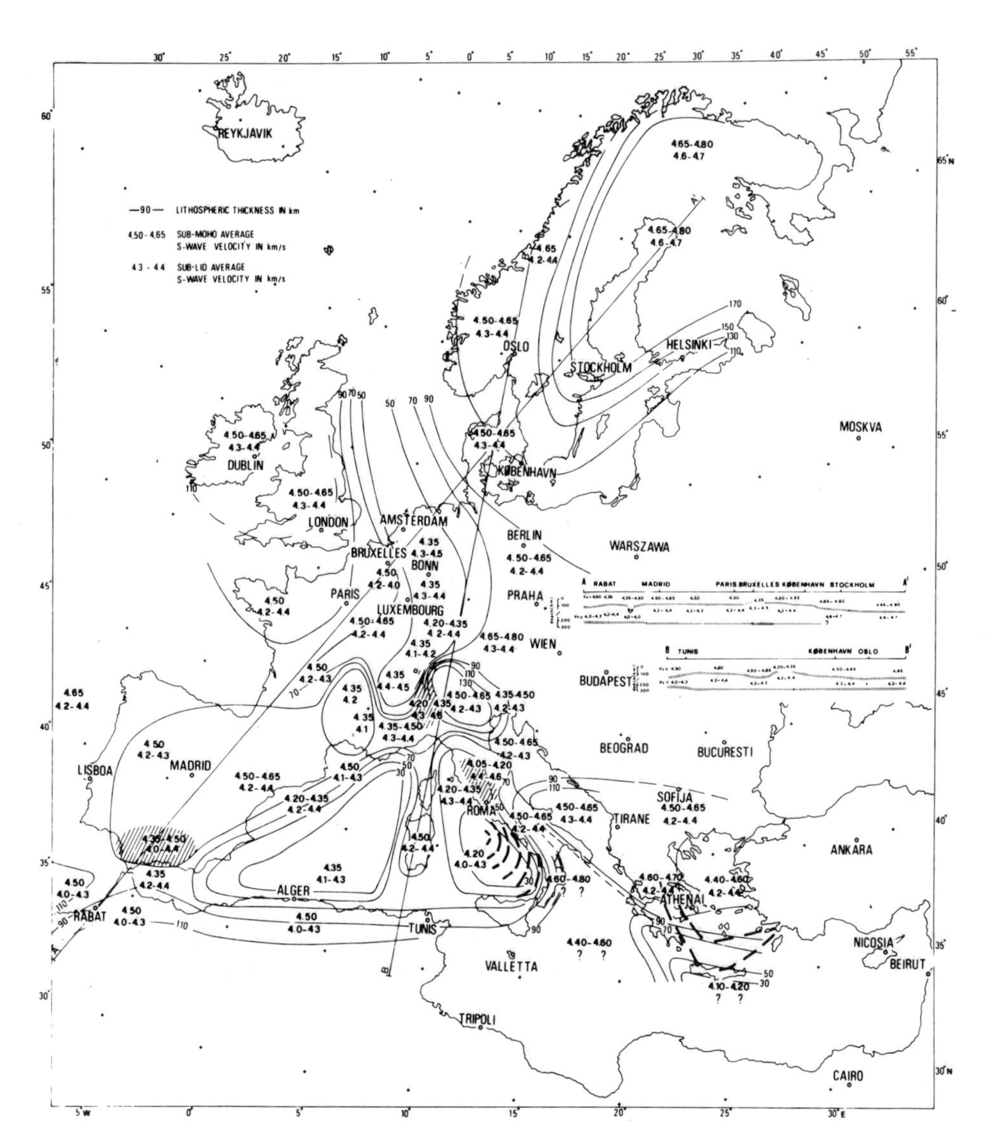

Figure 1

Map of lithospheric thickness (in km) in the European-Mediterranean region, deduced from the regional dispersion analysis of seismic surface waves (after PANZA *et al.*, 1980; CALCAGNILE, 1982; CALCAGNILE *et al.*, 1982; PANZA, 1985). Representative shear-wave velocities are given for the lower lithosphere or lid (row of upper numbers) and for the upper asthenosphere (row of lower numbers). The three shaded areas indicate the possible presence of 'lithospheric roots' to depth of about 200 km, while the thick dashed lines define the areas of intermediate and deep focus earthquakes in the Tyrrhenian and Aegean sea.

schematic map of the elastic properties of the lithosphere-asthenosphere system (PANZA *et al.*, 1980; CALCAGNILE, 1982; CALCAGNILE *et al.*, 1982; PANZA, 1985). When interpreting this map, it must be realized that it represents only an approximate solution of the inverse problem and is subject to inherent uncertainties (e.g., 15 to 20 km in lithospheric thickness). The first row of each set of numerals refers to the possible range of the average *S*-wave velocity, v_S, from the Moho to the depth indicated by isolines (lower lithosphere or lid), while the second row describes the possible range of the average *S*-wave velocity below that depth (upper asthenosphere or low velocity channel). Only in areas where a marked contrast exists between the values in the two rows, can the isolines be considered as representative of the lithosphere thickness. From the figure, the presence of strong lateral variations, both in thickness and in *S*-wave velocity, is evident.

A particularly striking structure delineated in the map is the lithospheric thinning in correspondence of the central European rift system, which extends from the Western Alps (Golfe du Lion) to the North Sea; the lithosphere has an average thickness of about 50 km and markedly lowered *S*-wave velocities. On the other side a very thick lithosphere is found beneath the Baltic Shield—it exceeds 150 km as an average—characterized by small, if any, contrast between shear wave velocities of the layers separated by the isolines.

Preliminary results of the new data inversion—very long period fundamental and higher modes dispersion values—point out a significant lateral heterogeneity not only in the uppermost 200 km but also down to depths of 400–500 km, where shield roots could be found, according to possible models for the central part of the shield (CALCAGNILE, 1984). Were these models confirmed, the classic idea of plate tectonic should be reviewed, with a consequent corroboration of the concept of tectosphere (JORDAN, 1975).

Deep investigations, to more than 400 km, have been carried out in the area by using *P*-wave travel time data but on a larger scale. Velocity-depth distributions down to several hundred kilometers beneath Europe have been obtained by several authors using well-known techniques. These distributions agree with the aforementioned dispersion results (Figures 2a, b). Lateral variations in the velocity distribution versus depth are present in the upper mantle beneath Europe, both in the uppermost 100–300 km and to depths of about 600–800 km. At depths of 400–500 km a velocity discontinuity is found for *P*-wave; this discontinuity raises at 320–360 km in some areas. A second discontinuity in *P*-wave velocity, v_P, is present at a depth of about 600–700 km, where v_P, is about 10.5–10.7 km/s. In the depth range 100–300 km *P*-wave velocity in the East European Platform (EEP) is larger than in central Europe. Finally, the presence of significant lateral variations in the elastic properties of the upper mantle is clearly confirmed by *P*-wave travel time residuals (CALCAGNILE and SCARPA, 1985; BABUSKA *et al.*, 1984).

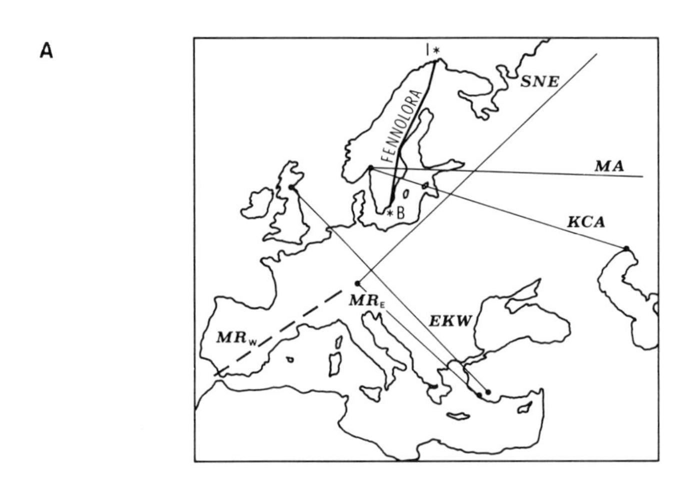

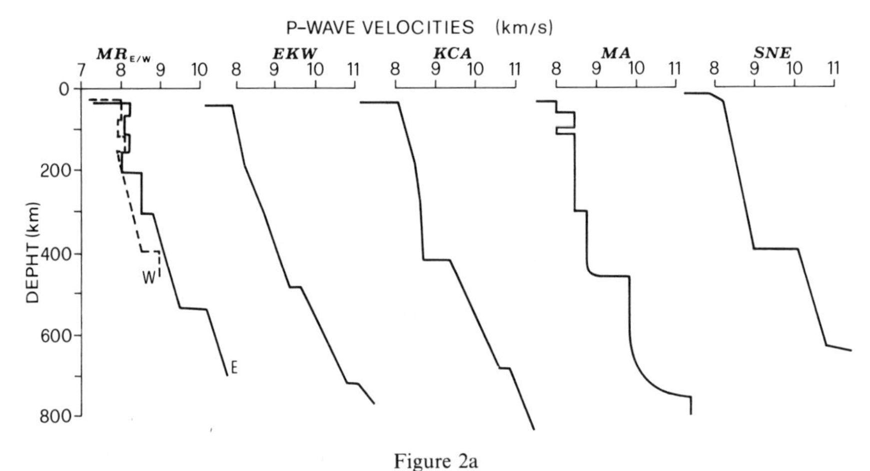

Figure 2a
P-wave velocity-depth functions in the European area (modified after CALCAGNILE and SCARPA, 1985).

3. Conductivity Models

There is a variety of different electromagnetic methods which sometimes appear difficult to compare with one another. Different data sets, as well as different mathematical tools applied to the same data sets, are often found to provide significantly different results for the same geographical location. The range of errors and uncertainties from such analyses often appear large and not adequately understood. This makes it sometimes difficult to interpret the different results, nevertheless the induction methods are very sensitive to discover a conducting

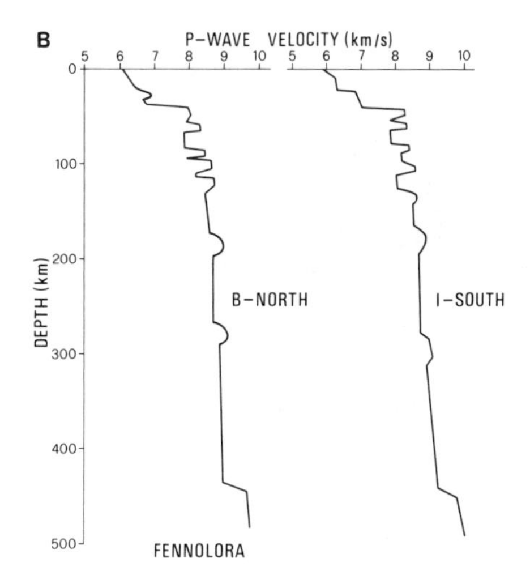

Figure 2b
P-wave velocity-depth function in Fennoscandia (redrawn from GUGGISBERG *et al.*, 1984).

anomaly. Concerning the depth of the conducting layers, the different inversion techniques generally give the same values.

An example is given in Figures 3a and 3b, taken from BERDICHEVSKY and ZHDANOV (1984), wherein geoelectrical models for the Baikal region are shown. The models were suggested by various authors using the same magnetotelluric data. The differences in the models are accounted for, according to the quoted authors, by differing degrees of confidence in the results of formal, i.e., one dimensional, interpretation additional to the subjectivity in neglecting the distortions. A dissolution of this difficulty entails abandoning interpretation of individual resistivity curves and a resort to statistical analysis, but this may cause loss of lateral variation detectability.

However, in spite of the aforementioned difficulties, some reliable conclusion already has been pointed out. Precambrian shields, old stable platforms, old oceans appear to be underlain by a thick low conductivity layer. High conductivity layers appear to occur at several hundred kilometers depth underneath shields, even if significant differences among them are reported, e.g., for the Australian shield the conductivity increase is approximately 500 km deep (LILLEY *et al.*, 1981) whereas for the Baltic shield it is in the range 100–300 km (e.g., JONES, 1983; HJELT, 1984). The high conductivity layers rise to about 100 km underneath tectonically perturbed continental areas and to a few (to several) tens of kilometers beneath rifts and grabens.

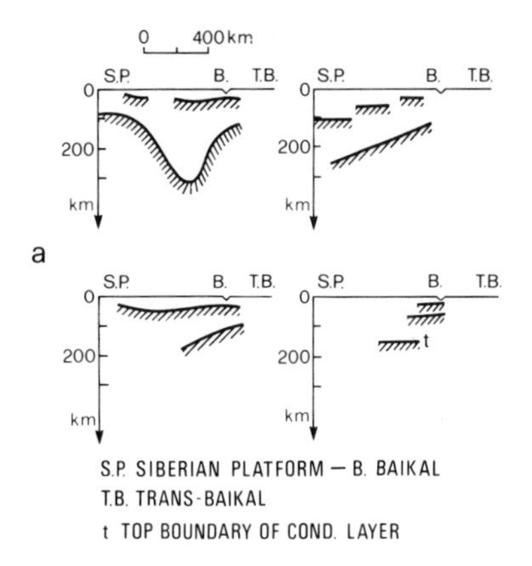

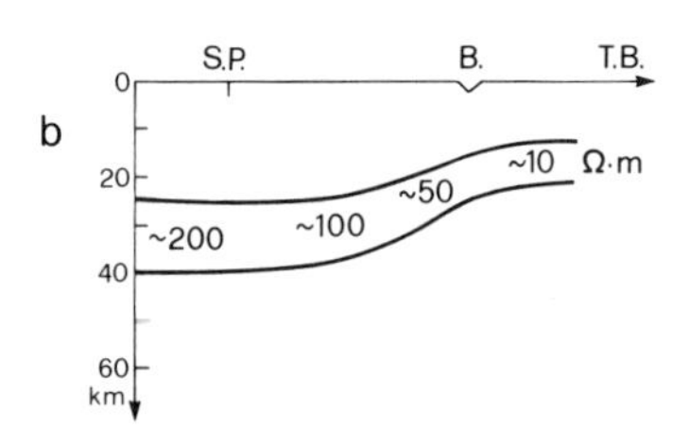

Figure 3
(a) Different geoelectrical models for the Baikal area, based on the same magnetotelluric data; (b) re-interpretation of the data used to construct Figure 3a by BERDICHEVSKY and ZHDANOV (1984).

4. *Relations between Seismological, Geothermal and Electrical Conductivity Models*

The shear-wave velocity, like electrical conductivity, is rather sensitive to thermal conditions. It is a good indicator of temperature in the upper mantle; in fact, the closer the temperature in the mantle is to the melting point, the lower is the *S*-wave velocity, which vanishes at temperatures above the melting point. Actually, anisotropy effects cannot be disregarded but they are believed to be small compared to the more important thermal effects. What is persuasive about this—with respect to both velocity anisotropy as well as chemical inhomogeneity—is a correlation of surface heat flow with the absence of the velocity anomaly (low velocity zone) or with the depth to the low velocity zone, if it is present. Further circumstantial evidence favouring this thermal model is the relationship between the geotherm for the younger stable regions, the melting curve for a peridotitic mantle in the presence of an excess of water (Figure 4) and the depth at which the low velocity zone begins; a similar relationship exists for oceans, tectonic regions and shields.

To corroborate with further evidence of the aforementioned thermal model, an

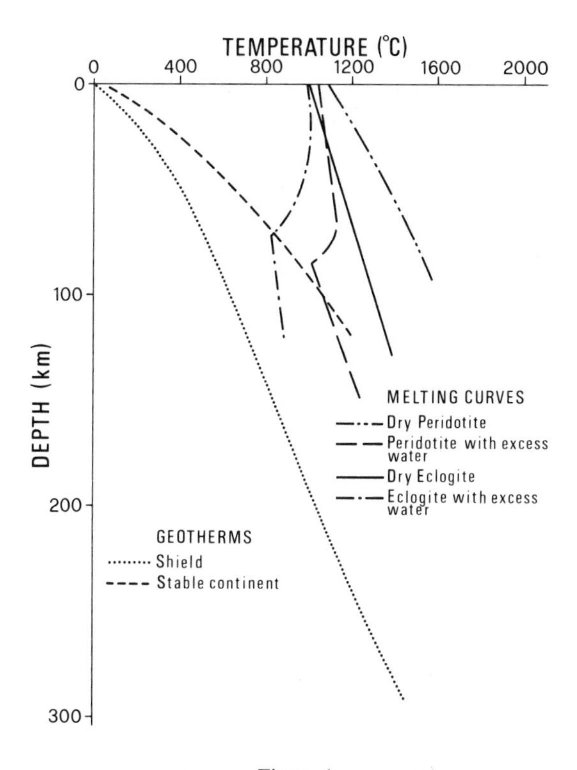

Figure 4
Geotherms for shields and stable continental areas, melting curves for peridotite and eclogite, dry and with excess water (redrawn from KNOPOFF, 1983).

attempt has been made to establish a relation between the elastic properties of the lithosphere-asthenosphere system and surface heat flow for the Mediterranean-European area (CALCAGNILE, 1983). A few years ago CHAPMAN *et al.* (1979) examined this relation for the European area on the basis of geothermal models, by computing geotherm-solidus crossing and checking the results with a few values of lithospheric thickness, h_{LIT}, available from seismological studies at that time, obtaining qualitative agreement. In the detailed analysis we discuss here, the adopted regionalization, consisting of about 50 subregions, is that used for the compilation of the map shown in Figure 1. A possible inverse linear relationship, in semilogarithmic scale, is found between lithospheric thickness and heat flow (Figure 5). If the lid shear-wave velocity is used as a weight in computing a 'corrected' lithospheric thickness, the result does not change significantly. Note that the subregion corresponding to the area of Crete, $h_{LIT} = 40$ km, is characterized by 'anomalous' (low) heat flow that could be explained by the blanketing effect of the thick sedimentary layer associated with the Hellenic Trench subduction zone. Also the Central Alps subregion, $h_{LIT} = 130$ km, is char-

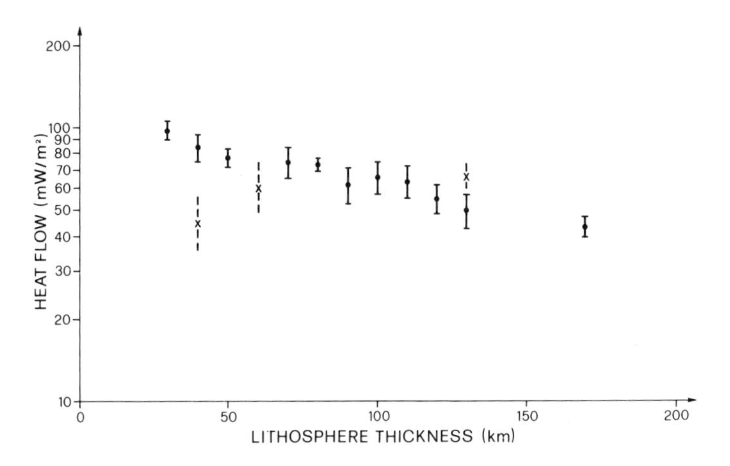

Figure 5
Surface heat flow versus lithospheric thickness for the European area. Data corresponding to the 'anomalous' regions described in the text are indicated by crosses. Error bars for lithospheric thickness are not reported since they can all be considered about 20 km, for heat flow bars length is a measure of the standard deviation.

acterized by 'anomalous' (high) heat flow, that however can be taken back to 'normal' values once it is corrected, about 30% reduction, for uplifting and denudation. The last 'anomalous' subregion, Central Apennines $h_{\mathrm{LIT}} = 60$ km, has low heat flow, qualitatively accounted for if one observes that this subregion has 'cool' lithospheric roots (PANZA *et al.*, 1980; CALCAGNILE and PANZA, 1981).

Further studies are necessary, however the previous results, if substantiated, are in favour of the thermal model when interpreting shear-wave velocity results.

Let us now consider the electrical conductivity of the crust and upper mantle. Conductivity data permit the detection of the presence of three layers with high conductivity in the uppermost 600 km of the Earth. ADAM (1978) has proposed quite convincing relations between regional heat flow and the depth of the crustal and upper mantle conducting layers (Figure 6). The more shallow ones are well correlated with heat flow values, while for the deepest layer, the scatter in the data makes the correlation rather weak.

The existence of the relationships shown in Figures 5 and 6 suggests that on a global scale the velocity of S-wave and the electrical conductivity in the Earth's mantle are both controlled by thermal conditions. This can be understood considering the geotherms reported in Figure 4. For instance, the geotherm for stable continental regions crosses the melting curve for a peridotitic mantle in the presence of an excess of water at a depth of about 100 km. A depth at which the low velocity zone in those regions begins as well as the intermediate conductive layer. On the other end the geotherm for shield areas is distant from the melting curves and

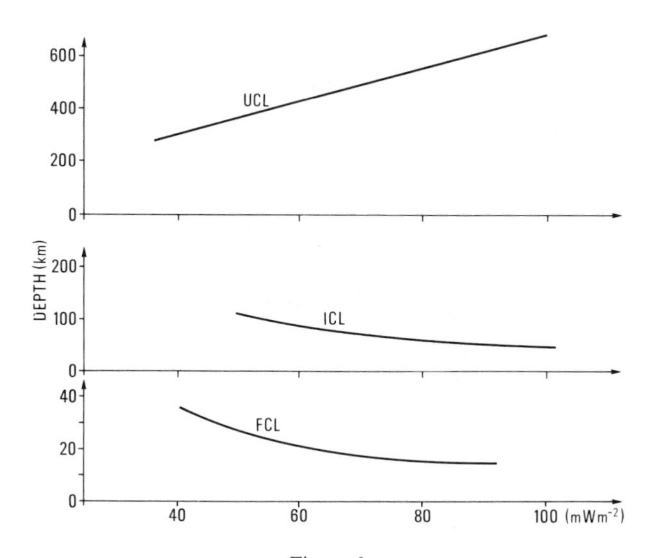

Figure 6
Connection between the regional heat flow and the depth of the conducting layers (simplified after ADAM, 1978). FCL indicates the conducting layer in the Earth's crust, ICL indicates the intermediate conducting layer (in the asthenosphere), UCL indicates the ultimate conducting layer.

this is in suitable agreement with the *S*-wave velocity distribution in the upper mantle of these areas, as well as with the absence of a pronounced conductive layer.

All these evidences may encourage a proposal that the seismic low velocity zone in the upper mantle coincides in fact with the intermediate conducting layer (ADAM, 1978). The amount of data available for parts of the European area allows the testing of this idea more in detail.

5. Structural Models Consistent with Seismological and Electromagnetic Data for Selected Areas

In Europe, a region fairly well covered by deep electromagnetic and seismological investigations is Fennoscandia. Many data have been collected in the area, some still in the stage of processing or interpretation.

A summary of electromagnetic results as well as conductivity models is given by HJELT (1984) (Figure 7). Namely, in northwestern Fennoscandia the top of the conductive layer is about 110–120 km deep for Kevo (KEV), 150–180 km deep for Kiruna (KIR) and about 200 km deep for Nattavaara (NAT), located almost on the seismic profile Fennolora. In north-central Finland for Sauvamaki (SAU) a conductive layer in the uppermost 150–200 km is not allowed; for the profile Sveka, in correspondence of KUK, a resistivity curve has been obtained that fits the 'normal' resistivity curve, that is the East European Platform resistivity curve (EEP), (VANYAN *et al.*, 1977). For the Kola peninsula no indication of any asthenospheric conductor is available, while for Karelia it seems roundly 150–200 km deep; however,

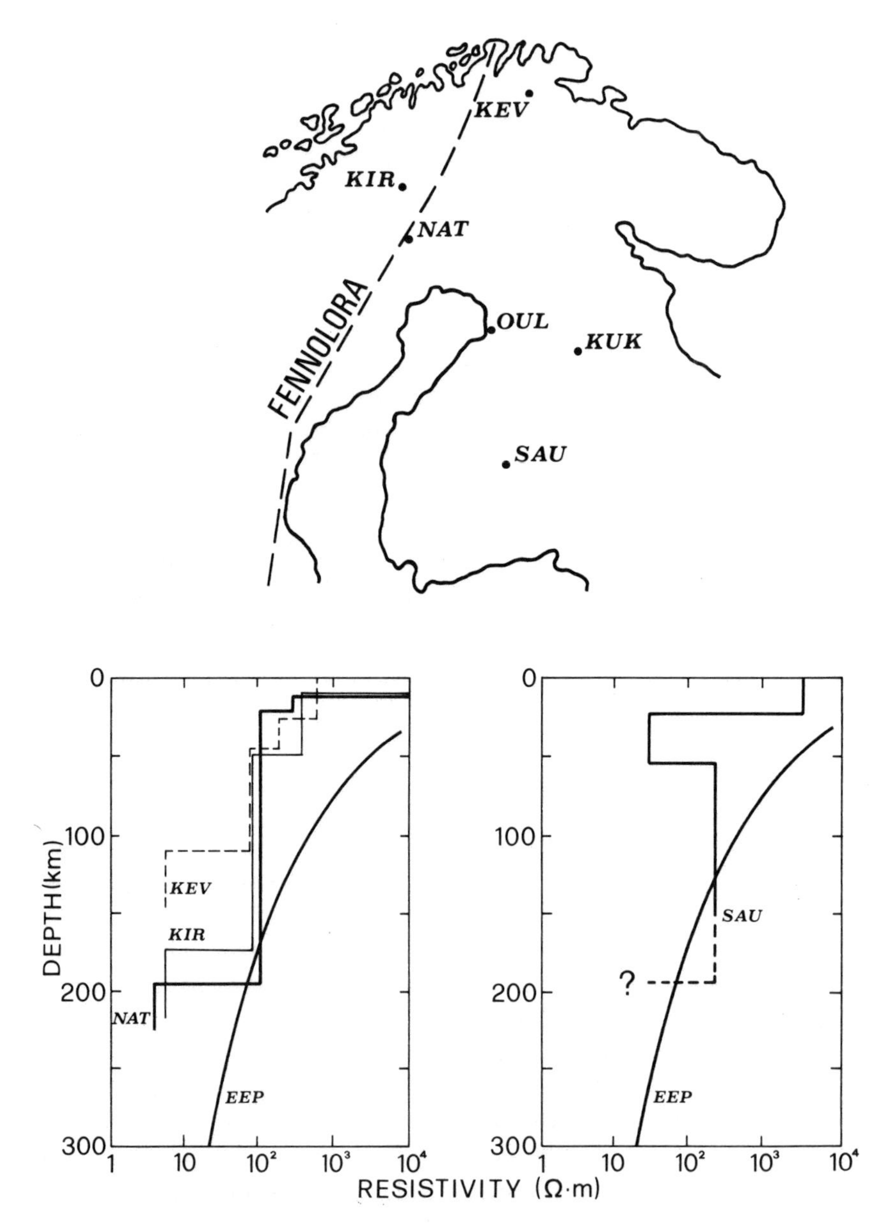

Figure 7
Conductivity models for Fennoscandia (after HJELT, 1984)

there is considerable distortion in the data.

If the hypothesis that a low velocity of elastic waves and low resistivity have the same physical origin in the thermal state of the upper mantle is accepted, the models of Figure 7 are in good agreement with the elastic properties of the lithosphere-asthenosphere system (Figure 1) and with the recent results obtained from the study of higher modes of Rayleigh waves dispersion (CALCAGNILE, 1984). In fact, it seems plausible that there is a decrease of electrical resistivity at depths where the shear wave velocities are low.

To further investigate this possibility, we have carried out a new inversion of the seismological data mindful of, particularly for the low velocity zone, the geometry obtained from electromagnetic data analysis.

The results of the inversion give for Kevo two possible groups of models. In the first, the top of the low velocity channel is about 110 km deep and the channel thickness is roughly 100–150 km; v_S is in the range 4.65–4.80 km/s, in the lower lithosphere or lid and in the range 4.35–4.50 km/s in the low velocity channel; the *S*-wave velocity in the layer below the channel—the 'subchannel'—is in the range 4.5–4.7 km/s. In the second, the top of the channel is 150–160 km deep and the subchannel *S*-wave velocity is only about 4.5 km/s, thus implying a rather thick (about 150 km) and deep low velocity zone. Based upon the models in Figure 7 the resistivity in the lid is about 95 Ω m while the one in the channel is between 5–10 Ω m.

In the model for Kiruna, the low velocity channel begins at a depth of about 160 km and is 100–150 km thick; v_S is about 4.65 km/s in the lid and about 4.35 km/s in the channel; in the subchannel v_S is in the range 4.5–4.9 km/s. In this case the resistivity in the lid is about 95 Ω m while the one in the channel is between 2–10 Ω m.

Corresponding to Nattavaara, the model is as follows: the low velocity channel begins at about 200 km of depth and is only 50 km thick; the *S*-wave valocity in the lid is about 4.65 km/s, in the channel is in the range 4.35–4.50 km/s and for the subchannel is between 4.6–4.8 km/s. For this area a second type of structural model is admitted, wherein the low velocity channel, about 80 km thick, begins at a depth of roundly 250 km. In this second type of models the depth of the low velocity zone coincides with that found by SACKS *et al.* (1979) using a totally independent method. The resistivity values for the lid and low velocity layer are respectively around 100 Ω m and 5 Ω m.

For Sauvamaki a set of models gives a low velocity channel only about 50 km thick starting at a depth of approximately 200 km; the *S*-wave velocity in the lid is around 4.65 km/s, in the channel around 4.35 km/s, in the subchannel around 4.9 km/s. Another set of models is consistent with a low velocity channel starting about 250–320 km deep. If the *S*-wave velocity in the subchannel is about 4.7 km/s, the channel is more shallow and about 100 km thick, however if the *S*-wave velocity in the subchannel is about 4.5 km/s, the channel is deeper and only about

50 km thick. The resistivity in the lid is about 200 Ω m, i.e. larger than in the previous cases, while the low velocity layer resistivity is again in the range 2–10 Ω m.

Another area reasonably covered by seismological and electromagnetic measurements is the Rhine Graben. In the summary of results obtained for this area (GREGORI and LANZEROTTI, 1982) two models are given (Figure 8): model *A*, with a conductive layer in the depth range 25–45 km, is consistent with surface waves dispersion data in the sense that low *S*-wave velocity corresponds to low resistivity. Indeed, in the layer with a low resistivity of about 50 Ω m also the *S*-wave velocity is rather low, in the range 4.10–4.25 km/s. Beneath this layer, i.e., for depths in the range 45–85 km, the resistivity is around 2000 Ω m and the *S*-wave velocity is around 4.4 km/s. At greater depths—approaching 150 km of depth, the *S*-wave velocity is again rather low (around 4.1 km/s) and the resistivity becomes as low as 25 Ω m (REITMAYR, 1975).

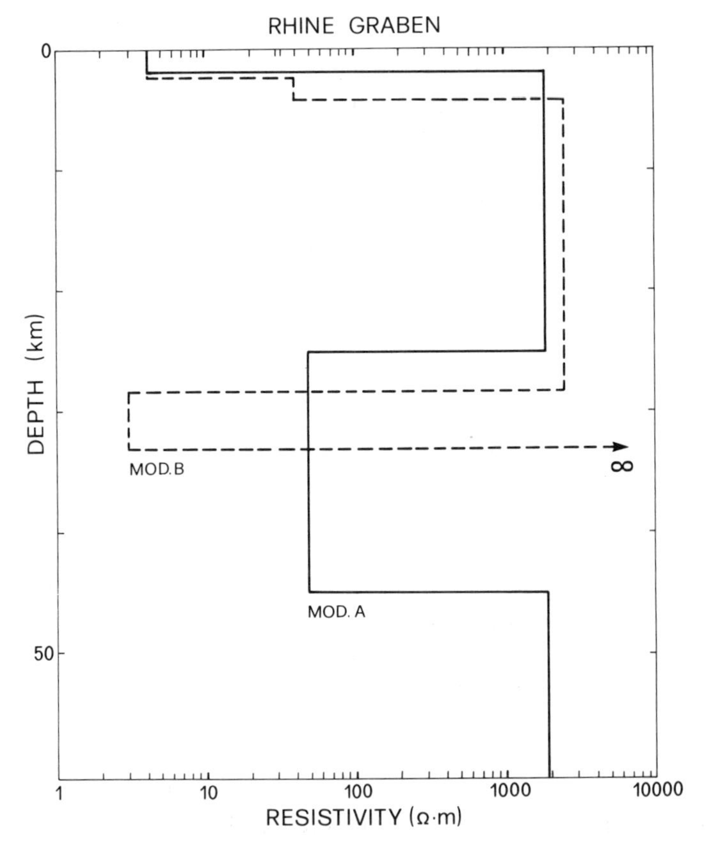

Figure 8
Conductivity models for the Rhine Graben (after GREGORI and LANZEROTTI, 1982)

6. Conclusions

The great importance of combining seismological and electromagnetic data for upper mantle studies has been pointed out. In general it is possible to conclude that to a decrease in resistivity occurring in some depth range a decrease in shear-wave velocity can be associated; the evidence concerning *P*-wave velocity is markedly less convincing (ADAM, 1980; RYABOY and D'ERLYATKO, 1984). This constitutes a rather important argument in favour of the presence of processes of partial melting in the upper mantle; in fact partial melt must affect *S*-wave propagation and resistivity decidedly more than *P*-wave propagation. The great importance of these arguments for the understanding of the physics of the Earth's interior requires an intensification of combined researches on the elastic and electrical properties of the Earth mantle, mainly with a view toward the regionalization of the data gathered over tectonically heterogeneous areas.

Usually the crossing between geotherms and melting curves is used for determining the depth at which the asthenosphere begins. It must be observed however that corresponding to the solidus there is only the beginning of fusion. The amount of liquid phase and the conductivity increase as the temperature increases whereas the shear-wave velocity decreases. Therefore the combined use of resistivity and *S*-wave velocity values can allow estimation of the temperature in the upper mantle. In particular, if hypotheses are made concerning the type of rocks present in the asthenosphere and on their volatile content, it is possible to estimate both the temperature and the amount of liquid phase from conductivity values. The models obtained in this way can be tested against seismological data. By iterating such a procedure it could be possible to improve our knowledge of the physical and chemical properties of the Earth's mantle.

Acknowledgments

We are grateful to Prof. A. Adam for critically reading the manuscript and for pointing out the Russian reference.

We are grateful to Mrs I. Galante for the typing of the manuscript and to Mr G. Cavicchi for the drawings.

Research was carried out with financial support of the Italian Ministry of Education.

REFERENCES

ADAM, A. (1978), *Geothermal effects in the formation of electrically conducting zones and temperature distribution in the Earth.* Phys. Earth Planet. Int. *17*, P21–P28 (Letter section).

ADAM, A. (1980), *Relation of mantle conductivity to physical conditions in the asthenosphere.* Geophys. Surv. *4*, 43–55.

BABUSKA, V., PLOMEROVA, J. and SILENY, J. (1984), *Large-scale oriented structures in the subcrustal lithosphere of Central Europe.* Annales Geophysicae *2*, 649–662.

BERDICHEVSKY, M. N. and ZHDANOV, M. S. *Advanced theory of deep geomagnetic sounding. Methods in Geochemistry and Geophysics 19.* (Edited by Keller G. V.) (Elsevier, Amsterdam, 1984).

CALCAGNILE, G. (1982), *The lithosphere-asthenosphere system in Fennoscandia.* Tectonophysics *90*, 19–35.

CALCAGNILE, G. (1983), *Struttura del sistema litosfera-astenosfera e flusso di calore nell'area europea-mediterranea.* Atti *2°* Convegno Nazionale GNGTS, CNR, Rome, 323–333.

CALCAGNILE, G. (1984), *Nuove evidenze sismologiche sulla struttura degli scudi.* (Abstract) Atti *3°* Convegno Nazionale GNGTS, CNR, Rome, 115.

CALCAGNILE, G., D'INGEO, F., FARRUGIA, P., and PANZA, G. F. (1982), *The lithosphere in the central-eastern Mediterranean area.* Pure Appl. Geophys. *120*, 389–406.

CALCAGNILE, G. and PANZA, G. F. (1981), *The main characteristics of the lithosphere-asthenosphere system in Italy and surrounding regions.* Pure Appl. Geophys. *119*, 865–879.

CALCAGNILE, G. and SCARPA, R. (1985), *Deep structure of the European-Mediterranean area from seismological data.* Tectonophysics *118*, 93–111.

CHAPMAN, D. S., POLLACK, H. N. and CERMAK, V, *Global heat flow with special reference to the region of Europe.* In *Terrestrial heat flow in Europe.* (Edited by Cermak V. and Rybach L.) (Springer-Verlag, Berlin, 1979), 41–48.

GREGORI, G. P. and LANZEROTTI, L. J. (1982), *Electrical conductivity structure in the lower crust.* Geophys. Surv. *4*, 467–499.

GUGGISBERG, B., ANSORGE, J. and MUELLER, S., *Structure of the upper mantle under southern Scandinavia from Fennolora data.* In Proceed. *First Workshop European Geotraverse.* (Edited by Galson D. A. and Mueller S.) (European Science Foundation, Strasbourg, 1984), 49–52.

HJELT, S. E. (1984), *Deep electromagnetic studies of the Baltic Shield.* J. Geophys. *55*, 144–152.

JONES, A. G. (1983), *The electrical structure of the lithosphere and asthenosphere beneath the Fennoscandian shield.* J. Geomag. Geoelectr. *35*, 811–827.

JORDAN, T. H. (1975), *Lateral heterogeneity and mantle dynamics.* Nature *257*, 745–750.

KNOPOFF, L. (1983), *The thickness of the lithosphere from the dispersion of surface waves.* Geophys. J. R. Astr. Soc. *74*, 55–81.

LILLEY, F. E. M., WOODS, D. V. and SLOANE, M. N. (1981), *Electrical conductivity profiles and implications for the absence or presence of partial melting beneath central and southeast Australia.* Phys. Earth Planet. Int. *25*, 419–428.

PANZA, G. F., *Lateral variations in the lithosphere in correspondence of the southern segment of EGT,* In *Second EGT Workshop—The Southern Segment* (Edited by Galson, D. A. and Mueller, S.) (European Science Foundation, Strasbourg, 1985), 47–51.

PANZA, G. F., CALCAGNILE, G., SCANDONE, P. and MUELLER, S. (1980), *La struttura profonda dell'area mediterranea.* Le Scienze *141*, 60–71.

REITMAYR, G. (1975), *An anomaly of the upper mantle below the Rhine Graben, studied by the inductive response of natural electromagnetic fields.* J. Geophys. *41*, 651–658.

RYABOY, V. Z. and D'ERLYATKO, E. K. (1984), *Horizontal inhomogeneities of the asthenospheric layer of the upper mantle in Northern Eurasia on the basis of deep seismic and geoelectric investigations* (in russian), Dokl. A. N. SSSR *227*, 577–581.

SACKS, I. S., SNOKE, J. A. and HUSEBYE, E. S. (1979), *Lithosphere thickness beneath the Baltic Shield,* Tectonophysics *56*, 101–110.

VANYAN, L. L., BERDICHEVSKY, M. N., FAINBERG, E. B. and FISKINA, M. V. (1977), *The study of the asthenosphere of the East European Platform by electromagnetic sounding.* Phys. Earth Planet. Int. *14*, P1–P2 (Letter section).

(Received 18th March, 1986; revised 3rd June, 1986, accepted 4th June, 1986)

PAGEOPH, Vol. 125, Nos. 2/3 (1987)

0033-4553/87/030255-30$1.50+0.20/0

Temperature Profiles in the Earth of Importance to Deep Electrical Conductivity Models

VLADIMÍR ČERMÁK[1] and MARCELA LAŠTOVIČKOVÁ[1]

Abstract—Deep in the Earth, the electrical conductivity of geological material is extremely dependent on temperature. The knowledge of temperature is thus essential for any interpretation of magnetotelluric data in projecting lithospheric structural models. The measured values of the terrestrial heat flow, radiogenic heat production, and thermal conductivity of rocks allow the extrapolation of surface observations to a greater depth and the calculation of the temperature field within the lithosphere. Various methods of deep temperature calculations are presented and discussed. Characteristic geotherms are proposed for major tectonic provinces of Europe and it is shown that the existing temperatures on the crust-upper mantle boundary may vary in a broad interval of 350–1,000°C. The present work is completed with a survey of the temperature dependence of electrical conductivity for selected crustal and upper mantle rocks within the interval 200–1,000°C. It is shown how the knowledge of the temperature field can be used in the evaluation of the deep electrical conductivity pattern by converting the conductivity-versus-temperature data into the conductivity-versus-depth data.

Key words: Heat flow; crustal and upper mantle temperature; electrical conductivity (temperature dependence of); lithospheric conductivity pattern.

1. *Introduction*

The electrical conductivity σ of minerals and rocks is an extremely variable property and depends on a number of factors, e.g., porosity, moisture content, its nature and chemical composition, weathering, temperature, pressure, etc. At the near-surface conditions, the state and content of water in the material fully controls the rock conductivity. Slight changes in the saturation degree result in the wide variation of conductivity and thus disqualify the laboratory measurements for any practical use of defining the characteristic electrical constants of rocks on a regional scale. However, deeper in the Earth's crust and upper mantle, when the pores and micro-cracks are closed and when the geological structure is more uniform, it is the pressure and temperature that govern the physical parameters of rocks, among them also their electrical conductivity. Electrical conductivity is definitely the most temperature dependent parameter, varying up to 7–10 orders of magnitude for the temperature conditions existing in the uppermost 50–100 km.

[1] Geophysical Institute, Czechosl. Acad. Sci., 141–31 Praha, Czechoslovakia.

Basically, any temperature model of the continental lithosphere requires the solution of the heat conduction equation with the corresponding boundary conditions. For the extrapolation of the measured near-surface temperatures to a greater depth, the knowledge of the terrestrial heat flow, together with certain assumptions on the deep structure, usually based on the explosion seismology, and on the depth distribution of the heat sources and thermal conductivity are necessary.

There is a relationship between the surface heat flow and the tectonic age; it means that specific tectonic regions are characterized by specific deep temperatures. Following this relationship, characteristic geotherms were proposed for all major tectonic provinces in Europe and it was shown that the existing crustal temperatures might vary in a broad interval of 350–1,000°C. As the geothermal data, interpreted for this purpose, covered the entire geological span, the results obtained may have universal validity and can be applied to the general continental lithosphere.

2. *Temperature Within the Earth*

The temperature on the Earth's surface is determined by the position of the Earth in the solar system and is controlled by the solar radiation. Below the surface the temperature increases, the rate of the increase being referred to as the geothermal gradient. While the time and regional variations in the surface temperatures are entirely governed by the climatological conditions, below the depth of only a few metres, the diurnal and annual oscillations fade out and the geothermal gradient reflects the crustal structure. The range of the values of the observed near-surface geothermal gradients is 10 to 70 $mK.m^{-1}$ with typical values between 20–30 $mK.m^{-1}$. Direct temperature measurements can be performed in deep mines or holes and are thus limited by their maximum penetration, i.e., by a depth of few kilometres. Any deeper temperature estimates in the crust and upper mantle thus involve downward extrapolation of the near-surface observations.

The internal temperature field of the Earth can be described by the solution of the heat conduction equation (see e.g., CARSLAW and JAEGER, 1959):

$$\frac{\partial T}{\partial t} = \frac{1}{c\rho}[\nabla(k\nabla T) + A] - v\nabla T, \tag{1}$$

where ρ is the density, $kg.m^{-3}$, c—the heat capacity, $J.kg^{-1}\ K^{-1}$, k—the thermal conductivity, $W.m^{-1}\ K^{-1}$, A—heat generation, $\mu W.m^{-3}$, v—the velocity, $m.s^{-1}$. The term involving thermal conductivity is a measure of the heat flux in and out of a unit volume, A is the rate at which heat is produced within the unit volume. The term involving the velocity is the time rate of the temperature change resulting from the advection of heat into the unit volume. This term couples the thermal equation to the equation of motion. All individual parameters ρ, c, k, A are generally the function of position (x, y, z) and the solution of (1) further requires the knowledge of the initial and boundary conditions.

It is obvious that equation (1) cannot be solved generally, and depending on the scale, time span, heat transfer mechanism and other specific features, certain simplifications are necessary. If we limit ourselves to the investigation of the temperature field within the continental lithosphere, the convective term (v grad T) can be neglected. However, the problem can be more complicated below the lithosphere if large scale horizontal convection takes place in the viscous asthenoshpere, the convective term may eventually dominate the conductive transport of heat. The curvature of the Earth's surface can also be neglected, the geometry then becomes simpler and eventually a one-dimensional solution can well describe the main features of downward temperature extrapolation.

In the major part of the continental lithosphere, the present temperature field is stationary, i.e., $dT/dt = 0$. This condition is satisfied if the age of the area investigated (Δt) is greater than the time of thermal relaxation. It requires $\Delta t \sim h^2/\mathscr{H}$, where $\mathscr{H}$ is the temperature conductivity, $\mathscr{H} = k/c\rho$, and h is the depth to which the temperature is evaluated. For $\mathscr{H} \sim 0.5 - 1 \times 10^{-6}$ m^2 s^{-1} the steady-state solution is applicable in geological regions older than about 50 millions of years to a depth of $\sim$100 km.

As the proposed geotherms (see further) are based on the steady-state solution of the heat conduction equation, the data are to be taken more carefully in Tertiary terrains. However, a possible deviation, due to the departure from the steady-state solution, probably will be smaller than the uncertainty in some parameters used for the calculation.

Under the above simplifications, original equation (1) becomes

$$\frac{d}{dz}\left(k\frac{dT}{dz}\right) + A(z) = 0, \tag{2}$$

which enables a ready estimate of deep temperatures when compared to the geothermal structure of the individual tectonic provinces.

It is possible to propose a number of heat production distributions $A(z)$, though, usually only two are mentioned, both satisfying the experimental relationship between surface heat flow, Q, and the radiogenic heat production of the surface rocks, A_0:

$$Q = q_0 + DA_0. \tag{3}$$

q_0 is the so-called reduced heat flow, which characterizes the outflow of heat below a certain layer, in which the local variations of radioactivity produce the variations of surface heat flow. Reduced heat flow is presumed to be constant in large regions, which have undergone a similar tectonic evolution and which are called 'heat flow provinces' (ROY *et al.*, 1972). Parameter D has the dimension of length and characterizes the vertical distribution of heat sources. According to its meaning two models were proposed:

(i) The step model (ROY *et al.*, 1968) in which the crust is assumed to be composed of a series of blocks of vertically constant but horizontally varying radioactivity. q_0 is the heat flow from below the depth D, which bounds the depth range of these blocks. Within the depth interval $0 \div D$ the temperature can be expressed:

$$T = T_0 + (Q/k)z - (A_0/2k)z^2, \tag{4}$$

where T_0 is the surface temperature.

(ii) The exponential model (LACHENBRUCH, 1968) characterizes the Earth's crust as a product of magma solidification when the distribution of heat sources was governed by geochemical laws. The radiogenic heat production decreases exponentially with depth $A(z) = A_0 \exp(-z/D)$. The linear relationship between heat flow and heat production (3) preserves its validity even in the case of differential erosion of the surface rocks. The solution of equation (2) in this case is the following:

$$T = T_0 + \frac{q_0 z}{k} + \frac{A_0 D^2}{k}(1 - e^{-z/D}), \tag{5}$$

where $q_0 = Q - A_0 D$.

The latter model was used for the computation of the characteristic temperatures of all major tectonic units of Europe (ČERMÁK, 1982a), see Figure 1. The calculation

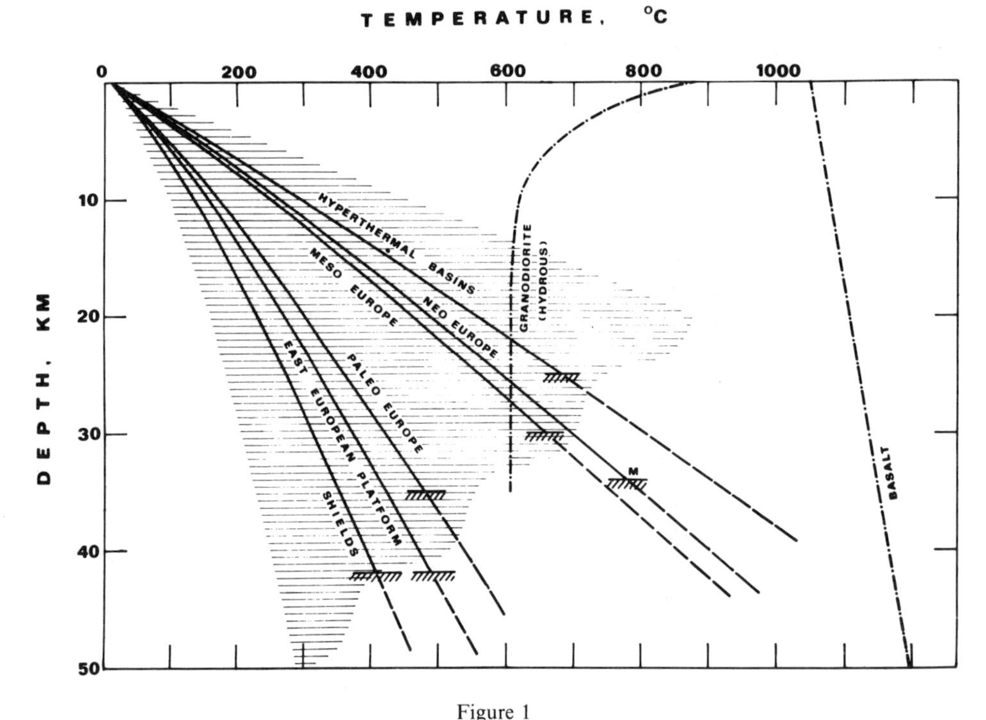

Figure 1
One-dimensional, steady-state temperature-versus-depth profiles for specific tectonic units in Europe.

Table 1

Deep temperatures at the Mohorovičić discontinuity and at 40 km depth for major tectonic features in Europe (model calculation) (ČERMÁK, *1982a*)

Tectonic feature	Heat flow ($mW.m^{-2}$)		Crustal thickness (km)		Temperature[2] at Moho		Temperature[3] at 40 km	
	Mean	Range[1]	Mean	Range[1]	Min.	Max.	Min.	Max.
Baltic and Ukrainian Shields	39	31–47	42	37–47	280	480	250	510
East European Platform	44	37–51	42	38–46	390	550	350	570
Caledonian Europe (Paleo-E.)	48	43–53	35	30–41	440	480	440	600
Western Siberia	54	48–60	38	35–42	540	640	520	720
Hercynian Europe (Meso-E.)	68	56–80	30	28–33	540	740	650	1,040
Ural	49	43–55	43	40–46	500	640	440	640
Alpine Europe (Neo-E.)	71	55–87	34	27–41	650	850	630	1,150
Hyperthermal Basins[4]	97	79–115	25	22–28	730	900	1,020	1,600[5]?

[1] Range calculated as the mean value ± standard deviation.
[2] For the calculation of the Moho-temperatures the following combinations were used: minimum heat flow and maximum crustal thickness and/or maximum heat flow and minimum crustal thickness, respectively. The range is thus narrower than in case of T_{40}.
[3] Minimum and maximum heat flow used together with the depth of 40 km.
[4] Pannonian Basin, Upper Rhinegraben.
[5] Extrapolated (unreal) value.

was made for the following values of the individual parameters: $A_0 = 2\ \mu W.m^{-3}$, $D = 10$ km, $k = 2.5\ W.m^{-1}\ K^{-1}$, which represent the mean values of a broad interval, but which can be used for the first approximation.

For each tectonic feature investigated, the limits of the computed temperatures were also determined, thus the probable minimum and maximum values, which may be expected in each case, were obtained. For this purpose, the mean surface heat flow and mean crustal thickness were complemented by their ranges and set as a mean value ± standard deviation. In view of the inverse relationship between heat flow and crustal thickness (ČERMÁK, 1979), the higher temperatures were attributed to the thinner crust and the lower temperatures to the thicker crust (a greater depth of the Mohorovičić discontinuity), see Table 1.

For higher values of the surface radioactivity A_0, we must presume lower values of D, to avoid unreliable heat production at the crustal base. $D = 10$ km is the mean value of the data found in various heat flow provinces the world over, D varies from 4 to 16 km (MORGAN and SASS, 1984), and can generally be deemed a typical value of the logarithmic decrement characterized by near-surface heat production distribution. For the above choice of parameters, the characteristic heat productions are $A = 0.7\ \mu W.m^{-3}$ at 10 km, $A = 0.3\ \mu W.m^{-3}$ at 20 km, and $A = 0.1\ \mu W.m^{-3}$ at 30 km.

With increasing temperature, the coefficient of thermal conductivity decreases (as does its lattice component which characterizes the heat transfer under the conditions existing in the bulk of the crust). This decrease of thermal conductivity takes place especially in the upper crust, while in the lower crust the conductivity is substantially independent on temperature. Constant thermal conductivity within the whole crust is thus only the first approximation, but is justifiable for the existing range of conductivities of the crustal rocks $1.5 \div 4$ W.m^{-1} K^{-1}. Lower mean crustal conductivity would produce higher temperatures at all depths.

If the temperature dependence of thermal conductivity in the form $k = k_0/(1 + CT)$, where k_0 is the conductivity at 0°C and C is an experimental constant, $C \sim 0.0005 - 0.0015$ K^{-1}, is included into the solution of equation (5), the temperature can be expressed by

$$\frac{k_0}{C} \ln (1 + CT) = k_0 T_0 + q_0 z + A_0 D^2 (1 - e^{-z/D}). \tag{6}$$

The geotherms shown in Figure 1 can be complemented by the regional distribution of deep temperatures. As an example, Figure 2 shows the regional distribution of the temperatures at a depth of 40 km in Europe (ČERMÁK, 1984) calculated for the mean surface heat flow values of the square coordinate grid 2° × 2°.

For a more detailed calculation of deep temperatures in the continental lithosphere (ČERMÁK, 1982b), a combination of both the above models was used coupled with another empirical relation between the mean surface heat flow in the heat flow province, $\bar{Q}$, and the reduced heat flow, q_0 (POLLACK and CHAPMAN, 1977):

$$q_0 = 0.6\,\bar{Q}. \tag{7}$$

If it is postulated that within a certain surface area the mean heat production $\bar{A}_0$ in the surface layer of thickness D is related to the mean surface heat flow $\bar{A}_0 = (\bar{Q} - q_0)/D$, then $\bar{A}_0 = 0.4\,\bar{Q}/D$. For exponentially decreasing radioactivity with depth, the mean heat production $\bar{A}_0$ in the layer bounded by depths z_1 and z_2 is given by

$$\bar{A}_0 = \int_{z_1}^{z_2} A_{01} \exp(-z/D)\, dz, \tag{8}$$

where A_{01} is the heat generation at depth z_1. For $z_1 = 0$ and $z_2 = D$ it follows that $A_{01} = \bar{A}_0/(1 - e^{-1})$ and $A_{02} = A_{01}\, e^{-1}$ is the heat production at depth $z_2 = D$.

The radioactivity of the lower crust is even less known, but to complete the model, the exponential function $A(z) = A_{02} \exp(-z/D_2)$ was applied with $D_2 = 5$ km. The mean heat production in the lower crust for the above values of A_{02} is in the interval of 0.1 to 0.7 μW.m^{-3}, which corresponds reasonably well to the real conditions. For the heat production in the upper mantle, the older model by CLARK and RINGWOOD (1964) was used with a slight depth scale modification as proposed by POLLACK and CHAPMAN (1977). The uppermost part of the mantle down to a depth of 120 km is characterized by a depleted dunite-peridotite composition [$A_{03} = 0.0084$ μW.m^{-3}] and is underlain by the layer of primitive pyrolite [$A_{04} = 0.042$ μW.m^{-3}].

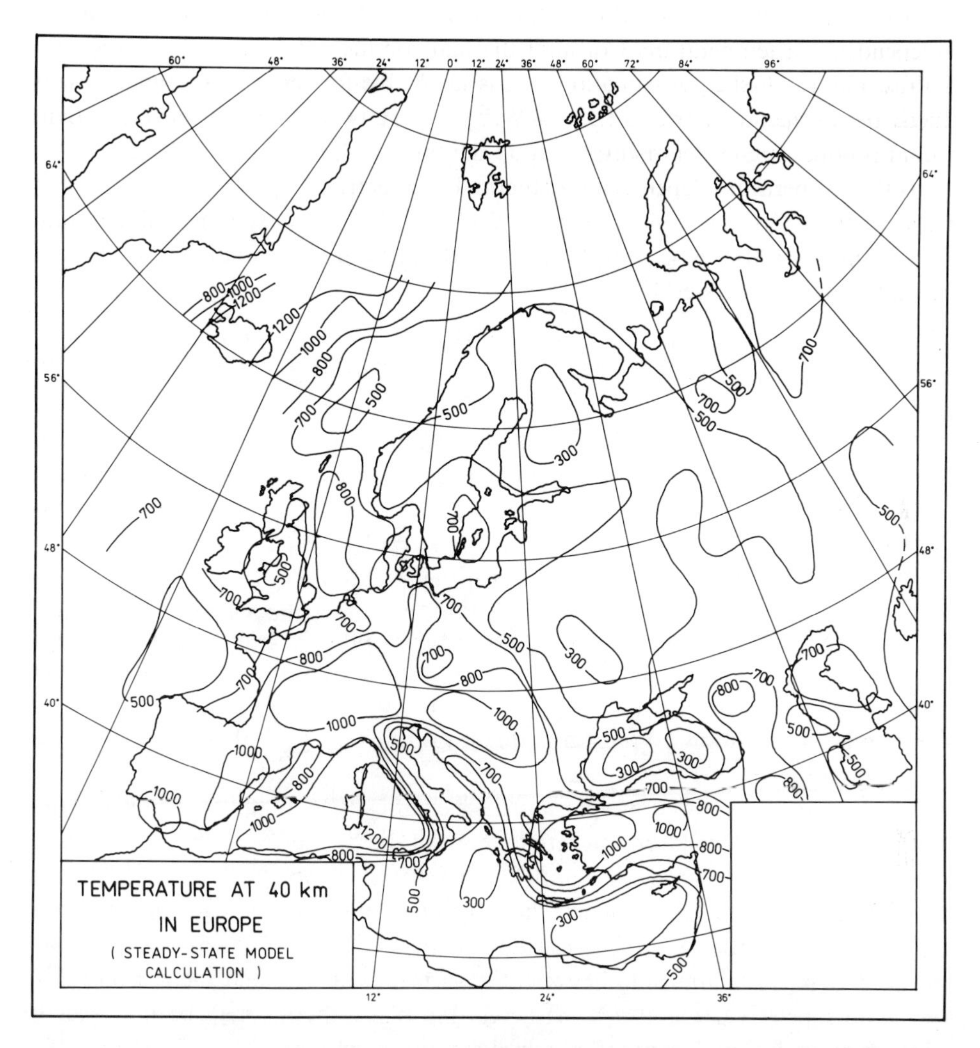

Figure 2
Regional distribution of calculated temperatures at 40 km depth in Europe.

For the vertical distribution of the thermal conductivity formula $k = k_0/(1 + CT)$ was used, and parameter C was selected according to the rock composition and heat transfer mechanism dominated at a corresponding depth. In the upper crust of granitic composition and dominating lattice conductivity, $C > 0$, and thermal conductivity decreases with increasing temperature; in the lower crust of predominantly granulite to basalt composition, $C = 0$, as conductivity is generally little dependent on temperature; in the upper mantle of ultrabasic composition where the radiative component of heat transfer starts to control the heat conduction at temperatures of over 500°C, $C < 0$, and the conductivity increases with increasing temperature (SCHATZ and SIMMONS, 1972).

The model of the two-layered crust and the two-layered upper mantle with temperature dependent thermal conductivity $k = k_0/(1 + CT)$ and the exponential heat production $A(z) = A_0 \exp(-z/D)$ gives the following steady-state one-dimensional solution $T(z)$ of equation (2) in each layer:

$$\frac{1}{C_i} \ln [1 + C_i T(z)] = \frac{1}{C_i} \ln [1 + C_i T_i] + Q_i(z - z_i)/k_{0i} - \frac{A_{0i} D_i}{k_{0i}} \left[D_i \exp \left(- \frac{z - z_i}{D_i} + z - z_i - D_i \right) \right],$$

with

$$Q_{i+1} = Q_i - \Delta Q_i,$$

where

$$\Delta Q_i = A_{0i} D_i \left[1 - \exp \left(- \frac{z_{i+1} - z_i}{D_i} \right) \right]. \tag{9}$$

In this expression T_i and/or Q_i are the temperature and the heat flow, respectively, at the depth z_i; k_{0i}, C_i, A_{0i} and D_i are the corresponding parameters in the i-th layer, bounded by depths z_i and z_{i+1}. For $k \neq k(T)$ and/or $A \neq A(z)$ (i.e., constant within a certain layer), solution (9) remains valid, if respectively, $C_i \rightarrow 0$ and/or $D_i \rightarrow \infty$ in the respective layer.

All the parameters used are presented schematically in Figure 3. The calculated geotherms slightly depend on the actual depth of the Mohorovičić discontinuity, z_3, for a thicker crust higher upper mantle temperatures are obtained and for a thinner crust lower temperatures are found. The calculated temperature-depth curves shown in Figure 4 correspond to standard crustal thickness $z_3 = z_M = 35$ km. Because of the observed negative relationship between the surface heat flow and the crustal thickness in Europe (ČERMÁK, 1979), the actual fan of geotherms may be narrower in comparison with that shown for standard crustal thickness.

The European continent is characterized by a large low heat flow zone covering most of northern and eastern Europe, surrounded by normal to high heat flow values spread in western, central, southern and southeastern Europe (ČERMÁK, 1979). The general 'northeast' to 'southwest' increase of the geothermal activity is the consequence of the geological evolution of the continent and the major European heat flow provinces correlate relatively well with the principal tectonic units of Ur-Europe, Paleo-Europe, Meso-Europe and Neo-Europe, according to STILLE (1924). The first order heat flow lows ($\sim$40 mW.m^{-2}) cover most of the East European platform, including both the Baltic and the Ukrainian shields, i.e., the greater part of the European craton, the oldest ($>1.5 \times 10^9$ yr) and the most stable portion of the whole continent. Other heat flow lows include smaller units such as the Bohemian

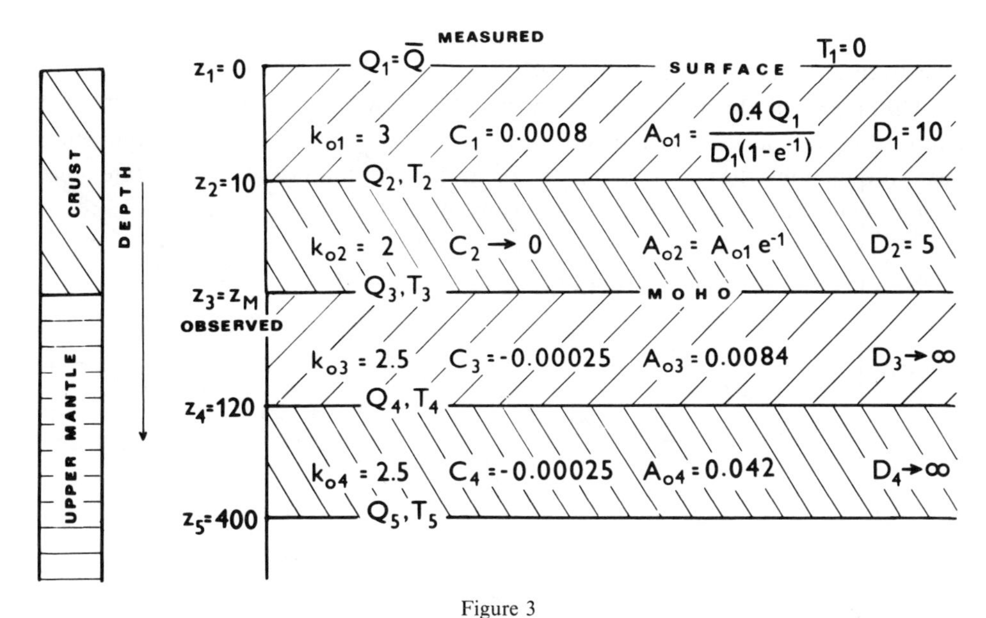

Figure 3
Schematic model and parameters used for the deep temperature calculation: z—depth, km; Q—heat flow, mW.m^{-2}; T—temperature, °C; k—thermal conductivity, W.m^{-1} K^{-1}; C—coefficient, K^{-1}; A—heat production, μW.m^{-3}; D—logarithmic decrement, km.

Massif, the Moesian plate, etc. representing stable, consolidated segments as well. Low heat flow is also typical of the Eastern Mediterranean Sea (probably part of the African plate), with a thick stable maffic crust (MORGAN, 1979), and the Black Sea. Geothermal highs (>80 mW.m^{-2}) are found in Iceland and its vicinity (part of the Mid-Atlantic ridge), the Rhinegraben, the Alps, the Pannonian Basin, several locations in the Balkans, Turkey and in the Caucasus, all geologically young structures, tectonically still active or recently active, and in the Western Mediterranean and the Aegean Sea, areas of thin subcontinental crust.

Using the characteristic mean surface heat flow as the main parameter for the deep temperature calculation (see Figures 1 and 4 and also Table 1) we can summarize the following conclusions. Relatively low temperatures (350–500°C) at the Mohorovičić discontinuity, at a depth of 45–50 km, are to be expected beneath the Precambrian shields and the East European platform (KUTAS and GORDIENKO, 1971; BALLING, 1976). Higher Moho temperatures of 500–600°C were calculated for the Paleozoic folded units such as the Bohemian Massif or the Ural Mountains at a depth of 35–45 km. Local temperature maxima of 600–700°C seem to exist in the zones of a weakened crust or in the areas of deep faults, which are characterized by a relatively higher heat flow on the surface (up to 70–80 mW.m^{-2}), (ČERMÁK, 1975). Very high crustal temperatures (800–1,000°C) are probable in hyperthermal regions,

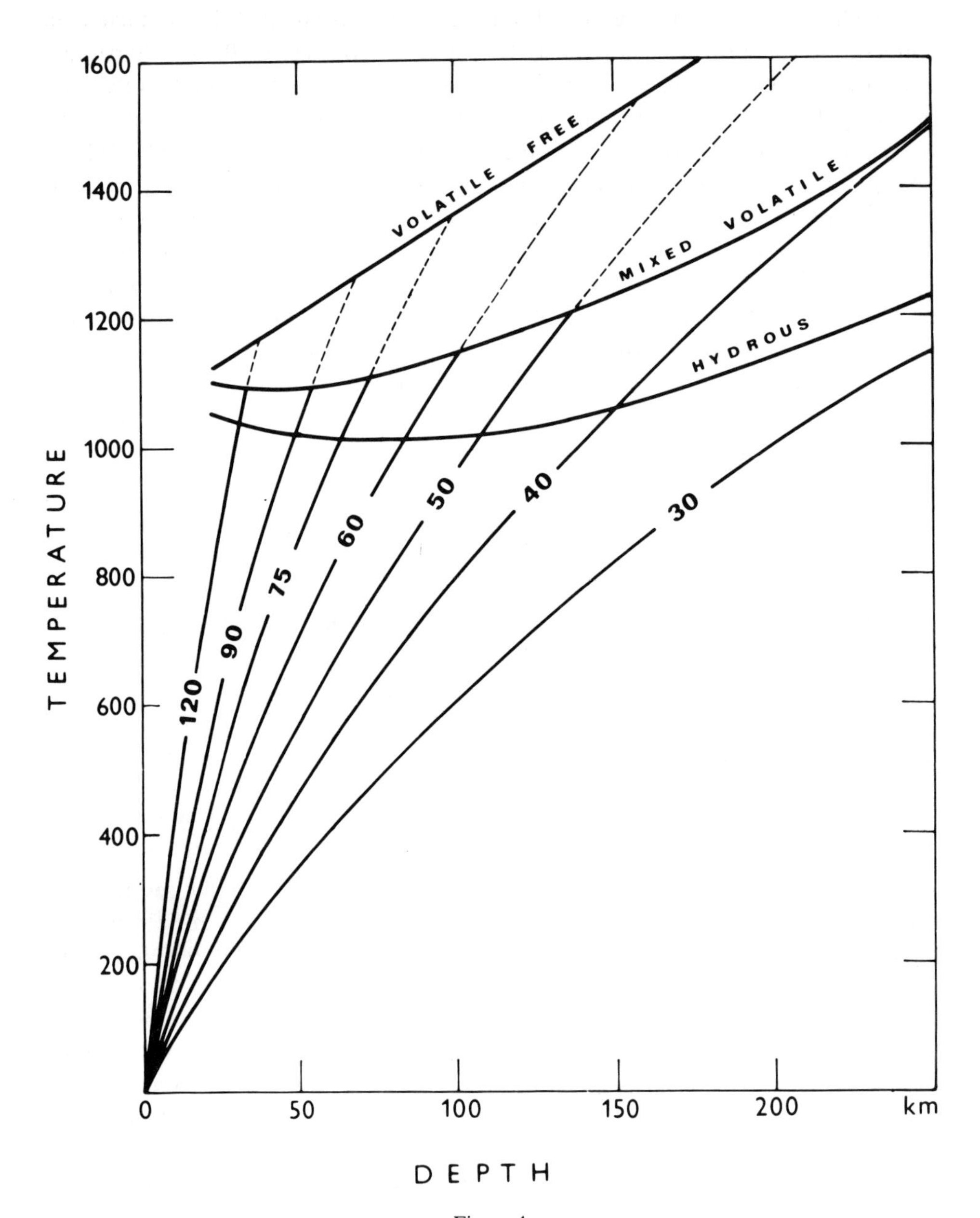

Figure 4
Geotherm family for continental lithosphere, individual curves are labelled in surface heat flow in mW.m^{-2}, melting relations after CHAPMAN *et al.* (1979).

such as the Pannonian Basin characterized by a relatively thin crust, 25 km, and a very high heat flow on the surface (BODRI, 1981; BUNTEBARTH, 1976). Similar high temperatures may exist in the Upper Rhinegraben (HAENEL, 1970). The horizontal crustal temperature changes on contacts between various tectonic units were computed by BALLING (1976) for the Baltic shield—Danish Embayment contact; by BUNTEBARTH (1973) for the Alps and the northern foreland; by ČERMÁK (1974) for a profile across the Western Carpathians and by HURTIG and OELSNER (1977) for several profiles across Europe. Using two-dimensional temperature modelling, STROMEYER (1984) has recently advanced several improvements and has discussed theoretical consequences of varying boundary conditions along the hypothetical profile from south to north Europe. Using seismological data on the actual crustal thickness and converting seismic velocities into heat production data, ČERMÁK and BODRI (1985) proposed deep temperature distribution adjoining five long-run East European geotraverses. A similar program is underway in Western Europe (GALSON and RYBACH, 1984).

3. *Electrical Conductivity*

The temperature dependence of electrical conductivity $\sigma(T)$ of minerals and rocks in the lithosphere can be expressed by

$$\sigma(T) = \sum_i \sigma_{0i} \exp(-E_i/kT) = \sum_i \sigma_i(T), \tag{10}$$

where T is the absolute temperature in K, k Boltzmann constant (0.862×10^{-4} eV.K^{-1}), E—activation energy (eV). The charge carriers are ions, ion-vacancies, electrons, or holes, and the total conductivity is thus a sum of the individual conductivity mechanisms $\sigma_i(T)$. Each of them is dominating in a certain temperature range and by plotting ln (or log) of the relative conductivity versus the inverse temperature the ln σ-values follow a curve consisting of several linear sections corresponding to these temperature ranges and each is characterized by a specific slope-value, $-E_i/k$.

The individual rock samples of the same rock type very often differ considerably, even if measured under the same conditions. It is therefore impossible to propose a single $\sigma(T)$-curve for a specific rock or for a certain crustal layer; only statistical material based on numerous measurements can yield representative data for the crustal structure on a regional scale. Endeavoring to judge the depth distribution of the electrical conductivity within the lithosphere, we have collected the data published for typical rocks in several reviews (PARKHOMENKO and BONDARENKO, 1972; VOLAROVICH and PARKHOMENKO, 1976; KELLER, 1982; HAAK, 1982; KARIYA and SHANKLAND, 1983; LAŠTOVIČKOVÁ, 1983) which we completed by some new results (see Appendix).

All the above compilations represent a number of laboratory measured DC conductivities. PARKHOMENKO and BONDARENKO (1972) and VOLAROVICH and PARKHOMENKO (1976) published data obtained for the territory of the USSR, their measurements were performed in dry air, not in the inner atmosphere, and formally their data are arranged as the interval of conductivity observed (minimum-maximum value) at the specific temperature value. KELLER (1982) compiled data for dunites and olivinites, which he presented in the form of graphs, from which we took values for our purposes. HAAK (1982) summarized the data obtained by numerous authors and presented characteristic data for basalts and ultrabasic rocks. KARIYA and SHANKLAND (1983) also summarized data obtained by several investigators and their data are presented as the mean value of log conductivity linked with the standard deviations. LAŠTOVIČKOVÁ (1983) gave a survey of DC conductivities of rocks from Central Europe, and also data in the Appendix mostly relate to this area.

Although the data used for our statistics are generated world wide, Central and Eastern Europe and the USSR results are represented most extensively. It is also obvious that such a set of data obtained by various authors and in various laboratories cannot be homogeneous regarding the experimental conditions. However, we believe that for our purpose, i.e., to have a relevant characteristic conductivity-versus-temperature curve for major rock types, the degree of unhomogeneity is moderate and that such a set can be used for general conclusions.

As the petrological composition of the lithosphere is complex, certain simplification was necessary and we have grouped the above data into several specific groups which may better characterize the individual crustal and upper mantle layers within the depth interval of approximately 0–100 km. Rocks were divided into four main groups which should represent the upper, intermediate and lower crust and the upper mantle (subcrustal lithosphere). We have followed partly, but not exactly, the division as proposed by KARIYA and SHANKLAND (1983), but as data based on different subdivisions by other authors were added, their grouping could not be managed. Furthermore, they proposed their gabbro and basalt groups rather to distinguish the grain size, coarse-grained versus fine-grained rocks, while our gabbro group represents intermediate to lower crust and our basalt group represents lower crust. Our division thus reflects the effect of composition rather than the texture.

(i) Granite group represents the rocks of the upper crust, with density smaller than 2.7 g.cm^{-3}, rich in quartz ($>10\%$): granites, granodiorites and also gneisses and slates from VOLAROVICH and PARKHOMENKO's (1976) compilation.

(ii) Gabbro group represents the rocks typical of intermediate to lower crust, $\rho > 2.7$ g.cm^{-3}, with less than 10 percent of quartz: gabbros, diabases and diorites, and also granulites.

(iii) Basalt group covers mostly rocks of lower crust, with a density over 2.8 g.cm^{-3}, less than 10 percent of quartz: andesites, basalts, dolerites, and amphibolites. From HAAK (1982) we have also included synthetic basalts in addition to natural ones. Our data (Appendix) contributed to this group various alkaline and calc-alkaline volcanic rocks (some of them are of lower density than 2.8 g.cm^{-3}) of basic composition: nephelites, trachytes, trachyandesites, rhyolites, dacites and teschenites.

(iv) Ultrabasite group encompasses rocks of mafic to ultramafic composition: ultramafic rocks and eclogites (HAAK, 1982), olivinites, pyroxenites, dunites, peridotites and eclogites (VOLAROVICH and PARKHOMENKO, 1976), and dunites and olivinites (KELLER, 1982). Our data (Appendix) relate to peridotites, eclogites, pyroxenites and also serpentinites.

The distribution of log conductivity versus temperature for all above groups of rocks are shown in Figures 5 through 9 and the values of log σ at selected temperatures in the interval 200–1,000°C are summarized in Table 2. Basalts are the most conductive rocks, then follow gabbros, granites, and the less conductive are ultrabasites. KARIYA and SHANKLAND (1983) found the same distribution of conductivity, as well as quite similar absolute values of conductivity for temperatures over 500°C. However, one must realize that the difference in characteristic conductivities among the groups are sometimes comparable or even smaller than the range of the existing conductivities within the respective groups. This underlines the extreme variability of rock conductivity, and unfortunately, may even question our formal division into the individual groups.

4. *Conductivity Versus Depth*

If we know the temperature dependence of the electrical conductivity and the depth distribution of temperature, it is possible to evaluate the probable depth dis-

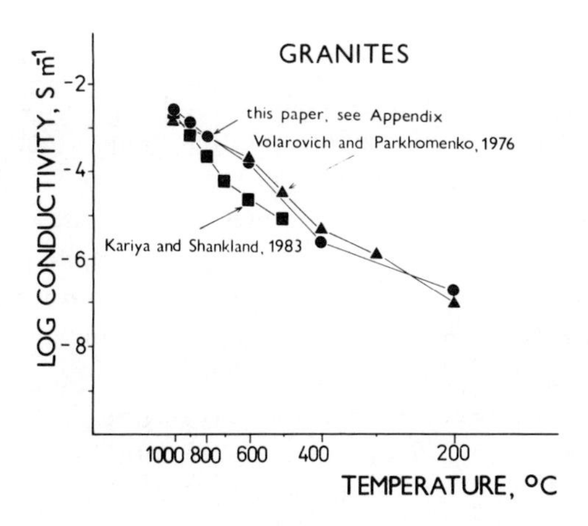

Figure 5
Log conductivity-versus-temperature for granite-group.

tribution of the electrical conductivity. This procedure is formally shown in Figure 10. In the left-hand side of the figure the generalized curves of the log conductivity-versus-temperature were adopted for four rock groups from Figure 9, in the right-

Table 2

Mean log conductivity, log (S/m), and its standard deviation, of rocks in individual groups at selected temperatures within the interval 200–1,000°C

Rock type	Volarovich and Parkhomenko (1976)	Haak (1982)	Keller (1982)	Karia and Shankland (1983)	This paper	Total[1]
Granite group						
Number of samples	25			27	28	80
200°C	−6.96 ± 2.08			—	−6.71 ± 0.95	−6.83 ± 0.18
300°C	−5.92 ± 1.87			—	—	−5.92
400°C	−5.35 ± 1.69			—	−5.62 ± 1.02	−5.49 ± 0.19
500°C	−4.49 ± 1.54			−5.07 ± 0.98	—	−4.78 ± 0.41
600°C	−3.72 ± 1.35			−4.64 ± 0.84	−3.74 ± 0.63	−4.03 ± 0.52
700°C	—			−4.23 ± 0.80	—	−4.23
800°C	—			−3.62 ± 0.82	−3.20 ± 0.69	−3.41 ± 0.30
900°C	—			−3.17 ± 0.80	−2.90 ± 0.89	−3.04 ± 0.19
1,000°C	−2.81 ± 0.64			−2.63 ± 0.79	−2.57 ± 0.78	−2.67 ± 0.12
Gabbro group						
Number of samples	40			37	31	108
200°C	−7.09 ± 1.11			—	—	−7.09
300°C	−6.36 ± 0.95			—	−6.94 ± 0.85	−6.65 ± 0.41
400°C	−5.41 ± 1.16			—	−6.09 ± 0.70	−5.75 ± 0.48
500°C	−4.79 ± 1.23			−4.77 ± 0.75	—	−4.78 ± 0.01
600°C	−3.92 ± 1.12			−3.95 ± 0.99	−4.29 ± 0.60	−4.05 ± 0.20
700°C	−3.83 ± 0.88			−3.54 ± 0.91	—	−3.68 ± 0.20
800°C	−3.40 ± 1.15			−3.25 ± 0.96	−3.25 ± 0.73	−3.30 ± 0.09
900°C	−3.05 ± 1.08			−2.75 ± 0.86	−2.43 ± 0.33	−2.74 ± 0.31
1,000°C	−2.43 ± 0.81			−2.09 ± 0.71	—	−2.26 ± 0.24
Basalt group[2]						
Number of samples	60	23		58	204	345
200°C	−6.44 ± 0.97	−5.90		—	−6.07 ± 1.43	−6.14 ± 0.28
300°C	−5.51 ± 0.87	−5.18		—	—	−5.35 ± 0.23
400°C	−4.51 ± 0.80	−4.65		—	−5.05 ± 0.93	−4.74 ± 0.28
500°C	−3.74 ± 0.66	−3.85		−3.93 ± 0.83	—	−3.84 ± 0.09
600°C	−3.44 ± 0.67	−3.23		−3.43 ± 0.87	−3.63 ± 0.57	−3.43 ± 0.16
700°C	−3.38 ± 0.44	—		−2.80 ± 0.74	—	−3.09 ± 0.41
800°C	−3.03 ± 1.32	−2.28		−2.29 ± 0.75	−3.00 ± 0.43	−2.65 ± 0.42
900°C	—	—		−1.84 ± 0.66	−2.66 ± 0.50	−2.25 ± 0.58
1,000°C	−1.60 ± 0.71	−1.70		−1.39 ± 0.69	—	−1.56 ± 0.16

Ultrabasite group[2] Number of samples	80	32	10	40	162
200°C	−8.77 ± 1.23	−6.3	—	−6.74 ± 1.51	−7.27 ± 1.32
300°C	—	−6.3	−7.80 ± 0.57	−6.24 ± 1.41	−6.78 ± 0.88
400°C	—	−5.2	−6.65 ± 0.78	−5.51 ± 0.76	−5.79 ± 0.76
500°C	−5.73 ± 1.14	−4.4	−5.60 ± 0.71	−4.59 ± 0.47	−5.08 ± 0.68
600°C	−5.05 ± 1.14	−3.7	−4.73 ± 0.25	−4.19 ± 0.36	−4.42 ± 0.59
700°C	−4.56 ± 0.90	—	−3.65	—	−4.10 ± 0.59
800°C	−3.91 ± 0.97	−2.9	—	−3.29 ± 0.40	−3.60 ± 0.51
900°C	−3.29 ± 0.95	—	—	−3.01 ± 0.51	−3.15 ± 0.20
1,000°C	−2.77 ± 1.07	−2.4	—	−2.71 ± 0.42	−2.63 ± 0.20
1,100°C	−2.15 ± 1.16	—	—	—	−2.15

Note:

[1] Standard deviation has formal meaning only and reflects the differences among individual authors, although the reliability of this value depends on numbers of samples and quality of data sources available.

[2] HAAK's (1982) data taken from figures, s.d. cannot be accounted for.

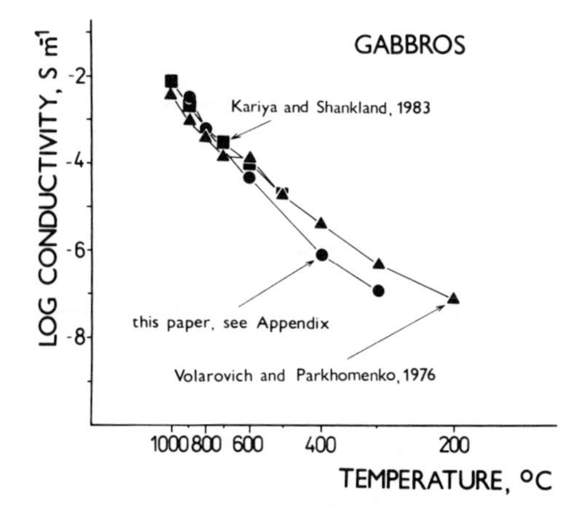

Figure 6
Log conductivity-versus-temperature for gabbro group.

hand bottom side three temperature-versus-depth curves for surface heat flow 40, 60 and 80 mW.m^{-2} were adopted from Figure 4. Right-hand upper side of the figure then shows the converted log conductivity-versus-depth for the surface heat flow 60 mW.m^{-2}. Similar log $\sigma(T)$/log $\sigma(z)$ conversion curves for heat flow 40 and/or 80 mW.m^{-2}, respectively, are shown in Figures 11 and 12.

As mentioned above, deep temperatures and surface heat flow are closely connected and chosen heat flow value thus controls the calculated temperature field. The global continental mean heat flow is 62.3 ± 40.1 mW.m^{-2} (JESSOP *et al.*, 1976) slightly lower than the European mean of 64.4 ± 28.3 mW.m^{-2} (ČERMÁK, 1979). The

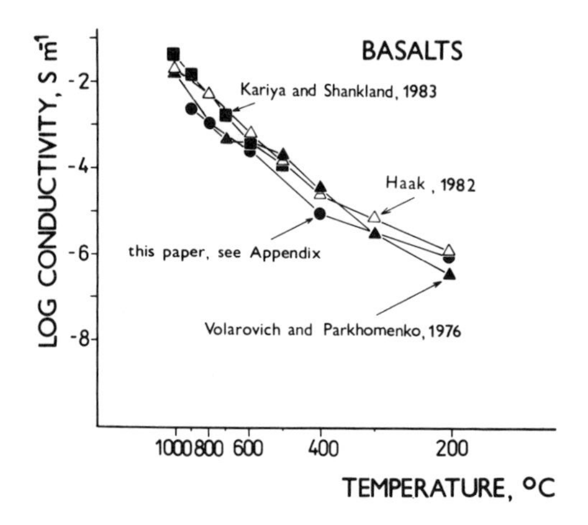

Figure 7
Log conductivity-versus-temperature for basalt group.

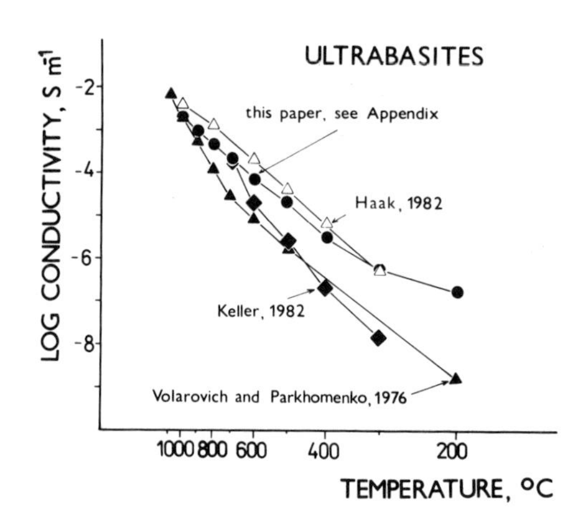

Figure 8
Log conductivity-versus-temperature for ultrabasite group.

value of 60 mW.m^{-2} characterizes the typical continental areas, usually of Paleozoic to Early Mesozoic consolidation, of normal crustal thickness of 30–40 km, such as e.g., Eastern North America, Central Europe, etc. Other two selected heat flow values 40 and/or 80 mW.m^{-2}, respectively, characterize (i) old shield or platform areas of Precambrian basement with low geothermal activity, thick crust (more than 40 km), such as e.g., the Canadian shield, East European platform, northern Asia, and (ii) young Tertiary orogenes and intramontane basins with high crustal temperatures, such as most Alpine–Carpathian ranges in Europe, Basin and Range provinces in

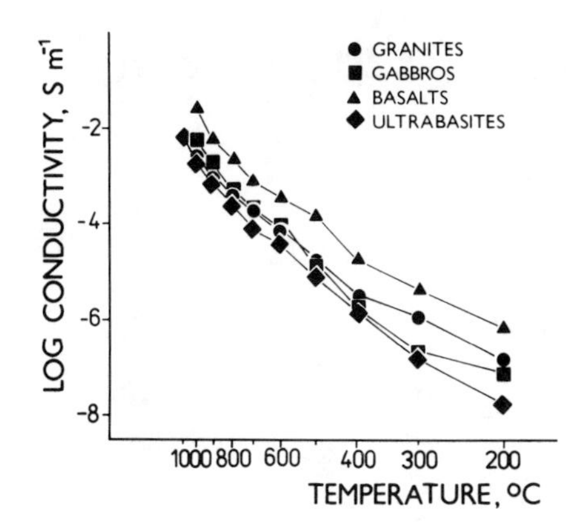

Figure 9
Log conductivity-versus-temperature for four individual groups of characteristic rocks for crustal and upper mantle structure.

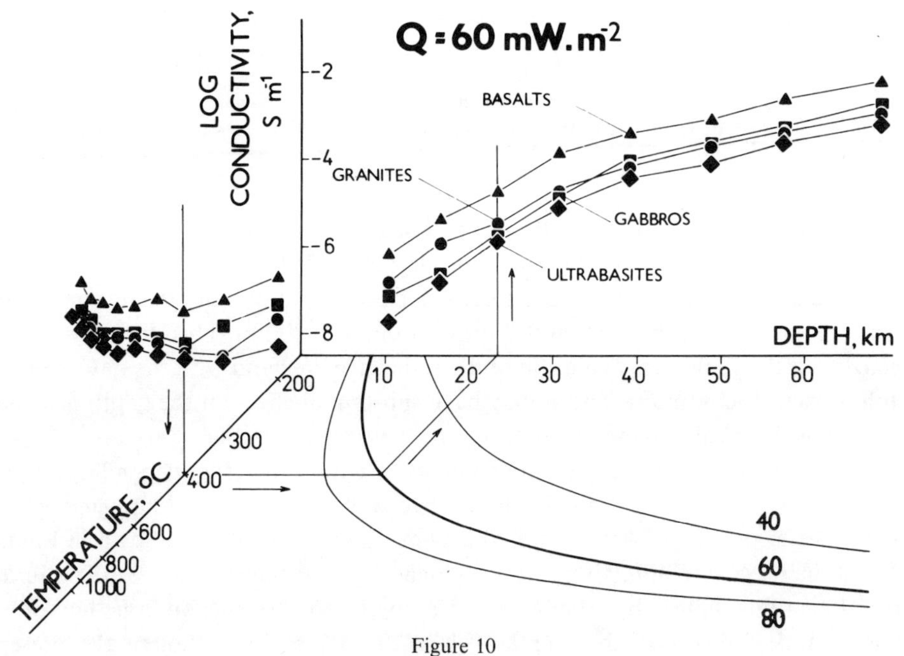

Figure 10
Conversion of log conductivity-versus-temperature into log conductivity-versus-depth curves for crustal temperature corresponding to surface heat flow of 60 $mW.m^{-2}$.

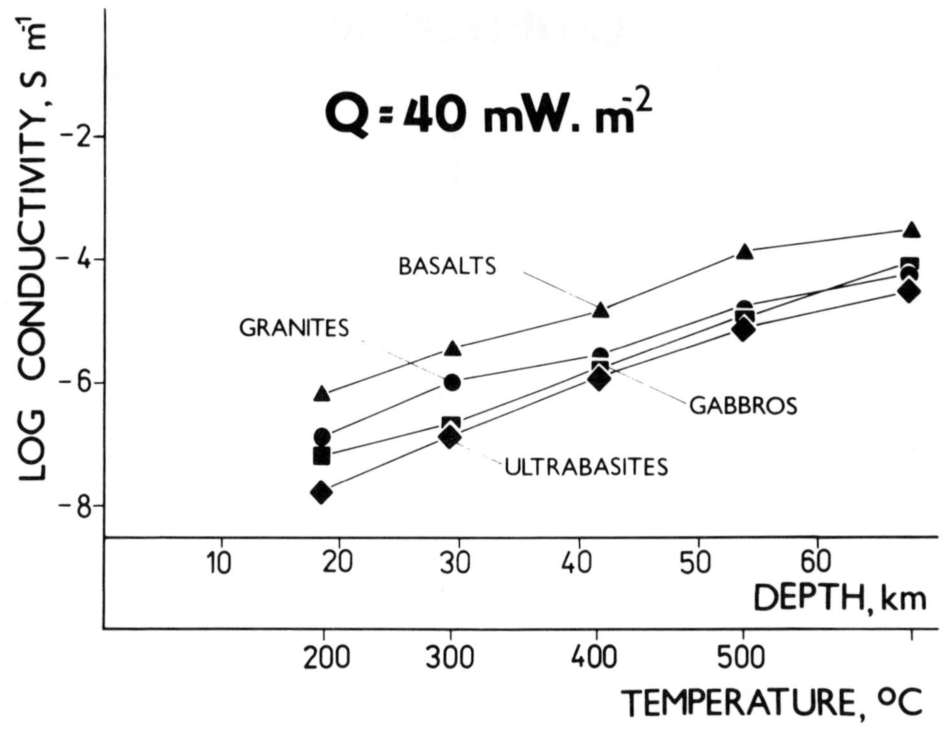

Figure 11
Converted log conductivity-versus-depth curves for surface heat flow of 40 $mW.m^{-2}$.

the western USA, the Pannonian Basin, etc. The latter tectonic units which all display high surface heat flow, however, may considerably differ in crustal thickness, while young Cenozoic orogenic belts constitute a thick crust (40–50 km), the intramontane basins or rift zones form a weakened crust of only 25–30 km. This diversity is connected with the geological evolution of the respective area and with the nature of the Mohorovičić discontinuity, and it may have substantial effect on the depth distribution of the electrical conductivity.

The probable range of electrical conductivity within the crust depending on the existing temperature conditions in the diverse tectonic provinces, characterized by the surface heat flows of 40, 60, and 80 $mW.m^{-2}$, are schematically shown in Figure 13. The increase of conductivity, due to higher crustal temperature existing in an area of heat flow higher by 20 $mW.m^{-2}$, can number one or one and one half order of magnitude for comparable depths below 10–15 km. Even though the present results are preliminary and correspond to the rather simple approximation, they clearly demonstrate the more conductive strata of the lower crust and the upper mantle in areas of high heat flow.

Basalt-type rocks are the most conductive of all investigated groups by one

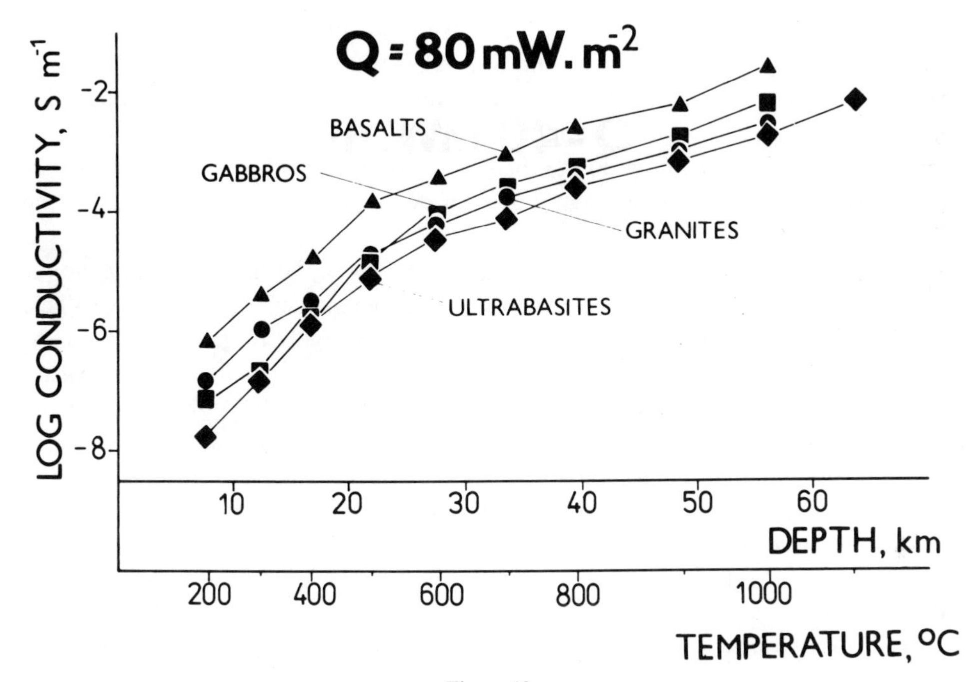

Figure 12
Converted log conductivity-versus-depth curves for surface heat flow of 80 mW.m^{-2}.

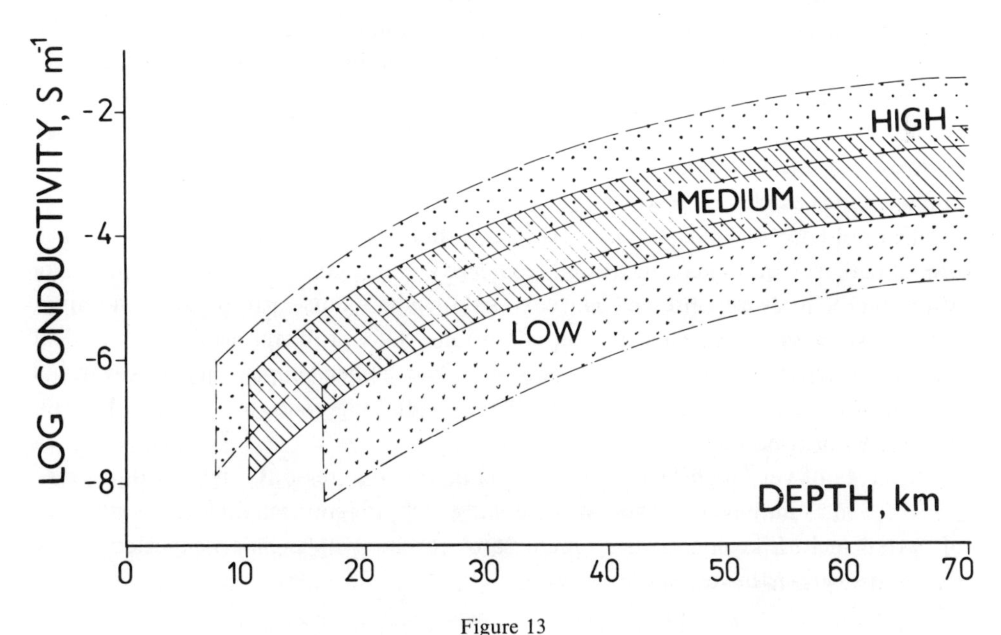

Figure 13
Probable range of electrical conductivity-versus-depth in the lithosphere for areas of low ($\sim$40 mW.m^{-2}), medium ($\sim$60 mW.m^{-2}), and high ($\sim$80 mW.m^{-2}) heat flow.

to two orders of magnitude, which agree with the findings of VOLAROVICH and PARKHOMENKO (1976) and KARIYA and SHANKLAND (1983), who explained this fact by smaller activation energy for their glassy and fine-grained components than for crystalline rocks. If this can be applied to the whole lower crust (which must be confirmed by future investigations), then we have to expect a certain increase of conductivity by one order of magnitude at a depth of 15–20 km, i.e., close to the surmised Conrad discontinuity.

KARIYA and SHANKLAND (1983) noted a statistically significant difference of about half to one order of magnitude between the conductivity of less conductive silicic rocks (granite group) and more conductive mafic rocks (gabbro group) for the temperature interval of 500–1,000°C. A clear decrease of conductivity with increasing silica content was experimentally proved by ALVAREZ *et al.* (1978) by measurements on powdered samples. Our data support this trend but give a smaller difference of only half the order of magnitude above 500°C, and no difference, or even a small opposite relation, was observed at temperatures below 500°C. Regardless, as granatic rocks constitute the upper crust and gabbro type rocks form the intermediate crust, this difference has little application to crustal studies of conductivity distribution.

The ultrabasite group has the lowest conductivities in the whole temperature interval studied by roughly half to one order of magnitude compared to granite or gabbro groups, and approximately two orders of magnitude lower than basaltic rocks. Principally the basalt-ultrabasite difference may be of significance for interpreting magnetotelluric data at the crust-upper mantle boundary. Passing from the lower crust to ultrabasite upper mantle, the conductivity may drop and the rapid change of conductivity should be observable. As the range delimited by the individual curves belonging to basalt and ultrabasite groups is slightly narrower for higher temperature, the above drop of conductivity should be more prominent in 'cold' areas, such as shields or platforms.

The presence of fluids or volatiles in the rock increases the conductivity. Since the laboratory rocks were measured under dry conditions, their conductivites are lower than if fluids were present. The curves shown here thus correspond to minimum conductivities to be expected in real conditions. This problem, however, is critical only at lower temperatures up to $\sim$200–300°C, at higher temperatures less water is present in rock, and there is probably no substantial difference if we apply 'dry' (no free water) or 'wet' data.

In areas of very high heat flow, such as in volcanic regions, high conductivities ($>10^{-2}$ S.m^{-1}) may be observed in the field, which are connected with the presence of partial melt. Likewise, a sharp increase of conductivity must be expected below the lithosphere-asthenosphere boundary.

References

ALVAREZ, R., REYNOSO, J. P., ALVAREZ, L. J. and MARTINEZ, M. L. (1978), *Electrical conductivity of igneous rocks: composition and temperature relations*. Bull Volcanol. *41*, 317–327.

BALLING, N. P. (1976), *Geothermal models of the crust and uppermost mantle of the Fennoscandian shield in South Norway and the Danish Embayement*. J. Geophys. *42*, 237–256.

BODRI, L. (1981), *Geothermal model of the Earth's crust in the Pannonian Basin*. Tectonophysics *72*, 61–73.

BUNTERBARTH, G. (1973), *Model calculation on temperature-depth distribution in the area of the Alps and the foreland*. Z. Geophys. *39*, 97–107.

BUNTERBARTH, G. (1976), *Temperature calculations on the Hungarian seismic profile-section NP-2*, In *Geoelectric and Geothermal Studies* (East-Central Europe, Soviet Asia) (ed. A. Ádám). KAPG Geophys. Monograph, Akadémiai Kiadó, Budapest, pp. 561–566.

CARSLAW, H. S. and JAEGER, J. C. (1959), *Conduction of Heat in Solids*. Clarendon Press, Oxford, 510 pp.

ČERMÁK, V. (1974), *Deep temperature distribution along the deep seismic sounding profile across the Carpathians* (*Model calculation*). Acta Geol. Sci. Hung. *18*, 295–303.

ČERMÁK, V. (1975), *Temperature depth profiles in Czechoslovakia and some adjacent areas derived from heat-flow measurements, deep seismic sounding and other geophysical data*. Tectonophysics *26*, 103–119.

ČERMÁK, V. (1979), *Heat flow map of Europe*, In *Terrestrial Heat Flow in Europe*. (eds V. Čermák and L. Rybach), Springer-Verlag, Berlin, Heidelberg, New York, pp. 3–40.

ČERMÁK, V. (1982a), *Crustal temperature and mantle heat flow in Europe*. Tectonophysics *83*, 123–142.

ČERMĂK, V. (1982b), *Regional pattern of the lithospheric thickness in Europe*, In *Geothermics and Geothermal Energy* (eds V. Čermák and R. Haenel). E. Schweizerbartsche Verlagsbuchhandlung, Stuttgart, pp. 1–10.

ČERMÁK, V. (1984), *Regional distribution of heat flow in Europe: Derived deep temperature and Moho heat flow patterns*, In *Ermittlung der Temperaturverteilung im Erdinnern* (ed. G. Bunterbarth). Inst. f. Geophysik, TU Clausthal, pp. 4–7.

ČERMÁK, V. and BODRI, L. (1985), *Temperature structure of the lithosphere based on 2-D temperature modelling applied to Central and Eastern Europe*, In *Proc. Int. Conf. on Modelling the Thermal Evolution of Sedimentary Basins* (ed. J. Burrus). Inst. Francais du Petrol, Rueil Malmaison, pp. 7–39.

CHAPMAN, D. S., POLLACK, H. N. and ČERMÁK, V. (1979), *Global heat flow with special reference to the region of Europe*, In *Terrestrial Heat Flow in Europe* (eds V. Čermák and L. Rybach). Springer-Verlag, Berlin, Heidelberg, New York, pp. 41–48.

CLARK, S. P. and RINGWOOD, A. E. (1964), *Density distribution and constitution of the mantle*. Rev. Geophys. Space Phys. *2*, 35–88.

GALSON, D. A. and RYBACH, L. (1984), *The European geotraverse project: general and geothermal aspects*, In *Ermittlung der Temperaturverteilung im Erdinnern* (ed. G. Bunterbarth). Inst. f. Geophysik, TU Clausthal, pp. 26–28.

HAAK, V. (1982), *Electrical conductivity of minerals and rocks at high temperatures and pressures*, In Landolt-Börnstein Monograph, *Physical Properties of Rocks*, Vol. V1b (ed. G. Angenheister). Springer-Verlag, Berlin, Heidelberg, New York, pp. 291–307.

HAENEL, R. (1970), *Interpretation of the terrestrial heat flow in the Rhinegraben*, In *Graben Problems* (eds J. H. Illies, S. Mueller). E. Schweizerbartsche Verlagsbuchhandlung, Stuttgart, pp. 116–120.

HURTIG, E. and OELSNER, CH. (1977), *Heat flow, temperature distribution and geothermal models in Europe: some tectonic implications*, Tectonophysics *41*, 147–156.

JESSOP, A. M., HOBART, M. A. and SCLATER, J. G. (1976), *The World Heat Flow Data Collection 1975*. Geothermal Service of Canada, Geotherm. Ser., Ottawa, 125 pp.

KARIYA, K. A. and SHANKLAND, T. J. (1983), *Electrical conductivity of dry lower crustal rocks*. Geophysics *48*, 52–61.

KELLER, G. V. (1982), *Electrical properties or rocks and minerals*, In *Handbook of Physical Properties of Rocks*, Vol. 1 (ed. R. S. Carmichael). CRC Press, Inc. Boca Raton, Florida, pp. 218–290.

KUTAS, R. I. and GORDIYENKO, V. V. (1971), *Teplovoye polye Ukrainy* (in Russian). Naukova Dumka, Kiev, 140 pp.

LACHENBRUCH, A. H. (1968), *Preliminary geothermal model of the Sierra Nevada*. J. Geophys. Res. *73*, 6977–6989.

Laštovičková, M. (1983), *Laboratory measurements of electrical properties of rocks and minerals.* Geophys. Surveys *6*, 201–213.

Morgan, P. (1979), *Cyprus heat flow with comments on the thermal regime of the Eastern Mediterranean,* In *Terrestrial Heat Flow in Europe* (eds V. Čermák and L. Rybach). Springer-Verlag, Berlin, Heidelberg, New York 1979), pp. 144–151.

Morgan, P. and Sass, J. H. (1984), *Thermal regime of the continental lithosphere,* J. Geodynamics *1*, 143–166.

Parkhomenko, E. I. and Bondarenko, A. T. (1972), *Elektroprovodnost gornikh porod pri visokihk davleniyakh i temperaturakh* (in Russian). Nauka, Moscow, 279 pp.

Pollack, H. N. and Chapman, D. S. (1977), *On the regional variation of heat flow, geotherms, and lithosperic thickness.* Tectonophysics *36*, 279–296.

Roy, R. F., Blackwell, D. D. and Birch, F. (1968), *Heat generation of plutonic rocks and continental heat flow provinces.* Earth Planet. Sci. Lett. *5*, 1–12.

Roy, R. F., Blackwell, D. D., and Decker, E. R. (1972), *Continental heat flow,* In *The Nature of the Solid Earth* (ed. E. C. Robertson). McGraw-Hill, New York, pp. 506–543.

Schatz, J. F. and Simmons, G. (1972), *Thermal conductivity of Earth materials at high temperatures.* J. Geophys. Res. *77*, 6866–6983.

Stille, H. (1924), *Grundfragen der vergleichenden Tektonik.* Bornträger, Berlin, 443 pp.

Stromeyer, D. (1984), *Downward continuation of heat flow data by means of the least squares method.* Tectonophysics *103*, 55–66.

Volarovich, M. P. and Parkhomenko, E. I. (1976), *Electrical properties of rocks at high temperatures and pressures,* In *Geoelectric and Geothermal Studies* (East-Central Europe, Soviet Asia) (ed. A. Ádám). KAPG Geophys. Monograph, Ákadémiai Kiadó, Budapest, pp. 321–372.

Appendix

Laboratory data on DC electrical conductivity of rock obtained in the Geophysical Institute, Czechosl. Acad. Sci., in 1974–85 are summarized here. Rock samples were collected from several localities within the Bohemian Massif and the Western Carpathians, two major tectonic units of the Czechoslovak territory (Fig. 14), additional samples were collected in Upper Saxony (south GDR) and southwest Poland, both areas belonging to the northernmost part of the Bohemian Massif, and a few samples are from Bulgaria. Altogether 303 samples were investigated covering the wide interval of characteristic crustal and upper mantle rocks ranging from granites to ultrabasites.

Uniform cylindrical samples (8 mm in diameter, 9 mm in height) were prepared and the conductivity of rocks was measured within the temperature range of 200–1,000°C by two electrodes method using a Yokogawa (or Goertz) XY recorder with a stabilized voltage of 60 V. All measurements were performed in the argon atmosphere under partial pressure of oxygen of 10^{-1} Pa. The temperature during the experiment was measured by means of a NiCr-Ni thermocouple and the Thermoprogramik P-829 unit controlled the heating guaranteed the linearity of the heating rate at 10°C/min. The contact of the sample was treated with a solution of 'glossy Pt' and by annealing to a maximum of 200°C. More details on the laboratory technique and rock-samples preparation can be found in Laštovičková and Popule (1974). The method and data were checked by repeated measurements and by

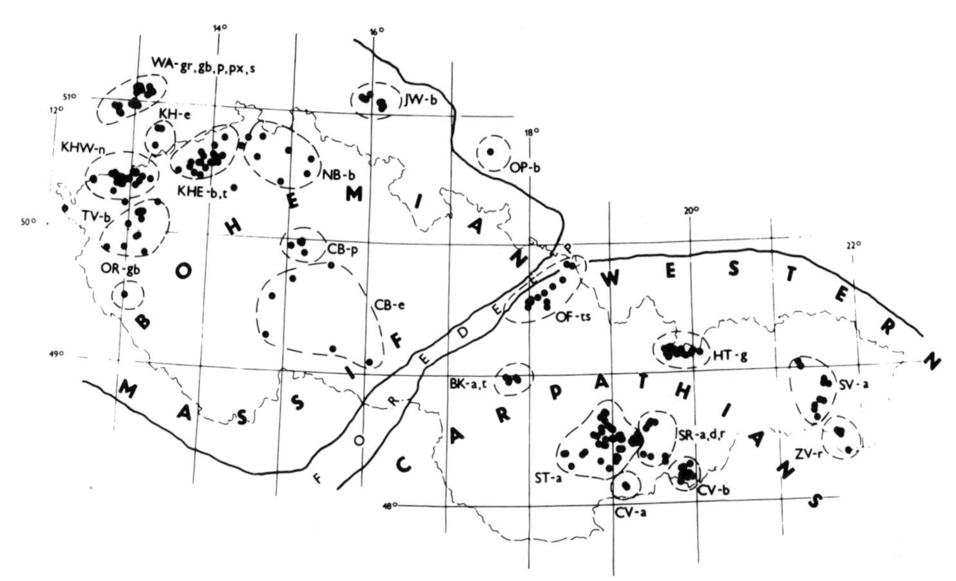

Figure 14
Simplified tectonic sketch of Czechoslovakia with localities where rock samples for laboratory conductivity measurements were collected. Explanations of codes: HT—High Tatras, WA—Waldheim, ST—Štiavnické vrchy, CV—Cerová vrchovina, SR—Slovenské Rudohorie, SV—Slánské vrchy, ZV—Zemplínské vrchy, JW—Jawor, OP—Opole, KH—Krušné Hory Mts, KHW—Krušné Hory west, KHE—Krušné Hory east, TV—Tepelská vrchovina, NB—Northern Bohemia, OF—Outer Flysch, BK—Biele Karpaty, CB—Central Bohemia; g—granites, gr—granulites, gb—gabbros, a—andesites, r—rhyolites, d—dacites, b—basalts, n—nephelinites, t—trachytes, ts—teschenites, p—peridotites, s—serpentinites, e—eclogites, px—pyroxenites.

comparing the results obtained by other methods (LAŠTOVIČKOVÁ, 1975, 1978a; BOCHNÍČEK and LAŠTOVIČKOVÁ, 1982).

The measured electrical conductivity data were completed with complex studies, which included the correlation of electrical properties of rocks with their composition, chemistry and structural parameters (PARKHOMENKO and LAŠTOVIČKOVÁ, 1978; LAŠTOVIČKOVÁ and PARKHOMENKO, 1978), combined examination of conductivity and magnetic properties (LAŠTOVIČKOVÁ and KROPÁČEK, 1976; LAŠTOVIČKOVÁ, 1978b), heat conductivity (CHANISHVILLI *et al.*, 1982, 1985), differential thermal analysis, thermogravimetry and X-ray analysis (KROPÁČEK and LAŠTOVIČKOVÁ, 1980) and the time factor studies of the temperature dependence of the electrical conductivity (LAŠTOVIČKOVÁ, 1982b, 1985).

All data are summarized in Table 3, which gives the mean log conductivities at selected temperatures of 200, 400, 600, 800, and 1 000°C together with corresponding standard deviations. Graphically the conductivity-temperature curves are shown in Figure 15.

The relatively highest conductivities were observed for the alkaline volcanic rocks of the Bohemian Massif (item No. 8), while similar rocks from the Western

Table 3

Temperature dependence of electrical conductivity of rocks, measured in argon atmosphere at partial pressure of oxygen $\sim 10^{-1}$ *Pa, n—number of samples*

No.[1]	Type of rock, locality and its code[2]	*n*	Mean log conductivity, log(S/m), and its standard deviation at selected temperatures, °C					Ref.[3]
			200	400	600	800	1,000	
1	Granite Western Carpathians HT-g	28	-6.71 ± 0.95	-5.62 ± 1.02	-3.74 ± 0.63	-3.20 ± 0.62	-2.57 ± 0.78	—
2	Acid granulite Upper Saxony WA-gr	17	-7.45 ± 0.77	-6.89 ± 0.70	-4.91 ± 0.80	-3.96 ± 0.54	-3.24 ± 0.29	1
3	Basic granulite Upper Saxony WA-gr	10	-6.50 ± 1.24	-5.80 ± 0.81	-4.24 ± 0.55	-3.29 ± 0.62	-2.78 ± 0.52	1
4	Gabbro Upper Saxony and Bohemian Massif WA-gb, OR-gb	4	-6.47 ± 0.83	-5.58 ± 0.91	-3.71 ± 0.34	-2.49 ± 0.36	-2.43 ± 0.33[4]	1
5	Calc-alkaline (andesite, rhyolite, dacite) and alkaline basaltic rocks West Carpathians ST-a, CV-b, SR-a, d, r, SV-a, ZV-r	56	-7.08 ± 1.03	-5.60 ± 1.08	-4.21 ± 0.85	-3.25 ± 0.87	-3.01 ± 0.82[4]	2,3
6	Basalt Bulgaria none	12	-5.06 ± 0.71	-3.97 ± 0.75	-3.22 ± 0.66	-2.58 ± 0.53	-2.31 ± 0.40[4]	—
7	Alkaline basalt Poland JW-b, OP-b	13		-5.15 ± 1.72	-3.75 ± 0.72	-3.22 ± 1.08		4
8	Alkaline volcanic rocks (basalt, nephelinite, trachyte) Bohemian Massif KHW-n, KHE-b, t, TV-b, NB-b	72		-3.94 ± 0.90	-3.00 ± 0.63	-2.42 ± 0.51		5

9	Teschenite Western Carpathians OF-ts	41		−5.26 ± 0.95	−3.22 ± 0.79	−2.49 ± 0.96		6
10	Trachyandesite Western Carpathians BK-a, t	10		−6.36 ± 0.59	−4.37 ± 0.61	−3.37 ± 1.08		6
11	Garnet peridotite Bohemian Massif CB-p	9	−9.41 ± 0.26	−6.88 ± 0.40	−4.11 ± 0.77	−3.53 ± 0.58	−3.12 ± 0.51	7
12	Peridotite and serpentinite Upper Saxony WA-p, s	9	−6.24 ± 1.55	−4.96 ± 0.83	−4.11 ± 0.85	−3.39 ± 0.81	−2.74 ± 0.67	1
13	Pyroxenite Upper Saxony WA-px	4	−5.70 ± 1.20	−5.07 ± 1.26	−4.22 ± 1.00	−3.43 ± 1.14	−2.86 ± 0.79	1
14	Eclogite Upper Saxony KH-e	6	−6.04 ± 1.03	−5.39 ± 1.01	−4.05 ± 1.03	−2.88 ± 0.89	−2.13 ± 0.65	1
15	Eclogite Bohemian Massif CB-e	8		−4.90 ± 0.85	−3.79 ± 0.81	−2.81 ± 0.38	−2.65 ± 0.42[4]	8
16	Miscellaneous, deep equivalents of volcanic rocks Western Carpathians none code	4	−6.33 ± 1.64	−5.84 ± 1.65	−4.88 ± 1.51	−3.85 ± 1.44	−3.37 ± 1.22[4]	2

[1] Number of item in Figure 15 showing log conductivity versus temperature distribution.
[2] Code showing locality in the schematic tectonic map, see Figure 14.
[3] Reference where detailed information on chemical analyses can be found: 1—Budzinski *et al.* (1985); 2—Kropáček *et al.* (1985); 3—Laštovičková (1984); 4—Jelenska *et al.* (1978); 5—Laštovičková and Kropáček (1983); 6—Laštovičková (1982a); 7—Laštovičková *et al.* (1985); 8—Dudek and Fediuková (1974). For code identifications see caption to Figure 14.
[4] At 900°C instead of 1,000°C.

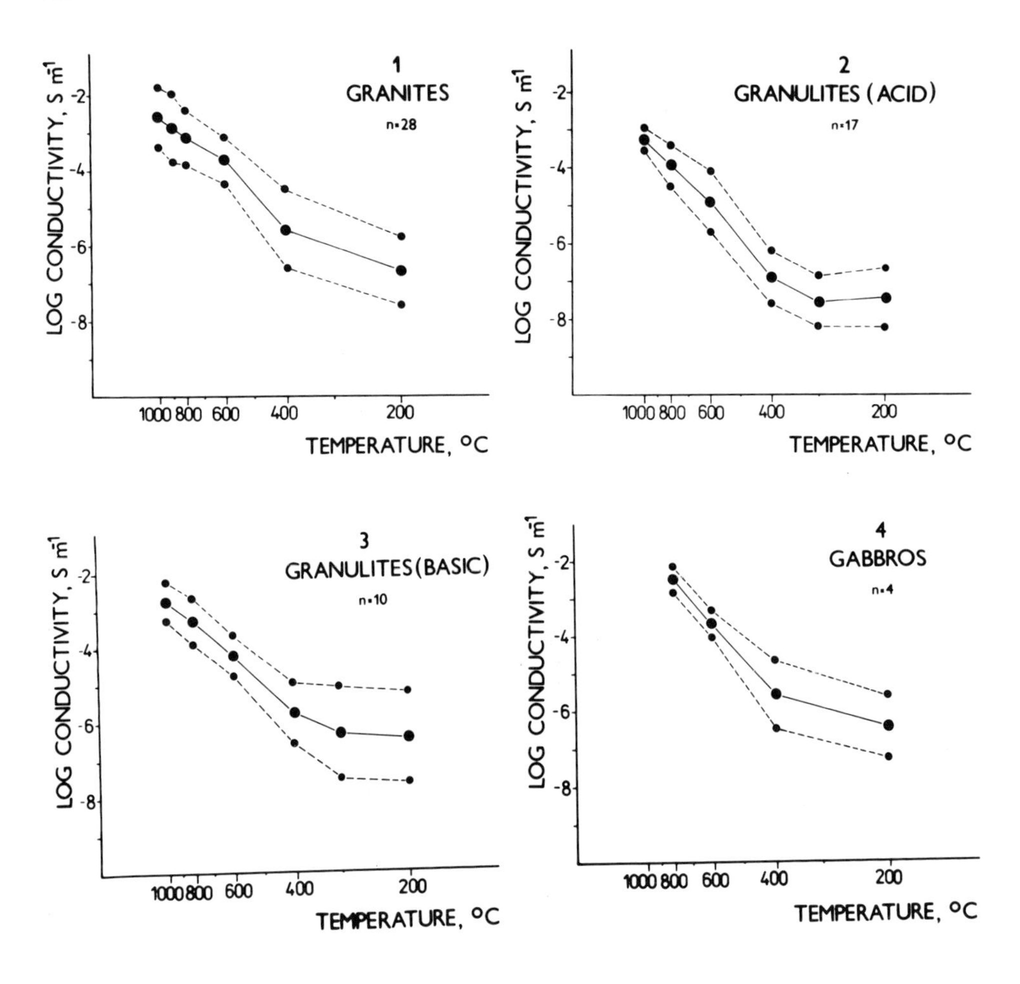

Figure 15a
Log conductivity-versus-temperature for various rocks, see Table 3.

Carpathians (item No. 5) revealed lower conductivities over the whole temperature range. There is a pronounced effect of admixture elements on the rock conductivity (Laštovičková and Kropáček, 1980; Laštovičková, 1984). Conductivity increases with rising content of metallic oxides: $\Sigma Fe = FeO + Fe_2O_3$, TiO_2 and/or with increasing content of *Ni*, *Cr* and *Co*, respectively. Experimental data were described numerically by applying the percolation theory to a set of 51 alkaline volcanic rocks from the Bohemian Massif and 16 calc-alkaline rocks from the Western Carpathians with a wide range of TiO_2 content, 0.1–6.0 wt%, and ΣFe, 2.5–17.0 wt% (Parkhomenko and Laštovičková, 1978). It was shown that the percolation theory might describe the behaviour of conductivity up to 500°C, but not above this

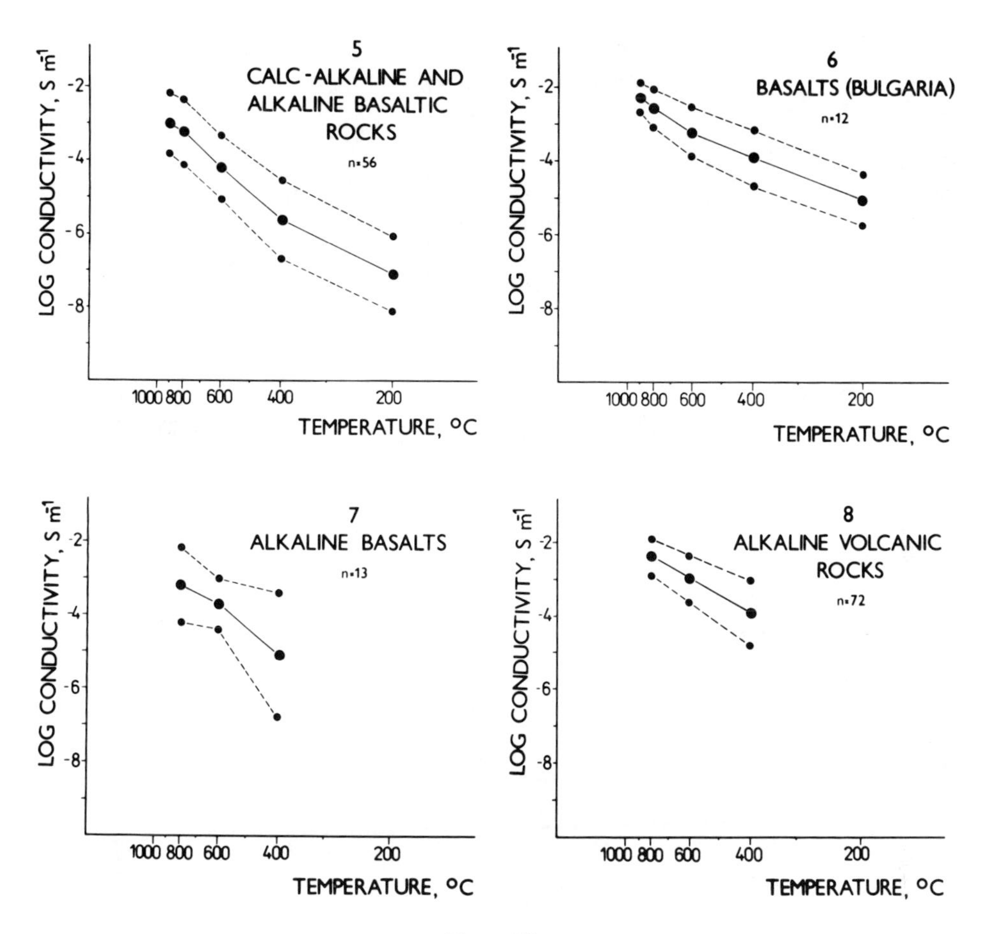

Figure 15b
Log conductivity-versus-temperature for various rocks, see Table 3.

temperature (Laštovičková and Kropáček, 1983). The electrical conductivity of *Fe-Ti-O* minerals, the main rockforming minerals of basaltic rocks, was investigated (Kropáček and Laštovičková, 1980), generally, the electrical conductivity was found to decrease with increasing degree of oxidation.

References for Appendix

Bochníček, J. and Laštovičková, M. (1982), *Time factor of polarization currents generating in measuring the electrical conductivity of rocks*. Stud. Geoph. Geod. *26*, 412–415.

Budzinski, H. Kopp, J., Kopf, M., Kropáček, V. and Laštovičková, M. (1985), *Elektrische und magnetische Eigenschaften der basischen Gesteine des sächsischen granulit Massif in Beziehung zur Petrologie*. Submitted to Gerlands Beitr. Geophys.

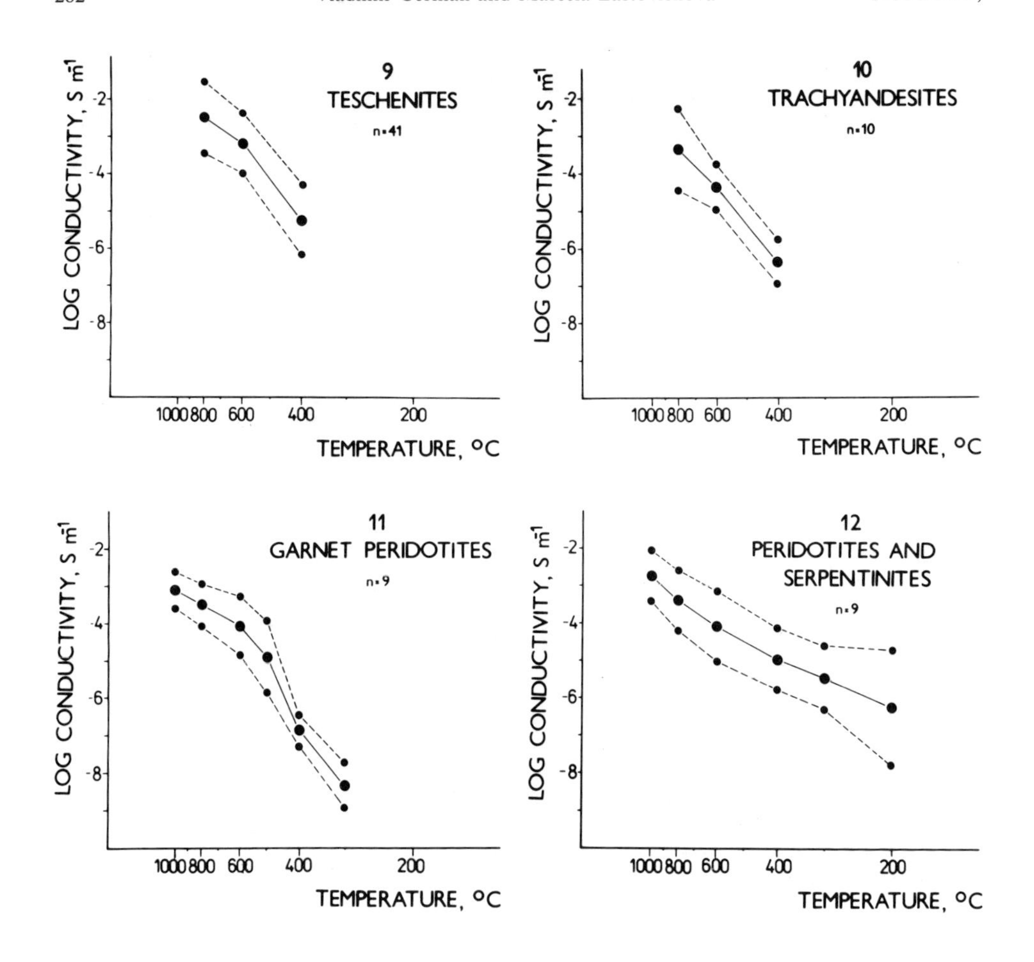

Figure 15c
Log conductivity-versus-temperature for various rocks, see Table 3.

CHANISHVILLI, Z. V., LAŠTOVIČKOVÁ, M. and KROPÁČEK, V. (1982), *Thermal and electric conductivity of three basaltic rocks of the Bohemian Massif under high pressure*. Stud. Geoph. Geod. *26*, 93–95.

CHANISHVILLI, Z. V., KROPÁČEK, V. and LAŠTOVIČKOVÁ, M. (1985), *Thermal and electric properties of basaltic rocks in the temperature interval of 20–900°C*, In *Physical Properties of the Mineral System of the Earth's Interior* (ed. A. KAPIČKA). Geoph. Inst., Czechosl. Acad. Sci., Prague, pp. 184–189.

DUDEK, A. and FEDIUKOVÁ, E. (1974), *Eclogites of the Bohemian Moldanubicum*. N. Jb. Miner. Abh., 127–159.

JELENSKA, M., KADZIALKO-HOFMOKL, M., KROPÁČEK, V., KRUCZYK, J. and POVONDRA, P. (1978), *Magnetic properties of basaltic rocks of Lower Silesia and their relationship to petrological characteristics*. *Čas. Min. Geol. 23*, 159–180.

KROPÁČEK, V. and LAŠTOVIČKOVÁ, M. (1980), *Changes of structure, phase composition and electrical conductivity under high-temperature oxidation of titano-magnetites*. J. Geophys. *48*, 40–46.

KROPÁČEK, V., LAŠTOVIČKOVÁ, M., POVONDRA, P. and KONEČNÝ, V. (1985), *Magnetic and electrical properties of volcanic rocks of the West Carpathians*. Trav. Inst. Géophys. Acad. Tchécosl. Sci., Geofys. sborník 1983, Academia, Praha (in press).

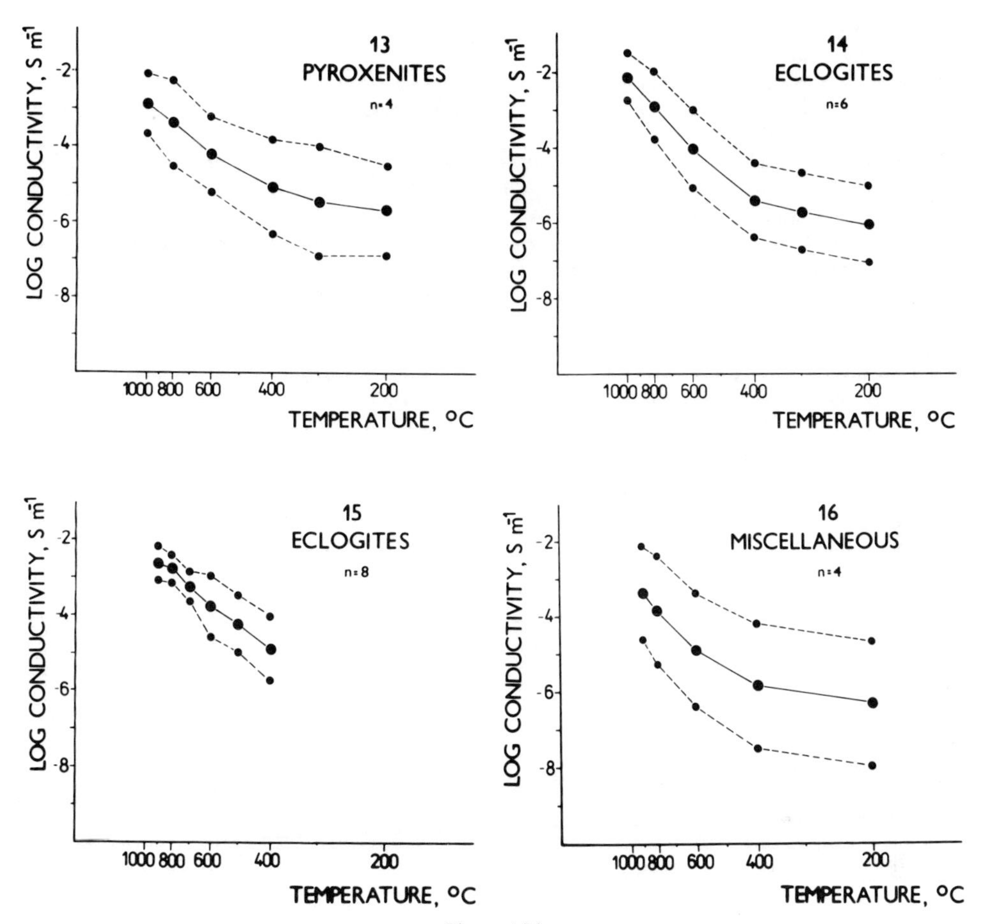

Figure 15d
Log conductivity-versus-temperature for various rocks, see Table 3.

LAŠTOVIČKOVÁ, M. (1975), *The electrical conductivity of eclogites measured by two methods*, Trav. Inst. Geophys. Acad. Tchécosl.Sci., Stud. Geoph. Geod. *19*, 394–398.

LAŠTOVIČKOVÁ, M. (1978a), *The electrical conductivity of rocks in the heating and cooling cycle.* Stud. Geoph. Geod. *22*, 184–189.

LAŠTOVIČKOVÁ, M. (1978b), *Changes of electrical conductivity in the vicinity of the Curie temperature of basaltic rocks—theoretical model.* Stud. Geoph. Geod. *22*, 98–102.

LAŠTOVIČKOVÁ, M. (1982), *Electrical conductivity of teschenites, trachyandesites and basalts under high temperatures.* Trav. Inst. Géophys. Acad. Tchécosl. Sci. *529*, Geofys. sborník 1979, Academia, Prague, pp. 233–250.

LAŠTOVIČKOVÁ, M. (1984), *Electric properties of neovolcanic rocks of Central and East Slovakia. 588*, Geofys. sborník 1981, Academia, Prague, pp. 217–233.

LAŠTOVIČKOVÁ, M. (1985), *Influence of the time-factor on laboratory measurements of high-temperature electrical conductivity of some rock-forming minerals*, In *Physical Properties of the Mineral System of the Earth's Interior* (ed. A. Kapička). Geophys. Inst., Czechosl. Acad. Sci., Prague, pp. 91–94.

LAŠTOVIČKOVÁ M. and KROPÁČEK, V. (1976), *Changes of electric conductivity in the neighbourhood of the Curie temperature of basalts.* Stud. Geoph. Geod. *20*, 265–272.

Laštovičková, M. and Kropáček, V. (1980), *Electrical conductivity of young basaltic rocks from Central and SE Slovakia.* Stud. Geoph. Geod. *24*, 389–399.

Laštovičková, M. and Kropáček, V. (1983), *Perkolyacionnaya model izmeneniya elektroprovodnosti shchelochnikh vulkanicheskikh porod Cheskogo massiva* (in Russian). Geof. Zhurnal Akad. nauk Ukr. SSR, Kiev 5, 77–81.

Laštovičková, M. and Parkhomenko, E. I. (1978), *Elekricheskiye svoystva bazaltov Cheskogo massiva pri visokikh termodinamicheskikh parametrakh.* Trav. Inst. Géophys. Acad. Tchécosl. Sci. *464*, Geofis. sborník 1976, Academia, Prague, pp. 289–304.

Laštovičková, M. and Popule, J. (1974), *A method of measuring the temperature dependence of the electrical conductivity of rocks.* Stud. Geoph. Geod. *18*, 370–380.

Laštovičková, M., Povondra, P., Fiala, J., Kropáček, V. and Pechala, F. (1985), *Electrical conductivity of pyropes and garnet peridotites in connection with their chemical analysis and other physical parameters.* Sb. Geol. věd ÚÚG *19*, 151–170.

Parkhomenko, E. I. and Laštovičková, M. (1978), *O vliyaniyi khimicheskogo i mineralnogo sostava na elektricheskoye soprotivleniye bazaltov i eklogitov* (in Russian), Fizika Zemli *11*, 79–85.

(Received 13th November, 1985, accepted 28th April, 1986)

PAGEOPH, Vol. 125, Nos. 2/3 (1987) 0033-4553/87/030285-05$1.50+0.20/0

Analyzing Electromagnetic Induction Data: Suggestions from Laboratory Measurements

A. G. Duba [1] and T. J. Shankland [2]

Abstract—Recent inversions of electrical profiles of the upper mantle beneath the oceans permit a variety of conductivity-depth profiles ranging from models with monotonically increasing conductivity to layered models having decreases of conductivity with depth. Laboratory data on possible mantle materials can physically explain high mantle conductivities in terms of a fluid phase (partial melt, hydrous fluid) or a good solid conductor (amorphous or graphitic carbon) and favor a profile having a high conductivity layer (HCL) underlain by a more resistive layer.

Key words: Mantle temperature profile, electrical conductivity.

Introduction

It is common practice in geophysics for researchers to use field observations to model quantities such as seismic velocities or electrical conductivity in an appropriate frequency range and then to use laboratory measurements of (presumably) the same quantities for interpretation. From these interpretations we obtain estimations of properties such as composition, porosity, or temperature. However, it can sometimes be helpful to turn the problem around and ask whether models based on laboratory data can constrain the analysis of field measurements. The issue that we address is whether electrical conductivity in the upper mantle rises monotonically with depth or whether the sharp rise in conductivity beginning about 40–80 km is followed by a decrease in conductivity at greater depth, thus producing a well-defined high conductivity layer (HCL). We show that present laboratory data favor the existence of an HCL having both a top and bottom to the region of high conductivity.

[1] Earth Sciences Department, Lawrence Livermore National Laboratory, Livermore, CA 94550, U.S.A.

[2] Earth and Space Sciences Division, Geophysics Group, Los Alamos National Laboratory, Los Alamos, NM 87545, U.S.A.

Conductivity profiles

Inversion of ocean bottom magnetotelluric data permits models to have either resistive or conductive layers beneath the HCL extending between 40 and 180 km depth (OLDENBURG *et al.*, 1984). To explain mantle conductivity profiles, we assume that the conductivity of the mantle above and beneath the HCL is controlled by solid silicates that compose the bulk of the mantle. Thus, we can use available data on the conductivity of relevant silicates to match the value of the electrical conductivity inferred for this region.

In contrast, the conductivity of the HCL is great enough that intercrystalline phases such as partial melt (CHAN *et al.*, 1973; WAFF, 1974; SHANKLAND and WAFF, 1977) or carbon (DUBA and SHANKLAND, 1982) have been invoked to supplement the low conductivity of silicates such as olivine.

If we consider the shallowest mantle layer above the HCL, the 'lithosphere' or 'lid' in seismic terms, this is the region of least uncertainty. The rock is primarily peridotite (WYLLIE, 1967; RINGWOOD, 1975) and eclogite (ANDERSON and BASS, 1984) whose mineralogies are dominated by olivine and pyroxene; petrological and heat flow studies indicate that temperatures range from 600–1,200°C. At the present time, the conductivity in this region is difficult to resolve, but it seems to be of the order 10^{-4} to 10^{-2} S/m (FILLOUX, 1982a, 1982b; YUKUTAKE *et al.*, 1983). Within these uncertainties there are no insuperable problems in reconciling the values of conductivities at plausible temperatures with experimental conductivity curves summarized in Figure 1. The curves of single crystal olivine (Red Sea Peridot or San Carlos are probable minimum conductivities. The effect of contact with other minerals and of grain boundaries is only likely to raise conductivities into the range indicated by the lherzolite (peridotite) curves.

Similar arguments apply to conductivities at depths greater than about 200 km. Again, mantle conductivities of the order of 0.01 to 0.1 (FILLOUX, 1982a, 1982b; YUKUTAKE *et al.*, 1983) can be explained by the laboratory data of Figure 1 for minerals. Plausible 'electrogeotherms' based on laboratory data have previously been published for global conductivity-depth profiles (DUBA, 1976).

The region that has been the most difficult to explain by single crystal or rock conductivities is the HCL. Conductivities of the order of 0.1 S/m at temperatures well below 1,400°C, and the relatively abrupt rise of conductivity with depth that seems associated with models proposed to date (OLDENBURG *et al.*, 1984), argue strongly for conduction taking place through a mechanism other than bulk conduction in mineral grains or in ordinary grain boundaries. Whether the enhanced electrical conduction takes place in a partial melt, elemental carbon in grain boundaries, or in a hydrous phase arising from amphibole dehydration (TOZER, 1979), is to first order less significant than the fact that these mechanisms are characterized by having both a top and a bottom. The top of an HCL occurs where the conducting fluid or carbon phase has sufficient volume to become interconnected.

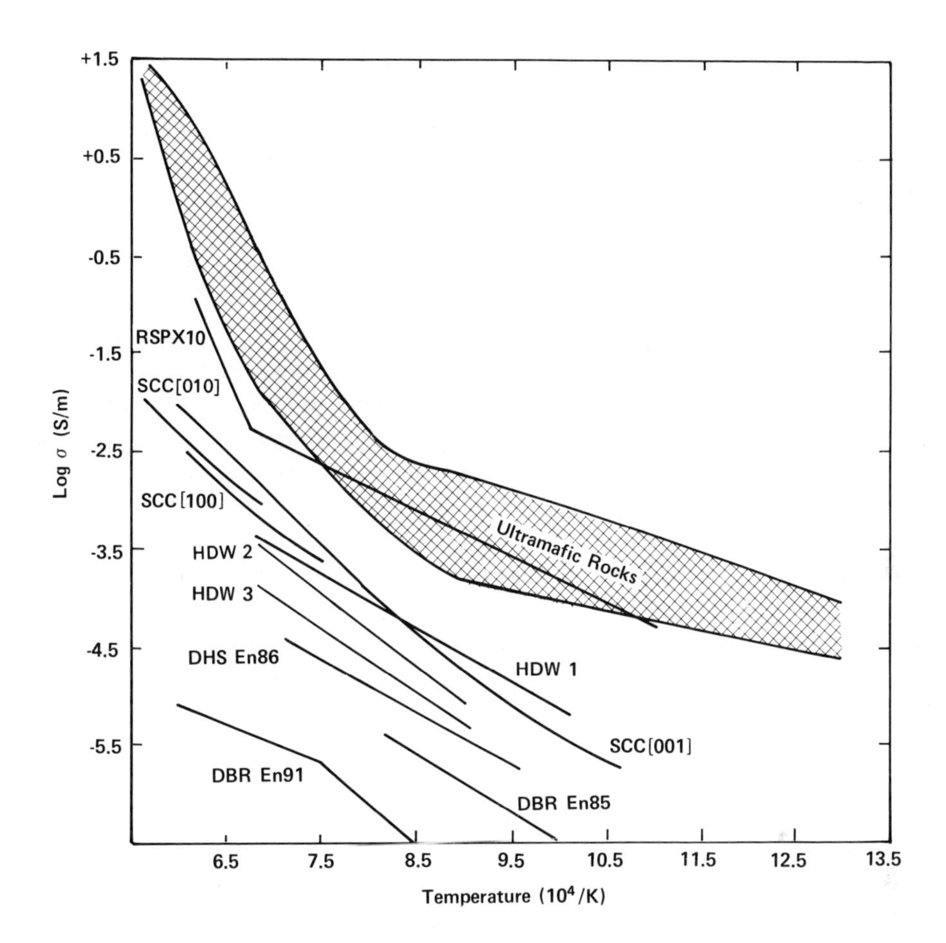

Figure 1

Electrical conductivity in olivine (DUBA and SHANKLAND, unpublished), pyroxene (DUBA, 1976; DUBA *et al.*, 1976; HUEBNER *et al.*, 1979), and a variety of ultrabasic rocks (RAI and MANGHNANI, 1978). RSPX10 is a peridot conductivity arbitrarily enhanced to allow for possible grain boundary contributions (SHANKLAND and WAFF, 1977).

At the bottom of the layer the conducting fluid or carbon phase is no longer present or has inverted to an insulator, i.e., carbon to diamond. Although all the models presented by OLDENBURG *et al.* (1984) have relatively abrupt conductivity rises at the top of the HCL, they show that the inversion procedures do not require a relatively abrupt drop of conductivity beneath the HCL. For their smoothest models (that have constant conductivity in the HCL and no bottom to the layer) the amount of the good conducting phase is continuously reduced as the single crystal conductivity rises with depth and temperature. However, from the point of view of a physical explanation involving a special conducting phase, the expectation is that the behavior of the bottom of the layer should resemble that of the top, namely a relatively abrupt change of conductivity with depth as volume fraction and interconnectivity decrease with depth. If the good conducting phase is amorphous or graphitic carbon, then the transition to a nonconducting diamond phase should also lead to a relatively sharp drop of conductivity beneath the HCL.

Conclusion

In this paper we have argued that physical explanations of the region of elevated conductivity in the region of 50–200 km in the mantle require the presence of a better conducting phase than the minerals thought to compose the uppermost mantle, as they are known from current laboratory measurements. We would expect that an HCL based upon such an explanation would have a conductivity drop beneath the HCL that would be as abrupt, if not so large, as that at the top of the HCL. From the point of view of those modeling electromagnetic induction profiles, while the existence of an abrupt bottom to the HCL may not be required by the inversion procedures, such a feature would be very consistent with a plausible physical explanation of upper mantle conductivity. Furthermore, incorporating such a feature into the model could aid in better definition of the conductivity of the HCL.

Acknowledgments

This work was performed under the auspices of the U.S. Department of Energy by the Lawrence Livermore National Laboratory under contract number W-7405-ENG-48 and by Los Alamos National Laboratory under contract number W-7405-ENG-36.

References

ANDERSON, D. L. and BASS, J. D. (1984), *Mineralogy and composition of the upper mantle.* Geophys. Res. Letters *11*, 637–640.

CHAN, T., NYLAND, E. and GOUGH, D. I. (1973), *Partial melting and conductivity anomalies in the upper mantle.* Nature Phys. Sci. *244*, 89–90.

DUBA, A. G. (1976), *Are laboratory electrical conductivity data relevant to the Earth?* Acta Geodaet., Geophys. et Montanist. Acad. Sci. Hung. Tomus. *11*, 484–495.

DUBA, A. G., BOLAND, J. N. and RINGWOOD, A. E. (1973), *Electrical conductivity of pyroxenes.* J. Geol. *81*, 727–735.

DUBA, A. G., HEARD, H. C. and SCHOCK, R. N. (1976), *Electrical conductivity of orthopyroxene to 1,400°C and the selenotherm.* Proc. Lunar Sci. Conf., 7th, 3173–3181.

DUBA, A. G. and SHANKLAND, T. J. (1982), *Free carbon and electrical conductivity in the Earth's mantle*, Geophys. Res. Letters *9*, 1271–1274.

FILLOUX, J. H. (1982a), *Magnetotelluric experiment over the ROSE Area.* J. Geophys. Res. *87*, 8364–8378.

FILLOUX, J. H. (1982b), *Seafloor magnetotelluric soundings in the Mariana Island Arc Area*, in *The Tectonic and Geologic Evolution of Southeast Asian Seas and Islands*, Part 2. AGU Geophysical Monograph Series *27*, American Geophysical Union, Washington, D.C., 255–265.

HUEBNER, J. S., DUBA, A. G. and WIGGINS, E. B. (1979), *Electrical conductivity of pyroxene which contains trivalent cations: Laboratory measurements and the lunar temperature profile.* J. Geophys. Res. *84*, 4652–4656.

OLDENBURG, D. W., WHITTAL, K. P. and PARKER, R. L. (1984), *Inversion of ocean bottom magnetotelluric data revisited.* J. Geophys. Res. *89*, 1829–1833.

RAI, C. S. and MANGHNANI, M. H. (1978), *Electrical conductivity of ultramafic rocks to 1820 kelvin.* Phys. Earth Planet. Interiors *17*, 6–13.

RINGWOOD, A. E. (1975), *Composition and Petrology of the Earth's Mantle.* McGraw-Hill, New York.

SHANKLAND, T. J. and WAFF, H. S. (1977), *Partial melting and electrical conductivity anomalies in the upper mantle.* J. Geophys. Res. *82*, 5409–5417.

TOZER, D. C. (1979), *The interpretation of upper-mantle electrical conductivities.* Tectonophys. *56*, 147–163.

WAFF, H. S. (1974), *Electrical conductivity in a partially molten mantle and implications for geothermometry.* J. Geophys. Res. *79*, 4003–4010.

WYLLIE, P. J. (1971), *The Dynamic Earth: Textbook in Geosciences.* Wiley, New York.

YUKUTAKE, T., FILLOUX, J. H., SEGAWA, J., HAMANO, Y. and UTADA, H. (1983), *Preliminary report on a magnetotelluric array study in the Northwest Pacific.* J. Geomag. Geoelectr. *35*, 575–587.

(Received 5th November, 1985, accepted 28th April, 1986)

PAGEOPH, Vol. 125, Nos. 2/3 (1987) 0033-4553/87/030291-28$1.50+0.20/0

Appreciation of Spherically Symmetric Models of Electrical Conductivity

J. PĚČOVÁ,[1] Z. MARTINEC[2] and K. PĚČ[2]

Abstract—A procedure is suggested of a more effective and faster computation of the impedance, the transfer function and amplitudes of the induced field in a spherically symmetric model of the electrical conductivity. The existing induction data have been supplemented by about 80 new values derived from the analysis of daily means. The fit of the existing 1-D models of the electrical conductivity of the mantle to the set of induction data is investigated. The characteristic equation for the free electromagnetic oscillations of a radially inhomogeneous Earth is derived and its possible importance in solving the inverse problem of electric conductivity is pointed out.

Key words: Electrical conductivity Earth's models, electromagnetic induction, free electromagnetic oscillations.

1. Introduction

In recent years, research has been devoting increased attention to the study of the spatial distribution of the Earth's parameters. Progress has been made in determining the lateral changes of the elastic parameters of the Earth (WOODHOUSE and DZIEWONSKI, 1984; DZIEWONSKI, 1984; NATAF *et al.* 1984) and of its density (PĚČ and MARTINEC, 1984; MARTINEC and PĚČ, 1986). Determining the distribution of the Earth's electrical conductivity is a more complicated problem than determining the distribution of the mechanical parameters. The reasons for this are the drastic changes of the conductivity in the Earth's mantle, as much as six orders of magnitude in the radial direction and five in the horizontal, whereas the lateral variations of, e.g., the density do not exceed 5% of the mean value for a particular depth. Especially the surface layers are strongly laterally inhomogeneous due to the contrast of the electrical conductivity between the oceans and continents. Finally, the frequency range of the variations of the electromagnetic field is very wide. In the last two decades, the theory of solving inverse geophysical problems has been widely developed (WIGGINS, 1972; PARKER, 1970, 1983; ANDERSSEN, 1979). Considerable progress

[1] Geophysical Institute, Czechosl. Acad. Sci., Boční II, 141 31 Prague 4.
[2] Department of Geophysics and Meteorology, Faculty of Mathematics and Physics, Charles University, V Holešovičkách 2, 180 00 Prague 8.

has also been made in the theory of electromagnetic induction, and effective procedures have been introduced for solving local 3-D problems, whereas the solution of the problem on a global scale is known only for the model of the 1-D radial distribution of the electrical conductivity (BERDICHEVSKY *et al.*, 1976; BANKS, 1981; BAILEY, 1970; ECKHARDT, 1963 and others).

A number of methods for solving inverse geophysical problems require an *a priori* estimate of the solution, which is then improved (TARANTOLA and VALETTE, 1982). From this point of view, a mean, spherically symmetric model of the Earth, consistent with the observations of the variations of the electromagnetic field, becomes important for solving the inverse 3-D problem of electrical conductivity. This paper deals with some of the aspects of this problem. This involves the modification of the theory of electromagnetic induction for a spherically symmetric Earth, which enables a fast solution of the direct problem for the impedance, the transfer function and for the depth dependence of the amplitudes of the induced field. Existing and additional induction data were used to appreciate and compare some of the models of electrical conductivity. Finally, thc equation of the free electromagnetic oscillations of an inhomogeneous Earth is derived, and the roots of the characteristic frequency equation are analysed.

2. *The Matrix Method and the Recurrent Formulas for the Electromagnetic Induction in a Spherically Layered Earth*

A number of authors has been concerned with treating the electromagnetic induction in the Earth, (e.g., LAHIRI and PRICE, 1939; PRICE, 1950, 1962; SRIVASTAVA, 1966). In the recurrent impedance formulas, the information on the radial dependence of the amplitudes of the electromagnetic field are suppressed, and many authors replace it with approximate formulas for the skin-effect which hold precisely for a homogeneous half-space. The recurrent amplitude formulas were derived by HVOŽDARA and SIRÁŇ (1975). A different approach to solving the impedance and amplitude problem, based on the matrix method, was presented by PĚČ *et al.* (1985), which we shall refer to as PMP. If only the impedance is being considered, both approaches are equivalent. PMP give more advantageous recurrent formulas for computing the impedance. The advantage of the matrix method over the recurrent formulas is mainly displayed in computing the amplitudes, as we shall show later.

Let us assume that the distribution of the electrical conductivity within the Earth is described by a spherically symmetric conductivity model. Let us divide this spherical conductor into a number of homogeneous and isotropic layers without field sources. Let the index m denote the m-th spherical layer ($m = 1, 2, \ldots, M$) with an inner radius r_m and outer radius r_{m-1}. The radius of the Earth's surface is r_0. Every layer is characterized by the constant electrical conductivity σ_m. Assume that the permeability μ_0 is constant within the model and equal to the value in vacuum. The

Maxwell equations in the quasistationary approximation for the m-th layer reduce to Helmholtz's differential equations for the magnetic intensity H_m and the electric intensity E_m:

$$\Delta H_m + k_m^2 H_m = 0, \quad \Delta E_m + k_m^2 E_m = 0, \tag{1}$$

where the wave number k_m is given as

$$k_m^2 = -i\omega\mu_0\sigma_m, \tag{2}$$

and the time dependence is assumed to have the form exp $i\omega t$. For the chosen model of the medium it holds that

$$\nabla \cdot H_m = 0, \quad \nabla \cdot E_m = 0. \tag{3}$$

The intensities H_m and E_m are related by the differential equation

$$\nabla \times E_m + i\omega\mu_0 H_m = 0. \tag{4}$$

The solution of Equations (1)–(4) in terms of spherical coordinates (r, θ, φ) can be found, e.g., in STRATTON, (1941). It can be proved that the electromagnetic field can be divided into two independent parts: (1) Into the part which contains the toroidal component of the electrical intensity and the poloidal component of the magnetic intensity—this is sometimes referred to as oscillations of the magnetic type, and (2) into the part which contains the poloidal component of the electrical intensity and the toroidal component of the magnetic intensity, sometimes referred to as oscillations of the electrical type.

We shall restrict ourselves to studying the oscillations of the magnetic type. In the m-th homogeneous layer $r \in (r_m, r_{m-1})$ the magnetic type is described by the relations

$$E_r = 0, \quad E_\theta = w_n(k_m r)\frac{1}{\sin\theta}\frac{\partial Y_{nl}(\theta, \varphi)}{\partial\varphi}, \quad E_\varphi = -w_n(k_m r)\frac{\partial Y_{nl}(\theta, \varphi)}{\partial\theta}, \tag{5}$$

$$\begin{aligned} H_r &= -\frac{n(n+1)}{i\omega\mu_0}\frac{1}{r}[w_n(k_m r)\ Y_{nl}(\theta, \varphi)], \\ H_\theta &= -\frac{1}{i\omega\mu_0}\frac{1}{r}\frac{d}{dr}[rw_n(k_m r)]\frac{\partial Y_{nl}(\theta, \varphi)}{\partial\theta} \\ H_\varphi &= -\frac{1}{i\omega\mu_0}\frac{1}{r}\frac{d}{dr}[rw_n(k_m r)]\frac{1}{\sin\theta}\frac{\partial Y_{nl}(\theta, \varphi)}{\partial\varphi}, \end{aligned} \tag{6}$$

where

$$w_n(k_m r) = \alpha_m j_n(k_m r) + \beta_m y_n(k_m r), \tag{7}$$

$j_n(k_m, r)$ and $y_n(k_m r)$ being spherical Bessel functions of the first and second kind, respectively, $Y_{nl}(\theta, \varphi)$ spherical harmonic functions, and α_m, β_m constants. We shall introduce the matrix

$$A_m(r) = \begin{pmatrix} j_n(k_m r) & y_n(k_m r) \\ -\frac{1}{i\omega\mu_0}\frac{1}{r}\frac{d}{dr}[rj_n(k_m r)] & -\frac{1}{i\omega\mu_0}\frac{1}{r}\frac{d}{dr}[ry_n(k_m r)] \end{pmatrix}. \tag{8}$$

The intensity components E_θ and H_φ for $r \in (r_m, r_{m-1})$ can then be expressed as follows:

$$\begin{pmatrix} E_\theta \\ H_\varphi \end{pmatrix} = A_m(r) \begin{pmatrix} \alpha_m \\ \beta_m \end{pmatrix}. \tag{9}$$

By eliminating the constants α_m and β_m, we obtain the relation between the intensities at the upper and lower boundary of the layer:

$$\begin{pmatrix} E_\theta \\ H_\varphi \end{pmatrix}_{r_{m-1}} = A_m(r_{m-1})A^{-1}(r_m) \begin{pmatrix} E_\theta \\ H_\varphi \end{pmatrix}_{r_m} = B_m \begin{pmatrix} E_\theta \\ H_\varphi \end{pmatrix}_{r_m}. \tag{10}$$

The elements of the matrix of layer B_m are (PMP)

$$\begin{aligned} B_{11} &= (k_m r_m)^2 q_n + k_m r_m p_n, \; B_{12} = \frac{(k_m r_m)^2}{\kappa_m} p_n, \\ B_{21} &= -\kappa_m (k_m r_m)^2 \left\{ s_n + \frac{r_n}{k_m r_m} + \frac{q_n}{k_m r_{m-1}} + \frac{p_n}{k_m^2 r_m r_{m-1}} \right\}, \\ B_{22} &= -(k_m r_m)^2 r_n - k_m \frac{r_m^2}{r_{m-1}} p_n, \quad \kappa_m = k_m/\omega\mu_0. \end{aligned} \tag{11}$$

Cross-products of spherical Bessel functions and their derivatives (the prime indicates a derivative with respect to the argument) have been introduced in Equations (11), i.e.,

$$\begin{aligned} p_n &= j_n(k_m r_{m-1})y_n(k_m r_m) - j_n(k_m r_m)y_n(k_m r_{m-1}), \\ q_n &= j_n(k_m r_{m-1})y_n'(k_m r_m) - j_n'(k_m r_m)y_n(k_m r_{m-1}), \\ r_n &= j_n'(k_m r_{m-1})y_n(k_m r_m) - j_n(k_m r_m)y_n'(k_m r_{m-1}), \\ s_n &= j_n'(k_m r_{m-1})y_n'(k_m r_m) - j_n'(k_m r_m)y_n'(k_m r_{m-1}). \end{aligned} \tag{12}$$

The continuity of components E_θ and H_φ at the boundaries between layers and Equation (10) enable the relation between E_θ and H_φ for the deepest boundary and the surface r_0 to be expressed as

$$\begin{pmatrix} E_\theta \\ H_\varphi \end{pmatrix}_{r_0} = P \begin{pmatrix} E_\theta \\ H_\varphi \end{pmatrix}_{r_M}, \tag{13}$$

where the (2×2) matrix P is the product of the layer matrices:

$$P = B_1 \cdot B_2 \cdot \cdots \cdot B_M. \tag{14}$$

Since the field must be regular at the origin, constant β_{M+1} must be equal to zero. The initial values for the matrix multiplication are

$$\begin{pmatrix} E_\theta \\ H_\varphi \end{pmatrix}_{r_M} = \alpha'_{M+1} \begin{pmatrix} 1 \\ [\chi_n(k_{M+1} r_M) - (n+1)]/(i\omega\mu_0 r_M) \end{pmatrix}, \tag{15}$$

where $\alpha'_{M+1} = \alpha_{M+1} j_n(k_{M+1} r_M)$ and

$$\chi_n(x) = \frac{x j_{n+1}(x)}{j_n(x)}. \tag{16}$$

The constant α'_{M+1} depends on the type of source exciting the induced field. The final values of the matrix multiplication (13) are

$$\begin{aligned} E_\theta(r_0) &= P_{11} E_\theta(r_M) + P_{12} H_\varphi(r_M), \\ H_\varphi(r_0) &= P_{21} E_\theta(r_M) + P_{22} H_\varphi(r_M). \end{aligned} \tag{17}$$

Equations (17) express the relation of the amplitudes of the components of the magnetic and electrical intensity at the surface and at the deepest boundary. They depend on the factor α'_{M+1}, i.e., in general on the type of source. However, if we are only interested in the relative changes of the amplitudes, the dependence on α'_{M+1} vanishes.

In induction problems, the concept of impedance is introduced to an advantage; it can be used to express the transfer function and the apparent resistivity. Let us define the radially dependent impedance $Z(r)$ in the same way as SRIVASTAVA (1966): $Z(r) = -E_\theta/H_\varphi = E_\varphi/H_\theta$. If we put $Z_m = Z(r_m)$ and $Z_{m-1} = Z(r_{m-1})$, (10) will yield the recurrent formula for Z_m:

$$Z_{m-1} = \frac{B_{11} Z_m + B_{12}}{B_{21} Z_m + B_{22}}, \tag{18}$$

where B_{ij} are given by (11). SRIVASTAVA (1966) derived Equation (18) in a different way, but he did not introduce the cross-products p_n, q_n, r_n and s_n (12). If, following BANKS (1969), we denote the ratio of the radial component H_r to the tangential component H_θ (on the surface) as $W_n(\omega)$, with a view to (6) we obtain

$$W_n(\omega) = n(n+1)\left\{w_n(k_1 r_0)/\frac{d}{dr_0}\,[r_0 w_n(k_1 r_0)]\right\}. \tag{19}$$

The transfer function (BANKS, 1969)

$$Q_n(\omega) = \frac{n - W_n(\omega)}{n + 1 + W_n(\omega)} \tag{20}$$

and the surface impedance

$$Z_0 = \frac{i\omega\mu_0 r_0}{n(n+1)}\,W_n(\omega) \tag{21}$$

are expressed in terms of function $W_n(\omega)$.

The derivation of Equation (18), which is the simple consequence of (10), indicates that the matrix method and the recurrent formulas are equivalent. However, the situation is different when the depth dependence of the amplitudes is computed.

3. *Amplitudes*

Equation (10) enables us to express the amplitudes E_θ and H_φ at any boundary, including the surface, with the aid of constant α_{M+1}, or, in other words, the matrix method makes use of the same matrix relations and the same combinations of Bessel functions in computing the impedances as in computing the amplitudes.

In determining the amplitudes by means of the recurrent formulas, one makes use of the continuity of E_θ and H_φ at the separate boundaries, e.g., at boundary $r = r_{m-1}$, in virtue of (5) and (6), the following system of two equations holds:

$$\begin{aligned}
&\alpha_m j_n(k_m r_{m-1}) + \beta_m y_n(k_m r_{m-1}) = \alpha_{m-1} j_n(k_{m-1} r_{m-1}) + \beta_{m-1} y_n(k_{m-1} r_{m-1}),\\
&\alpha_m \frac{d}{dr_{m-1}}\,[r_{m-1} j_n(k_m r_{m-1})] + \beta_m \frac{d}{dr_{m-1}}\,[r_{m-1} y_n(k_m r_{m-1})]\\
&= \alpha_{m-1} \frac{d}{dr_{m-1}}\,[r_{m-1} j_n(k_{m-1} r_{m-1}) + \beta_{m-1} \frac{d}{dr_{m-1}}\,[r_{m-1} y_n(k_{m-1} r_{m-1})].
\end{aligned} \tag{22}$$

For the deepest boundary, $r = r_M$ and $\beta_{M+1} = 0$, in particular

$$\begin{aligned}
&\alpha_{M+1} j_n(k_{M+1} r_M) = \alpha_M j_n(k_M r_M) + \beta_M y_n(k_M r_M),\\
&\alpha_{M+1} \frac{d}{dr_M}\,[r_M j_n(k_{M+1} r_M)] = \alpha_M \frac{d}{dr_M}\,[r_M j_n(k_M r_M)] + \beta_M \frac{d}{dr_M}\,[r_M y_n(k_M r_M)].
\end{aligned} \tag{23}$$

Equations (22) and (23) enable the recurrent formulas for α_m and α_{m-1}, and β_m and β_{m-1} to be expressed in terms of constant α_{M+1}. The relevant details can be found in the paper of HVOŽDARA and SIRÁŇ (1975).

As can be seen from Equations (22) and (23), the Bessel functions there have arguments different to those in the recurrent formulas for the impedance and, therefore, in computing the amplitudes by means of the recurrent formulas, further Bessel functions have to be computed without utilizing the spherical Bessel functions determined in computing the impedance. This duplicity can be avoided if the matrix method is used.

4. *Comments on the Numerical Computations*

The computation of the transfer functions and the impedances using the recurrent (18) or matrix formulas (13–15) is essentially simple and easy to program. From the numerical point of view, the only difficulty is determining the spherical Bessel functions with complex arguments. If the standard model of the electrical conductivity in the mantle is being considered, i.e., $10^{-3} \leqq \sigma \leqq 10^{3}$ $(\Omega m)^{-1}$, $4 \times 10^{6} \leqq r \leqq 6 \times 10^{6}$ m and periods ranging from 10^{2} s to 11 years, the real and imaginary parts of the arguments of the Bessel functions take values from the interval $(10^{-2}, 10^{4})$. The spherical Bessel functions can be computed numerically by expanding them into power series, or by means of recurrent formulas (ABRAMOWITZ, and STEGUN, 1970). It has been proved convenient to use the power series expansions if the absolute value of the argument of the Bessel functions is smaller than the order n; in the opposite case it is better to employ the recurrent formulas.

As regards the computer time, it is not advantageous to compute separately the spherical Bessel functions of the first and second kind for arguments relating to the top and bottom of the spherical layer, and their derivatives. A considerable amount of computer time can be saved by computing the crossproducts p_n, q_n, r_n and s_n (12) directly instead of the separate Bessel functions. The following recurrent formulas hold for these products (PMP):

$$
\begin{aligned}
p_{n+1} &= s_n + \frac{n^2}{z_1 z_2} p_n - \frac{n}{z_1} q_n - \frac{n}{z_2} r_n, \\
q_{n+1} &= \frac{n}{z_1} p_n - \frac{n+2}{z_2} p_{n+1} - r_n, \\
r_{n+1} &= \frac{n}{z_1} p_n - \frac{n+2}{z_2} p_{n+1} - q_n, \\
s_{n+1} &= p_n - \frac{(n+2)^2}{z_1 z_2} p_{n+1} - \frac{n+2}{z_1} q_{n+1} - \frac{n+2}{z_2} r_{n+1}.
\end{aligned}
\tag{24}
$$

To appreciate the accuracy of the numerical computations, one can use the identity

$$
p_n s_n - q_n r_n = 1/(z_1 z_2)^2. \tag{25}
$$

In Equations (24) and (25) $z_1 = k_m r_{m-1}$ and $z_2 = k_m r_m$. The initial values ($n = 0$) for the recurrent formula (24) are:

$$\begin{aligned} p_0 &= \sin(z_2 - z_1)/(z_1 z_2); \\ q_0 &= \cos(z_2 - z_1/(z_1 z_2) - \sin(z_2 - z_1)/(z_1 z_2^2); \\ r_0 &= -\cos(z_2 - z_1/(z_1 z_2) - \sin(z_2 - z_1)/(z_1^2 z_2); \\ s_0 &= \sin(z_2 - z_1)/(z_1 z_2) - \cos(z_2 - z_1)/(z_1^2 z_2) \\ &\quad + \cos(z_2 - z_1)/(z_1 z_2^2) + \sin(z_2 - z_1)/(z_1^2 z_2^2). \end{aligned} \tag{26}$$

It is recommended to start the matrix multiplication, or computation of the recurrent relations from the deepest boundary.

Function $\chi_n(x)$, defined by (16), occurs in Equations (15). The recurrent formulas for this function read

$$\chi_{n+1}(x) = (2n + 3) - \frac{x^2}{\chi_n(x)} \tag{27}$$

and the initial value is

$$\chi_0(x) = 1 - x \operatorname{cotan}(x). \tag{28}$$

A substantial advantage of recurrent formulas (24) and (26), as compared to the direct computation of the separate Bessel functions using standard recurrent formulas (ABRAMOVITZ and STEGUN, 1970), is that the difference of the arguments $[k_m(r_{m-1} - r_m)]$ occurs in the initial relations (26) instead of the arguments $k_m r_{m-1}$ and $k_m r_m$. This advantage will be appreciated especially for the large values of the arguments $|k_m r_m| > 5 \times 10^2$ because computer overflow occurs in the initial values if the standard recurrent formulas are used. Recurrent formulas (24) work with smaller arguments $|k_m(r_{m-1} - r_m)|$ and are, therefore, prone to overflow, especially if the layers are thin. If their thickness does not exceed tens of km, the argument reduces by as much as two orders of magnitude which decreases the overflow limit substantially. A large argument causes no difficulties in recurrent formulas (27) because the large values of the exponentials can be cancelled out in the initial function $\chi_0(x)$. However, this procedure is not possible, e.g., in computing $j_n(x)$, because the large exponential factors, which are responsible for the overflow, cannot be eliminated.

5. *Models of Electrical Conductivity Distribution in a Spherical Earth*

A large number of 1-D models of the distribution of the electrical conductivity along the Earth radius, derived by solving the plane or spherical inverse problem, have been published to date. These models and their combinations have been compared and treated in detail in a number of studies. The authors compared experimental results with the model curves, either for combinations of the values of the apparent resistivity and the impedance phase angle $\bar{\varphi}$, or for the response function $Q = i/e$ and its angle argument Φ_Q.

The newly derived matrix method of solving the direct problem enabled the fast computation of the theoretical curves of all four quantities ρ_T, $\bar{\varphi}$, Q and Φ_Q for a set of 20 selected models published between 1965 and 1982. The selection preferred global models which yielded the conductivity distribution pattern down to the depths of the lower Earth mantle. The models, adopted from ANDERSSEN *et al.* (1979), DEVANE (1978), BANKS (1972), KOVTUN (1980), ROKITYANSKY (1981), DUCRUIX *et al.* (1980), DIMITRIEV *et al.* (1977) and SCHMUCKER (1974) were modified in some cases, namely those which were published only in the form of a graph with no tables. The ten models, described in detail in this paper, are illustrated in Figures 1 and 2. Models EMAN 2 and EMAN 3 were derived and described by PĚČOVÁ and PĚČ (1982).

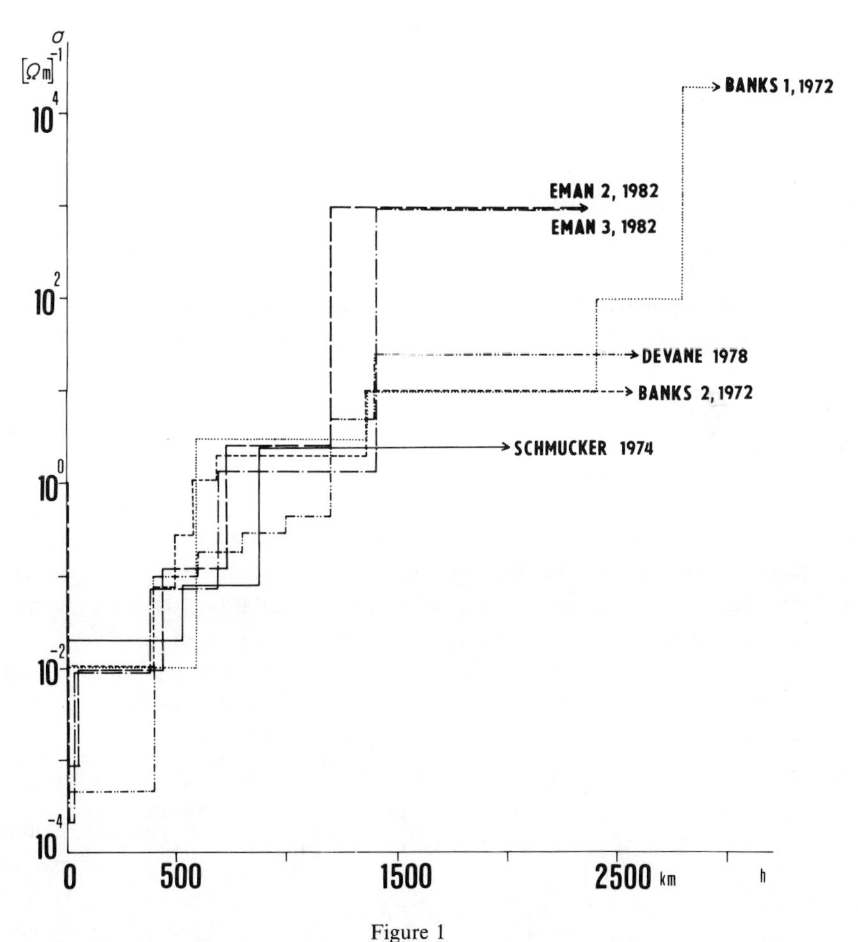

Figure 1
Models of radial electrical conductivity distribution of SCHMUCKER (1974), BANKS 1, 2 (1972), EMAN 2, 3 (PĚČOVÁ and PĚČ, 1982), and DEVANE (1978).

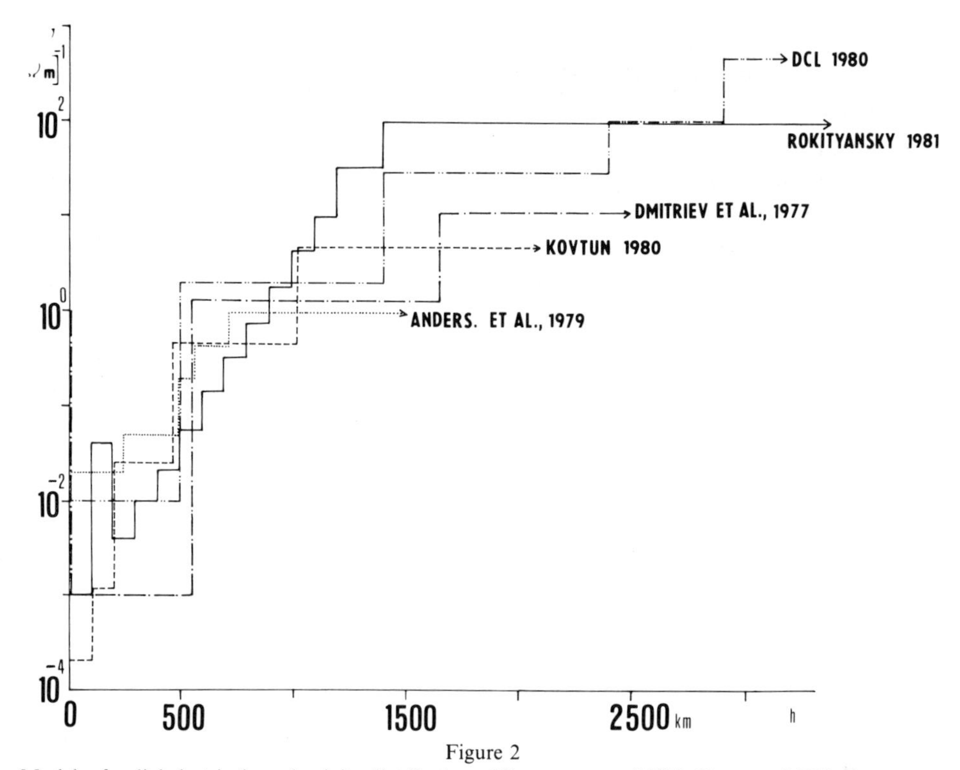

Figure 2
Models of radial electrical conductivity distribution of ROKITYANSKY (1981), KOVTUN (1980), DMITRIEV *et al.* (1977), DCL (DUCRUIX *et al.*, 1980), and ANDERSSEN *et al.* (1979).

The set of experimental data for quantities ρ_T, $\bar{\varphi}$, Q and Φ_Q for comparison with the theoretical curves were compiled from the literature, mostly from the publications of KOVTUN (1980) and ROKITYANSKY (1981), in which the results of various authors have been summarized, and from BANKS' paper (1972), on the one hand, and from the results of our own data analysis. We carried out the spectral analysis of series of daily average values of the geomagnetic components recorded at geomagnetic observatories worldwide and given in the table of PĚČOVÁ *et al.* (1980). We have used three-year intervals of daily averages from periods with different levels of solar activity, but also short three- to six-month intervals which contained well-defined variations with periods ranging from 30 to 8 days. Using the relations derived by RIKITAKE (1951) and BERDICHEVSKY *et al.* (1970), we computed the quantities Q, Φ_Q, ρ_T and $\bar{\varphi}$ for the spherical harmonic of degree $n = 1$ in the period range of 200 to 7 days from the values of the spectral densities and phases, determined by FFT with a standard filter (Tables 1 to 3). We omitted the data from observatories located at magnetic latitudes over 57° and in the neighbourhood of the equatorial electrojet, so that the number of stations N involved was decreased from 50 to 33–23. Finally, we reduced the whole set of our results to one third. The criterion for this reduction was

Table 1

Values of response function Q, *apparent resistivity* ρ_T *and their phases* Φ_Q, $\bar{\varphi}$

Three years' interval 1958, 59, 60													
$\sqrt{T_{s^{\frac{1}{2}}}}$	FIT A_H	$\varepsilon\%$	FIT A_Z	$\varepsilon\%$	FIT $\frac{A_Z}{A_H}$	$\varepsilon\%$	φ_H RAD	φ_Z RAD	N	Q	Φ_Q^0	$\rho_T[\Omega m]$	$\bar{\varphi}^0$
4206	4176.9	9.0	1860.1	16.0	0.4102	29.8	2.5115	5.339	25	0.244	16.7	0.902	72.0
3326	2509.7	13.7	1697.0	16.9	0.7551	30.5	6.084	2.881	23	0.122	8.1	3.329	86.5
3052	1962.6	7.2	1031.3	15.8	0.5177	26.0	5.468	2.116	23	0.198	15.2	2.387	78.0
2470	1798.3	13.4	880.9	15.2	0.5288	28.8	3.782	0.109	25	0.258	29.1	3.166	59.5
2429	1773.6	8.0	795.2	19.0	0.4489	27.0	1.232	4.102	24	0.238	14.8	2.744	74.4
2158	2226.4	10.7	973.2	12.2	0.4661	18.6	5.992	2.302	24	0.280	18.8	3.304	58.6
2130	2309.2	10.7	1033.5	14.0	0.4809	28.0	5.162	1.906	23	0.228	6.4	3.554	83.5
2103	2320.0	9.0	1049.5	15.0	0.4892	28.0	4.249	0.716	25	0.250	20.6	3.725	67.5
2005	1391.6	9.1	652.5	21.5	0.4917	25.6	3.593	6.044	26	0.296	32.3	4.402	50.4
1940	1646.6	13.6	706.2	22.2	0.4593	27.4	5.720	2.368	25	0.243	10.9	3.933	78.0
1845	1927.5	8.0	818.4	14.0	0.4383	27.6	4.844	1.307	26	0.263	19.0	4.266	67.3
1810	2847.0	6.2	1093.5	11.3	0.3926	17.4	4.174	0.680	25	0.276	14.9	3.625	69.8
1794	3807.7	6.6	1454.6	10.9	0.3972	15.4	2.595	5.429	25	0.273	13.1	3.652	72.3
1778	3387.1	7.7	1359.6	9.5	0.4309	11.7	1.064	3.989	26	0.257	10.1	4.106	77.6
1762	1899.9	9.8	904.4	11.0	0.5408	23.8	6.016	2.502	24	0.238	21.3	5.877	68.7
1676	2001.0	11.1	877.8	14.0	0.4904	28.8	0.999	3.744	25	0.257	19.9	5.517	67.2
1663	1960.0	5.8	804.1	13.9	0.4357	19.6	5.362	1.892	25	0.262	15.4	4.902	71.1
1637	2468.7	7.7	1012.2	9.8	0.4619	20.3	5.849	1.999	25	0.317	27.7	5.048	49.4
1625	2549.2	6.6	1129.8	10.9	0.4873	19.5	3.708	0.049	26	0.275	25.2	5.988	60.4
1613	2010.5	5.1	944.2	13.0	0.4967	20.9	1.861	4.556	26	0.251	24.2	6.825	64.4
1536	1905.1	4.2	707.2	10.7	0.3707	21.7	5.315	1.687	24	0.297	18.7	4.702	62.1
1516	3469.2	5.7	1377.2	12.8	0.4103	17.9	2.984	5.927	26	0.258	9.2	5.522	78.6
1506	3622.0	5.5	1392.9	12.6	0.3927	16.9	0.732	3.768	26	0.260	4.8	5.249	83.9
1497	2965.0	4.5	1086.5	12.0	0.3663	15.6	5.028	1.739	25	0.271	6.1	4.826	81.6

FIT A_H, A_Z-the best fitting spectral density amplitudes for spherical harmonic of degree n = 1 and mean phases φ_H and φ_Z. N is the number of the observatories, ε is the standard deviation.

Table 1 (*continued*)

Values of response function Q, *apparent resitivity* ρ_T *and their phases* Φ_P, $\bar{\varphi}$

Three years' interval 1958, 59, 60

$\sqrt{T_{s^2}^{\frac{1}{2}}}$	FIT A_H	ε%	FIT A_Z	ε%	FIT $\frac{A_Z}{A_H}$	ε%	φ_H RAD	φ_Z RAD	N	Q	Φ_Q^0	$\rho_T[\Omega m]$	$\bar{\varphi}^0$
1469	2598.3	2.98	1019.9	9.8	0.3880	12.4	3.061	5.881	26	0.269	14.18	5.7488	71.6
1410	2579.7	3.6	826.2	12.0	0.3114	17.6	4.850	1.488	25	0.299	7.56	4.1544	77.4
1402	2548.3	3.9	863.4	10.4	0.3339	16.1	3.287	0.051	26	0.284	3.56	4.7009	84.6
1394	2006.3	4.6	868.1	13.4	0.4437	19.6	3.5	0.24	25	0.238	4.0	7.7524	85.5
1274	2279.7	7.1	787.4	11.9	0.3600	15.2	3.704	0.155	26	0.300	14.7	5.9161	66.7
1268	2529.2	6.6	880.8	14.0	0.3585	19.0	2.275	5.276	25	0.280	5.47	6.0704	81.9
1262	2281.2	7.6	831.6	17.6	0.3799	27.4	1.649	4.549	24	0.277	9.8	6.7123	76.2
1230	2193.1	7.3	657.6	12.3	0.3091	16.0	3.238	5.982	26	0.322	17.26	4.7857	67.2
1225	2337.0	8.0	709.1	12.8	0.3137	17.2	1.168	3.911	26	0.320	12.2	4.9432	67.2
1158	1789.9	10.3	548.8	18.7	0.3276	19.0	2.721	5.623	25	0.307	7.75	5.6474	76.2
1153	2182.4	9.6	699.7	17.0	0.3430	16.5	0.872	3.728	25	0.302	9.68	6.2203	73.6
1090	3498.5	4.6	1033.8	12.0	0.2988	18.0	4.572	1.161	23	0.315	8.26	5.9202	74.6
1086	3600.4	4.3	1068.5	11.6	0.2972	18.0	3.046	5.750	25	0.327	12.93	6.0109	64.9
1082	2700.0	4.2	806.5	12.4	0.2989	18.6	0.676	3.250	25	0.339	16.16	6.1287	57.6
1079	2268.5	5.3	632.3	18.0	0.2743	25.0	3.540	0.161	26	0.322	6.78	5.3723	76.4
1075	2040.0	6.1	550.5	19.0	0.2623	27.0	1.814	4.646	26	0.331	8.40	5.0718	72.3
931	2108.2	4.6	585.8	17.0	0.2698	23.0	4.182	0.658	26	0.332	10.58	7.1663	68.1
929	2173.0	4.3	542.9	20.7	0.2404	28.0	2.707	5.456	26	0.347	9.54	5.8225	67.5
924	1954.0	5.0	596.7	9.0	0.3116	11.5	2.963	5.865	26	0.308	7.68	8.7833	76.3
859	1808.2	6.7	503.3	11.7	0.2842	23.1	5.127	1.639	24	0.329	9.70	8.4617	70.1
857	1669.6	7.9	512.2	10.3	0.3220	13.0	2.984	6.018	26	0.302	3.53	10.320	83.9
845	1337.6	9.0	421.5	18.6	0.3135	27.5	3.532	0.393	26	0.296	−0.08	11.204	90.2
843	1324.2	9.0	408.5	16.5	0.3105	24.2	2.924	5.666	25	0.317	12.52	10.78	67.1
810	1321.0	7.0	453.8	13.7	0.3540	30.4	1.394	4.433	26	0.281	3.92	14.501	84.1
808	1387.1	6.0	466.6	13.7	0.3463	22.1	5.7	2.6	25	0.284	−0.04	13.956	90.0
806	1026.3	8.9	337.6	17.5	0.3379	24.6	4.062	0.710	23	0.294	7.48	13.391	78.0
804	1303.0	8.9	403.8	14.4	0.3015	27.0	2.178	5.478	23	0.302	−5.18	11.972	99.0

Table 2

Values of response function Q, *apparent resistivity* ρ_T *and their phases* Φ_Q, $\bar{\varphi}$

$\sqrt{T_s}^{\frac{1}{2}}$	FIT A_H	ε%	FIT A_Z	ε%	FIT $\frac{A_Z}{A_H}$	ε%	φ_H RAD	φ_Z RAD	N	Q	Φ_Q^0	$\rho_T[\Omega m]$	$\bar{\varphi}^0$
Three years' interval 1963, 64, 65													
1082	1242.8	8.0	463.78	16.6	0.3843	28.0	3.243	5.9085	32	0.295	18.54	9.567	62.7
1079	1374.4	6.9	478.14	17.7	0.3528	27.0	2.063	4.6145	33	0.321	20.1	8.370	56.2
893	560.0	7.9	164.12	15.5	0.2939	26.7	0.805	4.0126	27	0.309	−2.03	8.677	93.8
Three years' interval 1967, 68, 69													
2130	1355.6	7.6	606.7	22.4	0.4353	28.2	4.011	0.5729	27	0.241	15.9	3.55	73.0
1579	1266.2	7.9	542.2	17.6	0.4148	20.4	1.051	3.7571	29	0.266	20.8	5.922	65.1
1568	1295.0	8.0	560.7	18.0	0.4358	21.0	5.991	2.3411	28	0.275	23.8	6.141	60.9
1536	1909.1	8.2	503.0	16.1	0.2605	27.4	5.415	1.7015	27	0.356	14.1	2.379	57.2
1100	924.9	12.0	223.6	22.7	0.2467	22.7	4.541	1.5764	23	0.341	−4.3	3.87	100.2
1090	1105.3	7.9	309.5	18.6	0.2745	18.6	3.835	1.0509	24	0.329	−10.0	5.32	110.4
934	916.2	5.1	309.6	14.1	0.3289	27.5	6.191	2.6835	28	0.300	13.0	10.55	69.1
924	915.7	5.1	247.0	17.8	0.2560	27.6	4.897	1.4072	27	0.333	9.4	6.85	70.0
922	977.3	5.0	297.4	14.6	0.2944	25.2	4.006	0.5366	28	0.314	10.3	8.76	71.2
901	1115.5	8.2	304.1	17.9	0.2731	32.0	1.755	4.7840	26	0.321	3.2	7.37	83.6
823	1224.4	5.0	386.1	14.1	0.3208	24.0	5.230	1.5975	29	0.323	15.4	11.81	61.9
822	1435.6	5.4	470.2	12.9	0.3302	17.0	4.020	0.5962	29	0.299	9..8	12.79	73.8
820	1155.0	6.4	440.4	13.2	0.3901	15.4	2.698	5.6468	28	0.265	8.4	17.40	79.0

FIT A_H, A_Z-the best fitting spectral density amplitudes for spherical harmonic of degree $n = 1$ and mean phases φ_H and φ_Z. N is the number of the observatories, ε is the standard deviation.

Table 3

Values of response function Q, apparent resistivity ρ_T and their phases Φ_Q, $\bar{\varphi}$

$\sqrt{T_{s^{\frac{1}{2}}}}$	FIT A_H	$\varepsilon\%$	FIT A_Z	$\varepsilon\%$	FIT $\frac{A_Z}{A_H}$	$\varepsilon\%$	φ_H RAD	φ_Z RAD	N	Q	Φ_Q^0	$\rho_T[\Omega m]$	$\bar{\varphi}^0$
Short intervals 1958, 59, 60													
1568	661.8	6.3	237.57	12.5	0.3672	26.6	0.478	3.267	27	0.288	13.64	4.22	69.8
1487	493.3	5.3	151.18	17.3	0.3029	23.0	5.460	1.965	25	0.315	11.12	3.42	69.8
1487	1038.7	3.5	430.12	11.5	0.4179	15.6	0.116	2.999	29	0.253	12.55	6.24	75.2
1127	416.9	7.8	151.77	11.2	0.3464	12.6	2.859	4.852	23	0.420	30.10	8.40	24.2
961	492.3	8.6	176.2	10.8	0.3815	18.0	0.782	3.737	26	0.277	7.45	11.16	79.3
922	707.4	6.9	233.68	12.5	0.3481	19.2	0.695	3.515	29	0.300	11.24	10.32	71.6
831	459.7	12.8	170.68	9.0	0.4299	14.5	4.327	1.389	26	0.271	−8.54	16.06	101.7
773	344.2	7.7	97.74	15.3	0.2784	19.0	2.445	5.333	22	0.320	7.42	10.85	75.5
Short intervals 1963, 64, 65													
2217	479.4	6.2	210.62	21.5	0.4227	31.2	1.794	4.988	25	0.230	2.92	3.15	93.0
1093	445.2	8.4	140.82	20.0	0.3127	31.0	2.462	5.260	29	0.309	11.27	7.97	107.4
Short intervals 1967, 68, 69													
1716	1111.5	5.3	388.8	9.2	0.3477	16.1	0.764	3.442	26	0.304	16.7	3.34	63.4
1524	830.1	4.3	272.6	14.6	0.3217	25.2	5.724	1.964	25	0.332	19.4	3.73	54.6
812	540.0	7.5	179.2	12.5	0.3192	27.5	2.733	5.576	29	0.297	10.6	13.39	72.9

FIT A_H, A_Z-the best fitting spectral density amplitudes for spherical harmonic of degree n = 1 and mean phases φ_H and φ_Z. N is the number of the observatories, ε is the standard deviation.

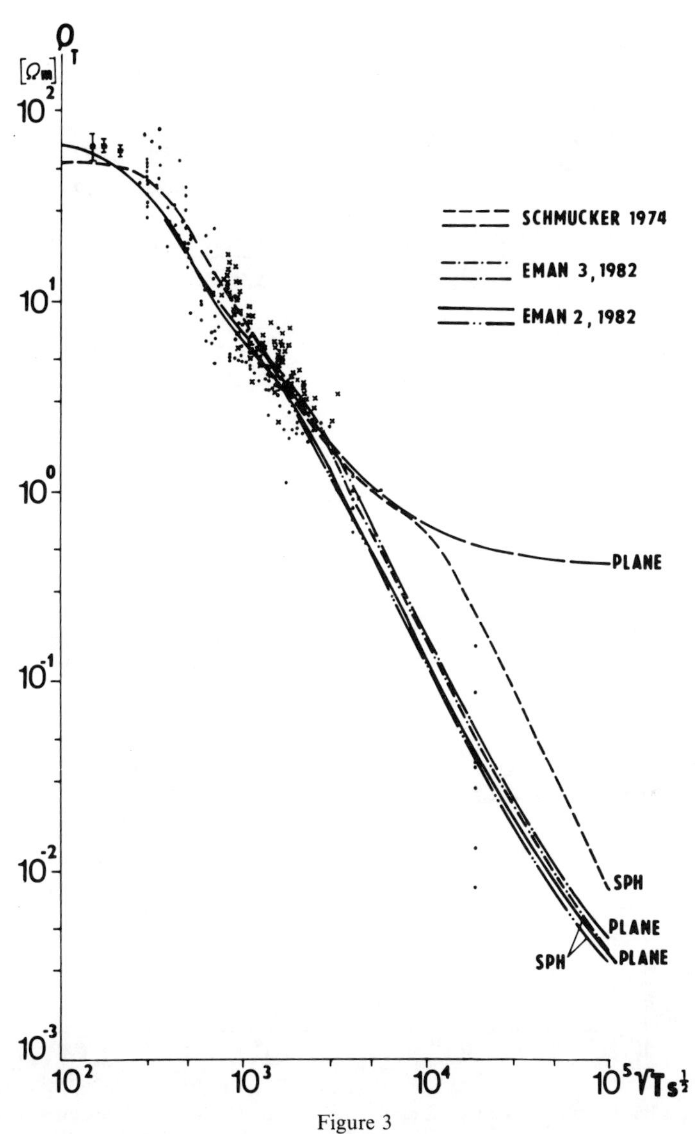

Figure 3

Curves of the apparent resistivity ρ_T of models SCHMUCKER (1974), EMAN 2, 3 for the spherical harmonic of degree $n = 1$. The curves of the same models in the plane configuration are also shown. The black dots represent the ρ_T-data from literature. The crosses represent the ρ_T-values from Tables 1, 2, 3 in this paper. The asterisks indicate averages of ρ_T from the analysis of S_q-variations for spherical harmonics $n = 5$, $m = 4$, $\sqrt{T} = 147\ s^{\frac{1}{2}}$; $n = 4$, $m = 3$, $\sqrt{T} = 170\ s^{\frac{1}{2}}$; $n = 3$, $m = 2$, $\sqrt{T} = 208\ s^{\frac{1}{2}}$, given with their standard deviations (ROKITYANSKY, 1981, Table 12) to illustrate the data distribution at shorter periods. The results of the analysis of the S_q-variations $n = 2$, $m = 1$, $\sqrt{T} = 294\ s^{\frac{1}{2}}$ (KOVTUN, 1980) are marked as dots.

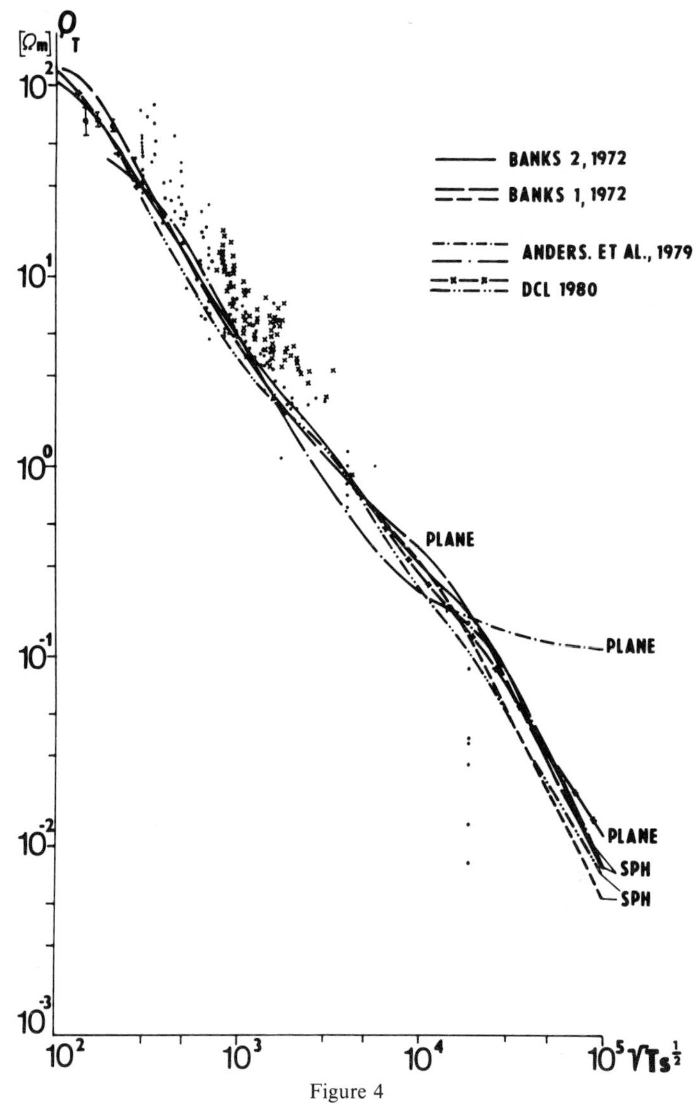

Figure 4
Same as Figure 3 but for models BANKS 1, 2 (1972), DCL (DUCRUIX *et al.*, 1980), and ANDERSSEN (1979).

the standard deviation of the spectral densities of the *H*- and *Z*-components to the first degree harmonic. On the whole, we added 80 values to each set of quantities ρ_T, $\bar{\varphi}$, Q and Φ_Q. In the following figures our results are marked with crosses, whereas the published data with black dots. As can be seen from Tables 1 to 3, the three-year interval of the solar activity maximum, 1958–1960, yielded the most suitable data.

Figures 3–14 give a qualitative idea of the degree to which the theoretical curves, computed for the selected models, fit the sets of experimental data. The ρ_T-curves

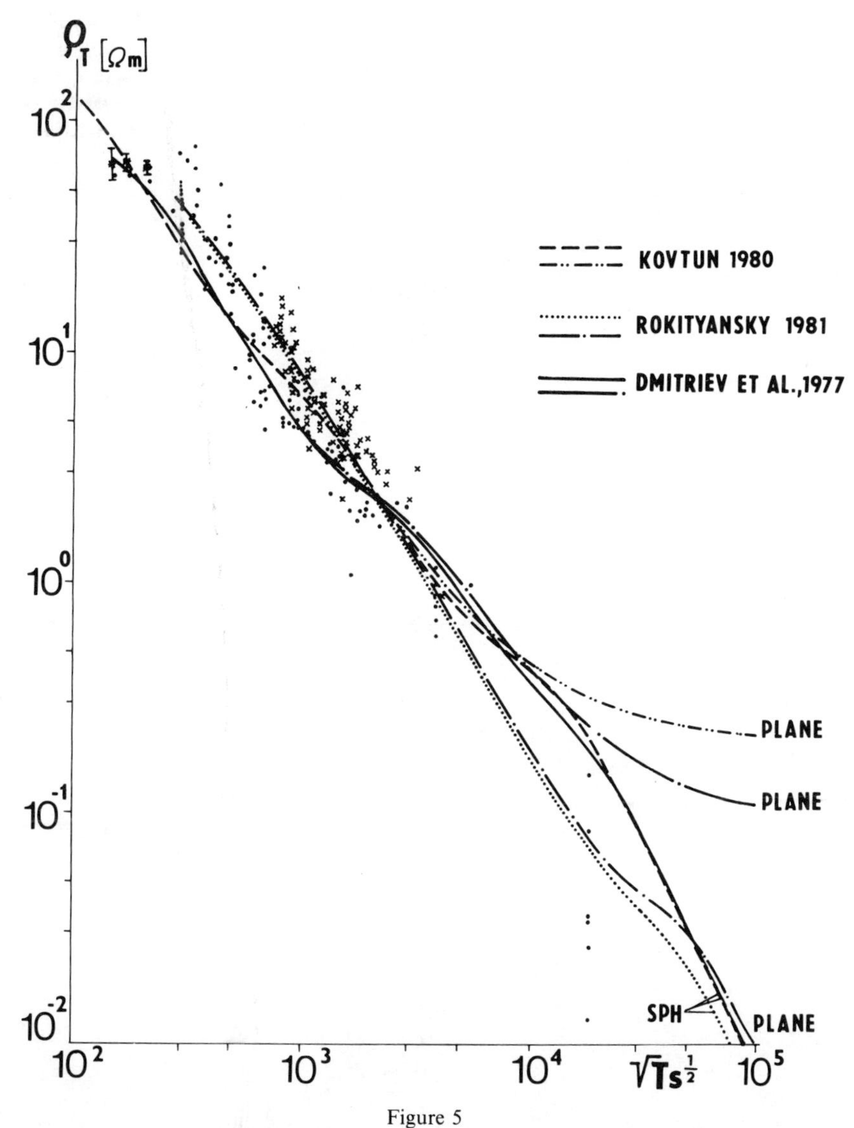

Figure 5
Same as Figure 3 but for models ROKITYANSKY (1981), KOVTUN (1980), DMITRIEV *et al.* (1977).

(Figures 3, 4, 5) for the Schmucker, Kovtun, Rokityansky, EMAN 2 and EMAN 3 models run through the middle of the data set, while the models of Dmitriev *et al.*, Banks 72, Anderssen *et al.* and DCL (DUCRUIX *et al.*, 1980) form a narrow pencil of curves slightly displaced from the middle of the set towards lower values of ρ for $\sqrt{T}$ ranging from 10^2 to 3×10^3 $s^{\frac{1}{2}}$. Equivalent information about the response function Q, plotted on a linear scale in the range 0 to 0.5 for $n = 1$, is shown in

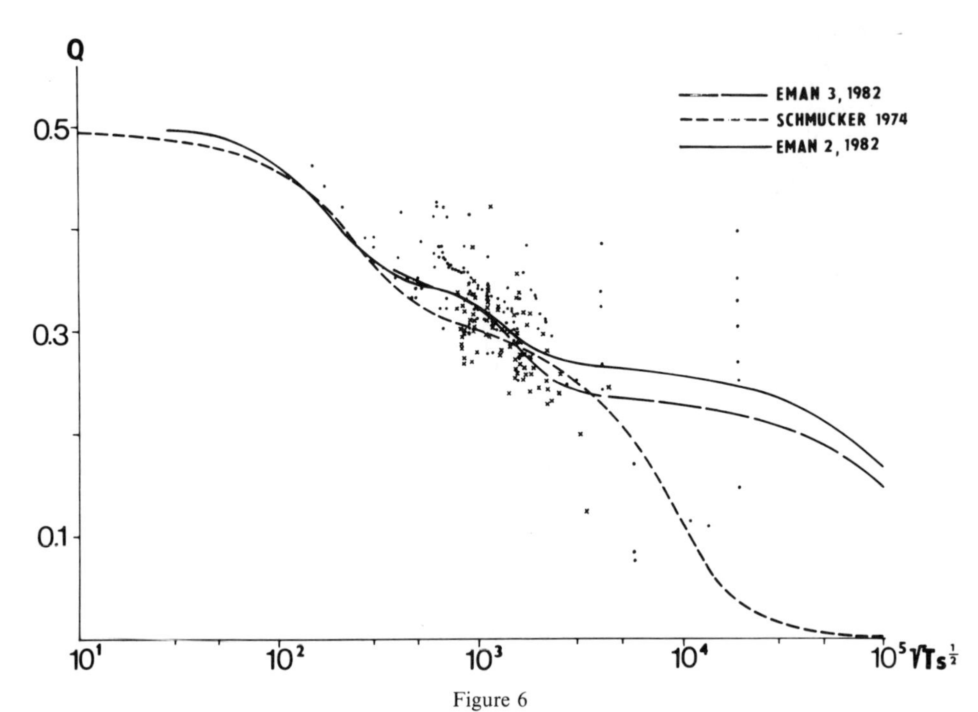

Figure 6
Curves of the response function Q for models in the Figure 3 as compared with results of the analysis of the variations for spherical harmonic $n = 1$.

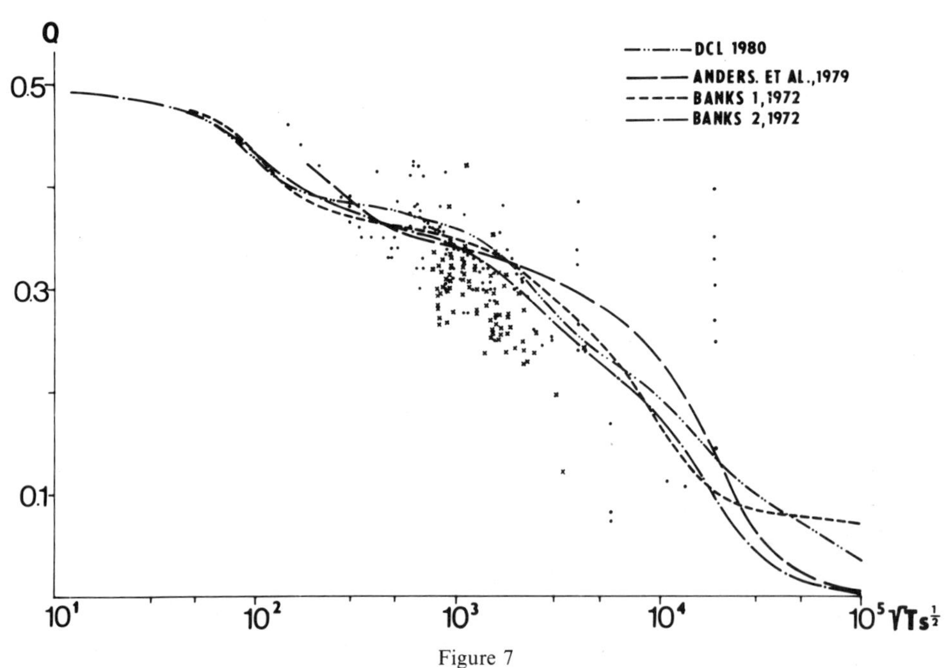

Figure 7
Response function Q for models in Figure 4.

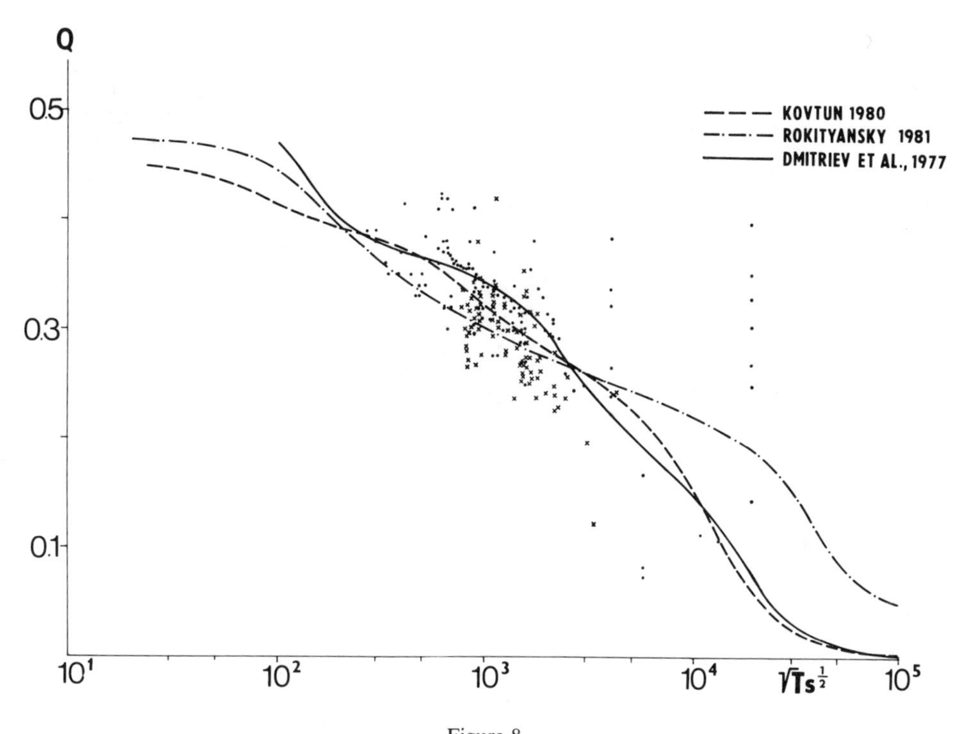

Figure 8
Response function Q for models in Figure 5.

Figures 6, 7 and 8, whereas for ρ_T the values are in the interval from 10^{-2} to 10^2 Ωm.

Figures 9–14 refer to the phase data, $\bar{\varphi}$ and Φ_Q, both displaying a large scatter. To emphasize the trend of the values, arithmetic averages with standard deviations in narrow period ranges were used for the phases. Some of the published results are depicted only as points. In spite of the large scatter of both the phase values, one can see that the curves for some models in the range of $\sqrt{T}$ between 3×10^2 to 2×10^3 run through the middle of the data sets, or at least are within the limits defined by the standard deviations. The curves of models Banks 72, DCL and Dmitriev *et al.* correspond to the $\bar{\varphi}$-values in the interval of $\sqrt{T}$ between 8×10^2 to 2×10^3, and the curves of Schmucker's and Anderssen's models in the interval of $\sqrt{T}$ between 10^2 and 8×10^2. The Schmucker, Kovtun, Dmitriev, Banks 72 and DCL models fit the values of Φ_Q satisfactorily in the range of $\sqrt{T}$ between 5×10^2 and 2×10^3.

The comparison of the curves of the selected models with the experimental data yields the following conclusions:

a) None of the models fits the data within the whole range of periods. The curves of the Schmucker, Rokityansky, Kovtun, Banks, Anderssen, EMAN 2 and EMAN 3 models fit the sets of ρ_T- and Q-data best. In the graphs of the two associated phases

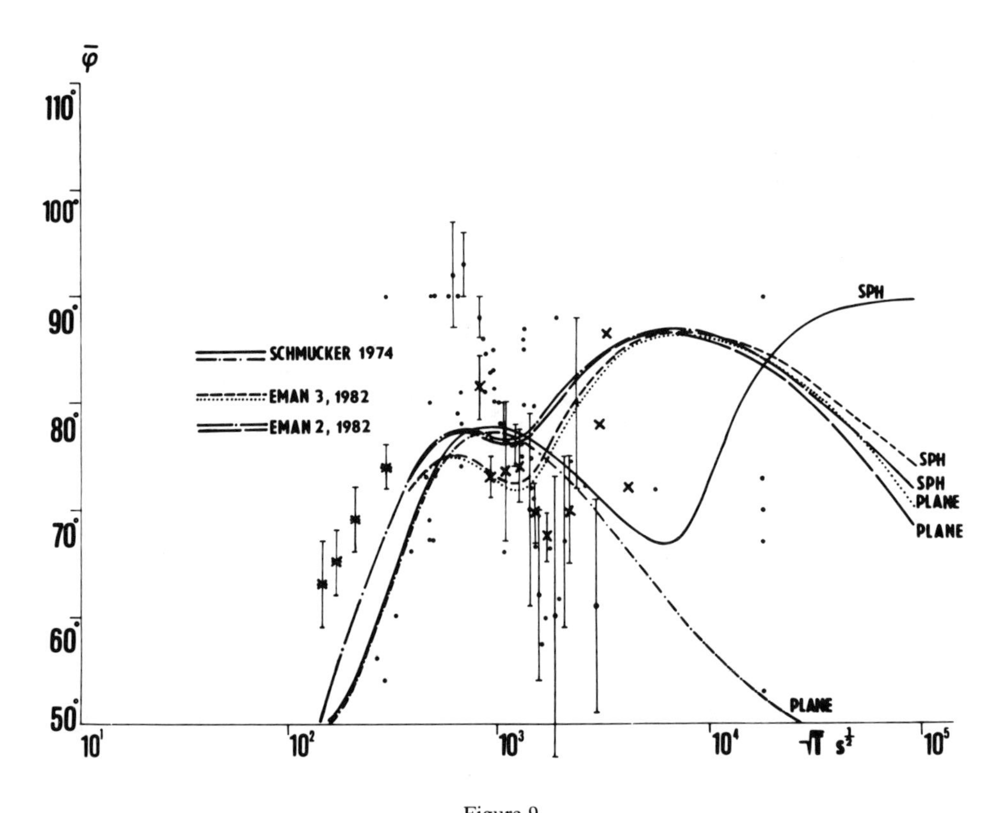

Figure 9
The functional dependence of the impedance argument $\bar{\varphi}(\sqrt{T})$ for models as in Figure 3. The data distribution is represented by averages with standard deviations. ROKITYANSKY's data (1981, Table 12) were again used for the period range of daily variations.

$\bar{\varphi}$ and Φ_Q, the fit is far worse than for the amplitudes, but the tendency of the curves to fit the data is evident in the Schmucker, Banks, DCL, Dmitriev and Kovtun models. The common features of these models, whose ρ-, Q-, $\bar{\varphi}$ and Φ_Q-curves fit the set of experimental data over a substantial part of the interval of long periods is the 'integrated conductance' of the upper layers, which was assumed for solving the inverse problem down to depths of, e.g., 400 km. According to BANKS (1981), 4000 S is a favourable starting parameter of the superficial layer for the inversion. The models in the present paper had conductances ranging from 4000 to 8000 S.

b) One of the conditions for successfully solving the inverse problem and producing a realistic model of the electrical conductivity, whose theoretical curves would fit all four data sets, is to compile data, derived by the same procedure with a high accuracy for the complete range of long periods, e.g., ANDERSSEN *et al.* (1979).

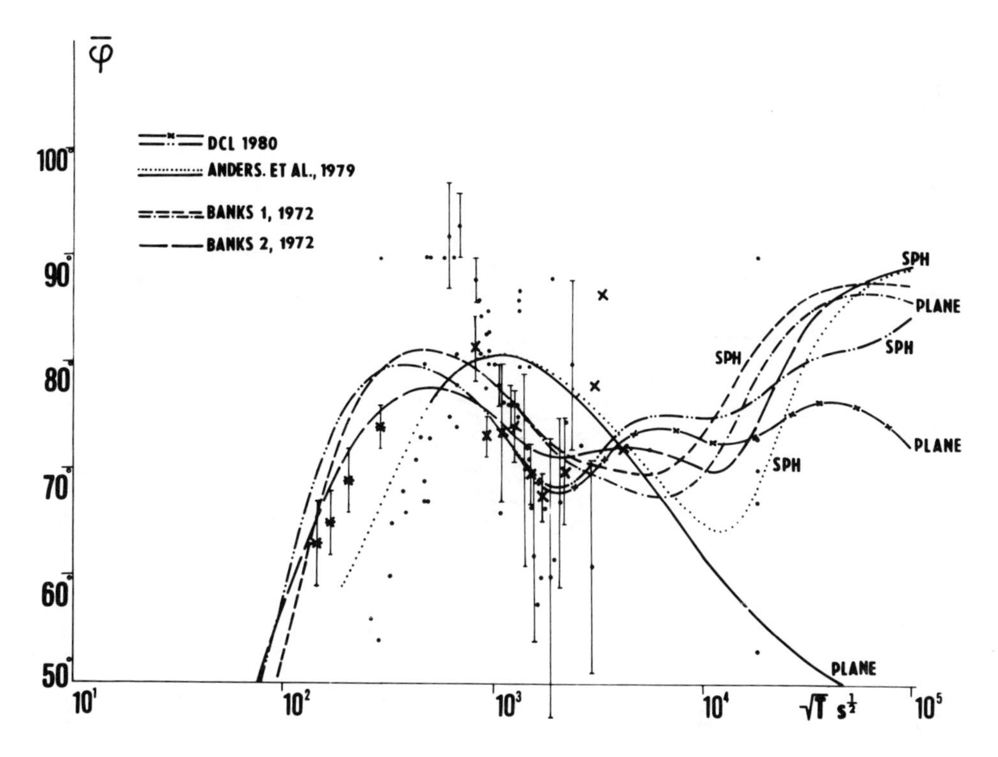

Figure 10
Same as Figure 9 but for models in Figure 4.

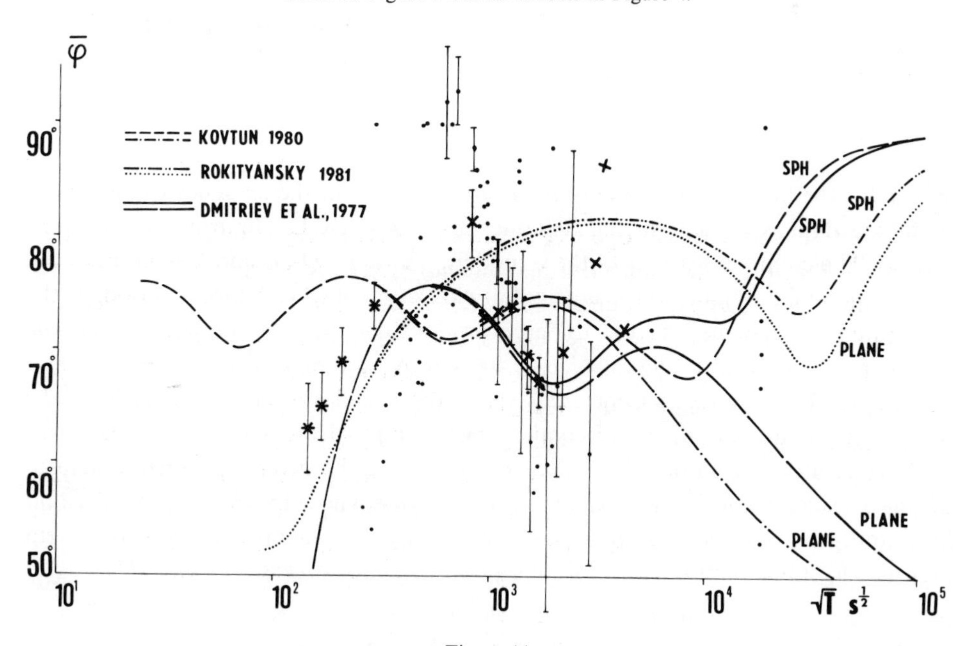

Figure 11
Same as Figure 9 but for models in Figure 5.

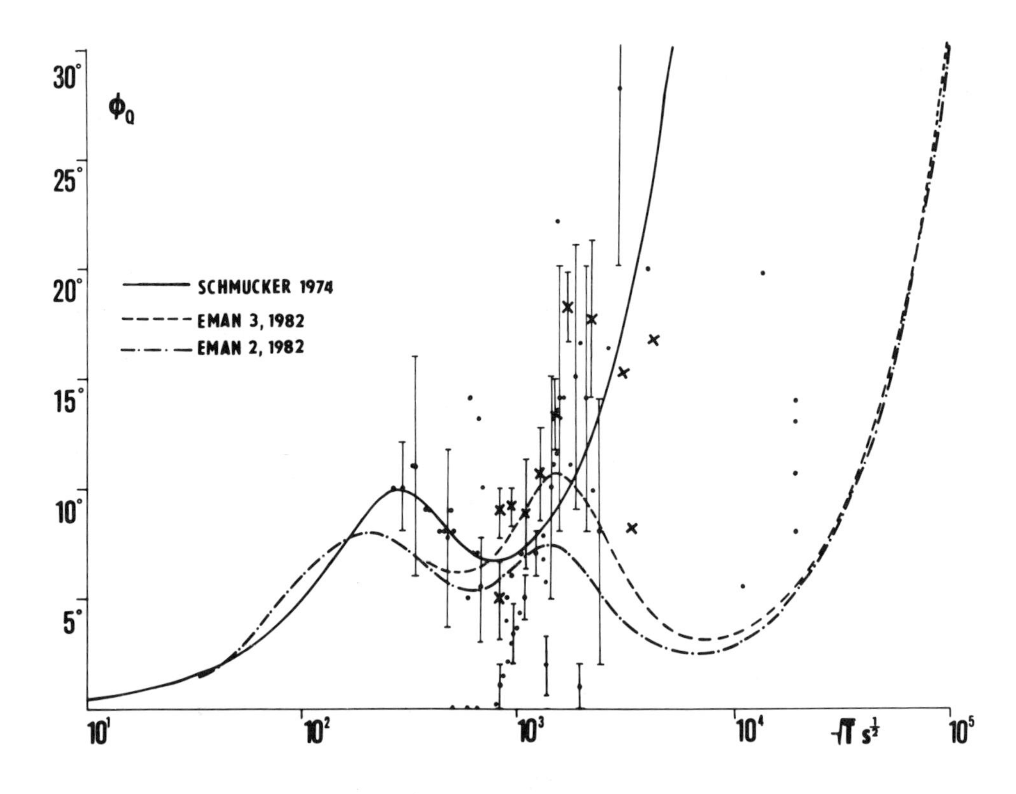

Figure 12
Comparison of the curves of functions arg $Q = \Phi_Q(\sqrt{T})$ for models in Figure 3.

6. *Free Decay and Damped Oscillatory Modes*

The external time-variable source induces an electromagnetic field and electrical currents within the conducting Earth. We shall study the physical situation which will occur, if all the external sources are switched off suddenly. The induced electrical field will decay due to ohmic losses and possibly radiation. The conditions of continuity of the vertical component of magnetic induction and of the horizontal components of the magnetic intensity must be satisfied on the Earth's surface, which is a conspicuous conductivity discontinuity. These boundary conditions will only be satisfied for particular discrete ${}_s\omega_n$, which may, in general, be complex. This problem was first treated by LAMB (1883). His results bear the sign of the contemporaneous ideas of the remote action of the electromagnetic field. The reader should be reminded that only in 1888 H. Hertz experimentally proved that electromagnetic phenomena propagated with a finite velocity. In 1909, P. DEBYE solved the problem of free oscillations of a homogeneous conducting sphere by applying the correct theory.

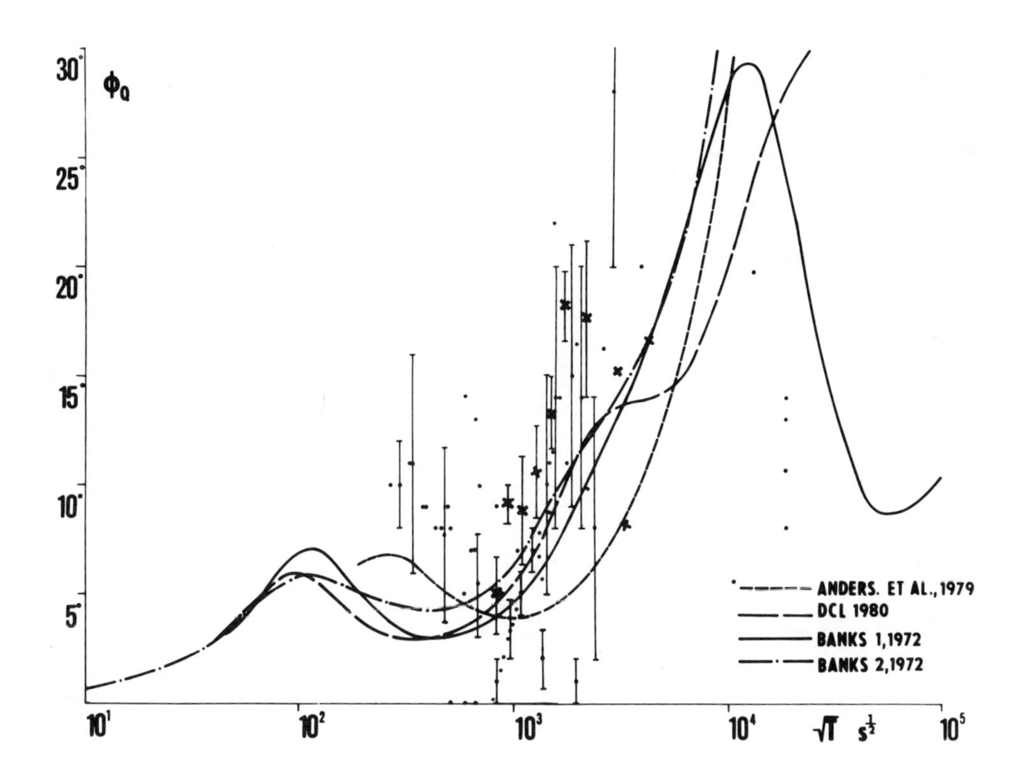

Figure 13
Same as Figure 12 but for models in Figure 4.

The result was that the roots of the characteristic equations for the frequencies were, in this case, purely imaginary, as well as generally complex, which means that, in the case of a homogeneous conducting sphere, pure decay modes as well as free damped oscillations exist.

The theory of magnetic diffusion of the field in a spherically symmetric mantle, generated by sources in the core, was presented by SMYLIE (1965). He considered the asymptotic case of Bessel's differential equation and came to the surprising conclusion that free oscillations were pure decay modes.

We shall now derive the characteristic equation for the free oscillations of a spherically symmetric model without the constraints imposed on spherical Bessel functions and, using DEVANE's current model of the mantle (1978), we shall demonstrate that the free oscillations are purely imaginary, in agreement with SMYLIE (1965), on the one hand, and that they have a nonzero real component, in agreement with DEBYE (1909), on the other.

The continuity of the horizontal components of the intensity of the electro-

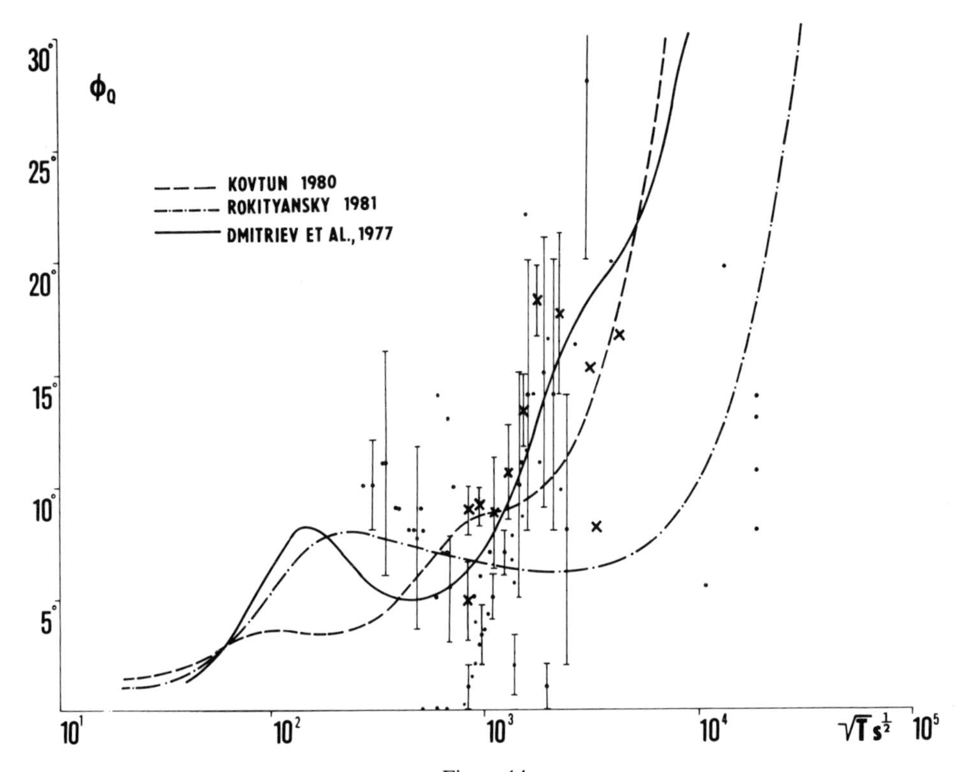

Figure 14
Same as Figure 12 but for models in Figure 5.

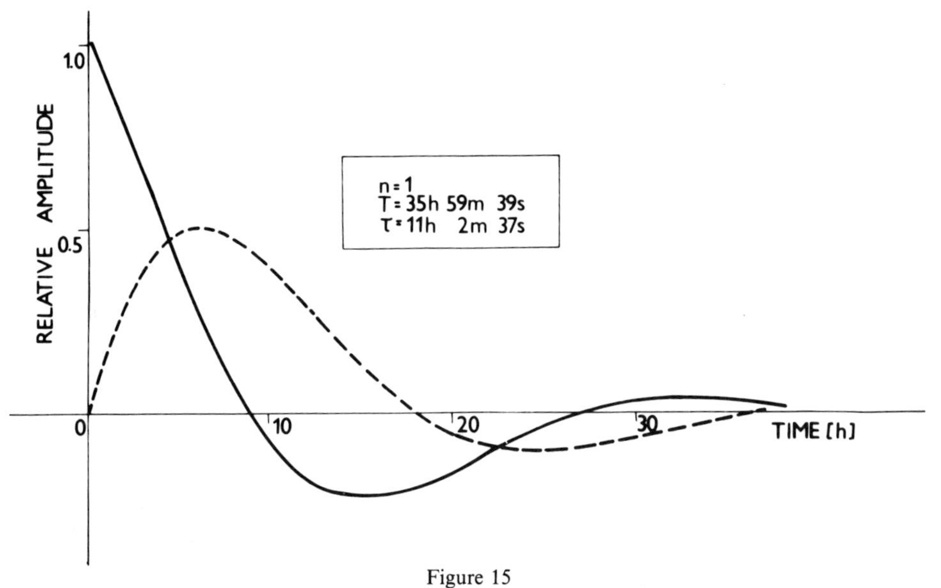

Figure 15
The time variation factor of amplitudes of free electromagnetic motion (damped oscillations) for frequency of $s = 3$, Table 4.

Table 4

Frequencies of free damped oscillations for electrical conductivity model DEVANE (1978); $n = 1$; ${}_s^0\omega_1 = 2\pi/{}_s^0T_1 + i/{}_s^0\tau_1$, *and quality factor* ${}_sQ_1$

s	${}_s^0T_1$	${}_s^0\tau_1$	${}_sQ_1$
1	24d 16h 59m 8s	2d 3h 21m 23s	0.28
2	21d 8h 55m 25s	15h 22m 18s	0.09
3	35h 59m 39s	11h 2m 37s	0.97
4	23h 40m 30s	3h 54m 58s	0.52
5	17h 22m 21s	2h 4m 37s	0.37
6	11h 16m 40s	53m 30s	0.25

Table 5

Relaxation time of free decay modes for electrical conductivity model DEVANE (1978); $n = 1$; ${}_s^D\omega_1 = i/{}_s^D\tau_1$.

s	${}_s^D\tau_1$	s	${}_s^D\tau_1$
1	2y 236d 13h 40m 44s	8	41d 21h 31m 21s
2	1y 45d 13h 33m 23s	9	30d 10h 46m 8s
3	198d 15h 27m 53s	10	27d 22h 52m 25s
4	138d 21h 38m 52s	11	21d 2h 3m 34s
5	85d 19h 27m 22s	12	19d 23h 24m 38s
6	69d 17h 15m 39s	13	15d 11h 3m 41s
7	47d 19h 11m 38s	14	14d 23h 39m 32s

magnetic field implies the continuity of the impedance. It is assumed that the external space is a vacuum, μ_0, ε_0, in which the impedance is determined as

$$Z_e(r) = i\omega\mu_0 r\, y_n(k_0 r)/\frac{d}{dr}\,[r y_n(k_0(k_0 r)],\quad r \geqq r_0, \tag{29}$$

where

$$k_0 = \omega\sqrt{(\varepsilon_0\mu_0)} = \omega/c \tag{30}$$

and c is the velocity of light.

The continuity at the surface, $r = r_0$, requires that $Z_e(r_0) = Z_0$, where Z_0 is the surface impedance of the Earth model computed as described in Section 2. In the frequency range of the observed variations, $|k_0 r_0| \ll 1$, and

$$\frac{y_{n+1}(k_0 r_0)}{k_0 r_0 y_n(k_0 r_0)} \sim 2n + 1. \tag{31}$$

With the aid of the asymptotic relation (31) we shall modify the formula for the surface impedance (29) to read

$$Z_e(r_0) = -i\omega\mu_0 r_0/n, \tag{32}$$

and the characteristic equation now becomes

$$nZ_0 - i\omega\mu_0 r_0 = 0. \tag{33}$$

7. *Values of the Characteristic Frequencies for the Devane Model of the Electrical Conductivity of the Mantle*

The characteristic equation (33) is a nonlinear transcendent equation for the characteristic frequencies in the complex domain. Since there is no special algorithm for determining the roots of such an equation, we adopted the method of successive approximations and chose DEVANE's (1978) model of the electrical conductivity of the mantle. The characteristic frequencies are marked with the superscript 0 for frequencies with a nonzero real component and the superscript D for purely imaginary frequencies, and represented in the following form:

$${}^0_s\omega_n = (2\pi/{}_sT_n) + (i/{}^0_s\tau_n); \quad {}^D_s\omega_n = i/{}^D_s\tau_n, \tag{34}$$

where ${}_sT_n$ is the period of the undamped oscillations and ${}_s\tau_n$ the relaxation period. The quality factor ${}_sQ_n$ is being introduced to appreciate the damping:

$${}_sQ_n = \pi({}_s\tau_n/{}_sT_n). \tag{35}$$

On the imaginary axis in the complex plane of frequencies ω, there is an infinite number of roots ${}^D_s\omega_n$, $s = 1,, 2, \ldots,$ for every mode $n = 1, 2, \ldots,$ which represent the decay modes and for which ${}_sT_n \rightarrow \infty$ and ${}_sQ_n = 0$. The first 14 roots of this type are given in Table 5.

The roots of the characteristic equation not on the imaginary axis, which we shall denote ${}^0_s\omega_n$, were determined by numerical mapping of the functional values of the l.h.s. of Equation (33) in the complex ω-plane. These roots and the corresponding periods ${}_sT_1$, relaxation periods ${}^0_s\tau_1$ and quality factors ${}_sQ_1$, related to the first mode $n = 1$, are given in Table 4.

Figure 15 shows the time variation of the intensity of the electromagnetic field for the initial phase 0 and $1/2\pi$ after the external source has been switched off.

8. *Conclusion*

Some of the problems related to the direct problem of electromagnetic induction and with the selection of representative data for solving the 1-D inverse problem have been discussed.

a) Recurrent and matrix methods for solving the direct 1-D problem of electromagnetic induction have been discussed. Cross-products of spherical Bessel functions, which make the numerical computation more effective, were introduced into the recurrent and matrix formulas for the impedance. It was proved that the matrix and recurrent methods were equivalent as regards impedance computations. In the problems of interest in the variation of the amplitudes of the induced field with depth, the matrix method is more economical because it uses Bessels functions of the same argument.

b) A set of 80 new values, obtained by analysing the daily averages of geomagnetic variation components, periods ranging from 7 to 200 days, recorded in the years 1958–1960, 1963–1965, 1967–1969, were added to the published induction data. The combined data set was compared with the data computed on the basis of the existing 1-D models of electrical conductivity. The qualitative comparison indicated that none of the models fits the data in the whole range of periods, especially regarding the phases. The conductivity models of SCHMUCKER (1974), BANKS (1972), ROKITYANSKI (1981), KOVTUN (1980) and DMITRIEV *et al.* (1977) fit the data set best in the period range of 7 to 50 days.

c) The characteristic equation was derived for the periods of the free electromagnetic oscillations of the model of the conductivity inhomogeneous Earth. It was proved that the roots of the equation were, on the one hand, purely imaginary and represented the decay modes, on the other, generally complex and represented damped oscillations. This method could be an alternative approach to solving the inverse conductivity problem, provided the free oscillation frequencies could be determined from variation observations.

REFERENCES

ABRAMOWITZ, M. and STEGUN, I. A. (1970), *Handbook of Mathematical Functions (Dover Publ., New York).*

ANDERSSEN, R. S. (1979), *The one-dimensional electromagnetic induction problem.* Geophys. J. R. Astr. Soc. *57*, 67.

ANDERSSEN, R. S., DEVANE, J. F., GUSTAFSON, S.-A. and WINCH, D. E. (1979), *The qualitative character of the global electrical conductivity of the Earth.* Phys. Earth Planet Inter. *20*, 15.

BANKS, R. J. (1969), *Geomagnetic variations and the electrical conductivity of the upper mantle.* Geophys. J. R. Astr. Soc. *17*, 457.

BANKS, R. J. (1972), *The overall conductivity distribution of the Earth.* J. Geomagn. Geoelectr. *24*, 337.

BANKS, R. J. (1981), *Strategies for improved global electromagnetic response estimates.* J. Geomagn. Geoelectr. *33*, 569.

BAILEY, R. C. (1970), *Inversion of the geomagnetic induction problem.* Proc. Roy. Soc. *315*, 185.

BERDICHEVSKY, M. N., FAINBERG, E. B., ROTANOVA, N. M., SMIRNOV, J. B. and VANYAN, L. L. (1976), *Deep electromagnetic investigations.* Ann. Géophys. *32*, 143.

BERDICHEVSKY, M. N., VANYAN, L. L., LAGUTINSKAYA, L. P., ROTANOVA, N. M. and FAINBERG, E. B. (1970), *The Experience of the Earth frequency sounding by the results of the spherical analysis of geomagnetic field variations*, Goemagn. Aeronom. *10*, 376 (in Russian).

DEBYE, P. (1909), *Der Lichtdruck auf Kugeln von beliebigen Material.* Annal. Phys. *30*, 57.

DEVANE, J. F. (1978), *Monte Carlo inversion of geomagnetic induction data.* Rep. 4 Workshop on EMI, Murnau 1978.

DMITRIEV, V. I., ROTANOVA, N. M., BALYKINA, O. N. and ZAKHAROVA, O. K. (1977), *About resolving power of curves of the deep magnetovariational sounding.* Geomagn. Aeronom. *17*, 1092 (in Russian).

DUCRUIX, J., COURTILLOT, V. and LEMOUËL, J. L. (1980), *The late 1960s secular variation impulse, the eleven year magnetic variation and the electrical conductivity of the deep mantle.* Geophys. J. R. Astr. Soc. *61*, 73.

DZIEWONSKI, A. M. (1984), *Mapping the lower mantle: determination of lateral heterogeneities in P velocity up to degree and order 6*, J. Geophys. Res. *89*, No B7, 5929.

ECKHARDT, D. H. (1963), *Geomagnetic induction in a concentrically stratified Earth.* J. Geophys. Res. *68*, 6273.

HVOŽDARA, M. and SIRÁŇ, G. (1975), *Penetration of longperiod geomagnetic variations to the core of the Earth.* Acta F.R.N. Univ. Comen., Astr. Geophys. *1*, 27.

KOVTUN, A. A. (1980), *The Use of the Natural Electromagnetic Field in Studying the Electrical Conductivity of the Earth.* (Leningrad. Univ.) (in Russian).

LAHIRI, B. N. and PRICE, A. T. (1939), *Electromagnetic induction in non-uniform conductors, and the determination of the conductivity of the Earth from terrestrial magnetic variations.* Phil. Trans. R. Soc. London *237*, 509.

LAMB, H. (1883), *On electrical motion in a spherical conductor.* Phil. Trans. R. Soc. London *174*, 519.

MARTINEC, Z. and PĚČ, K. (1986), *Normal earth density models.* Studia Geoph. Geod. *30*, 124.

NATAF, H.-C., NAKANISHI, I. and ANDERSON, Don L. (1984), *Anisotropy and shear-velocity heterogeneities in the upper mantle.* Geophys. Res. Lett. *11*, 109.

PARKER, R. L. (1970), *The inverse problem of electrical conductivity in the mantle.* Geophys. J. R. Astr. Soc. *22*, 14.

PARKER, R. L. (1983), *The magnetotelluric inverse problem.* Geophys. Surveys *6*, 5.

PĚČ, K. and MARTINEC, Z. (1984), *Constraints to the three-dimensional non hydrostatic density distribution in the Earth.* Studia Geoph. Geod. *28*, 364.

PĚČ, K., MARTINEC, Z. and PĚČOVÁ, J. (1985), *Matrix approach to the solution of electromagnetic induction in a spherically layered Earth.* Studia Geoph. Geod. *29*, 139.

PĚČOVÁ J., PĚČ, K. and PRAUS, O. (1980), *Remarks on spatial distribution of long period variations in the geomagnetic field over European area.* J. Geomagn. and Geoelectr. *32*, Supplement I, SI 171.

PĚČOVÁ, J. and PĚČ, K. (1982), *Models of electrical conductivity in the upper mantle.* Studia Geoph. Geod. *26*, 196.

PRICE, A. T. (1950), *Electromagnetic induction in a semiinfinite conductor*, Quart. J. Mech. Appl. Math. *3*, 385.

PRICE, A. T. (1962), *The theory of magnetotelluric method when the source is considered*, J. Geophys. Res. *67*, 4309.

RIKITAKE, T., (1951), *Electromagnetic induction within the Earth and its relation to electrical state of the Earth's interior Part III.* Bull. Earthq. Res. Inst. *29*, 61.

ROKITYANSKY, I. I., *Induction sounding of the Earth* (in Russian), (Naukova Dumka, Kiev 1981).

SCHMUCKER, U. (1974), *Erdmagnetische Tiefensondierung mit langperiodischen Variationen*, Protokoll über das Kolloquium: Erdmagnetische Tiefensondierung, Grafrath/Bayern, 313.

SMYLIE, D. E. (1965), *Magnetic diffusion in a spherically symmetric conducting mantle.* Geophys. J. Astr. Soc. *9*, 169.

SRIVASTAVA, S. P. (1966), *Theory of magnetotelluric method for a spherical conductor*, Geophys. J. R. Astr. Soc. *11*, 373.

STRATTON, J. A. (1941), *Electromagnetic Theory* (McGraw Hill, New York).

TARANTOLA, A. and VALETTE, B. (1982), *Generalized nonlinear inverse problems solved using the least-squares criterion*, Rev. Geophys. Space Phys. *20*, 219.

WIGGINS, R. A. (1972), *The general linear inverse problem: implication of surface waves and free oscillations for Earth structure.* Rev. Geophys. Space Phys. *10*, 251.

WOODHOUSE, J. H. and DZIEWONSKI, A. M. (1984), *Mapping the lower mantle: three-dimensional modelling of Earth structure by inversion of seismic waveforms.* J. Geophys. Res. *89*, No. B7, 5953.

(Received 4th November, 1985, accepted 29th April, 1986)

PAGEOPH, Vol. 125, Nos. 2/3 (1987) 0033-4553/87/030319-22$1.50+0.20/0

Upper Mantle Lateral Heterogeneities and Magnetotelluric Daily Variation Data

J. L. Counil,[1] M. Menvielle,[1] and J. L. Le Mouel[1]

Abstract—We use telluric and magnetic data of the diurnal variation recorded in Europe, Australia and North America to study the magnetotelluric tensor in the 6h–24h period range. We use associate directions and we eliminate the effects of deviation of telluric currents. We thus obtain for each observatory reliable phases and apparent resistivity values representative of the neighbouring stratified substratum. It appears that the values obtained in the four European observatories (Saint-Maur, France; Ebro, Spain; Toledo, Spain; Nagycenk, Hungary) give similar results and that these results are different from those obtained either in Tucson (USA) or in Watheroo (Australia).

Using Bostick transform we interpret these phase and apparent resistivity values in terms of conductivity of the upper mantle. We discuss then the conductivity heterogeneities in terms of change either in temperature, or partial melting or percentage of fluids of the upper mantle: at depths of about 300 km, the upper mantle appears to be 100 °C hotter under Australia than under Europe; the probable presence of fluids at depths about 100 km in the southwestern North America upper mantle appears to be responsible for the high observed conductivities. All these conductivity values are coherent with tomography results from Woodhouse and Dziewonsky: high (low) conductivities are coherent with low (high)seismic wave velocities.

Key words: Daily variation, upper mantle, lateral heterogeneities, conductivity, magnetotelluric.

I. Introduction

Within the past 10 years, geophysical studies have been performed in efforts to elucidate the 3-dimensional structure of the upper mantle. As expanded data have become available, lateral variations of increasingly complex nature were brought to light.

The inversion of seismological data allowed to map heterogeneity and azimuthal anisotropy in the upper mantle at a global scale (e.g., Nataf *et al.*, 1984; Woodhouse and Dziewonski, 1984) as well as at a local scale (e.g., Cara, 1978; Romanowicz, 1979, 1980; Montagner and Jobert, 1983; Montagner, 1986). At a global scale, these 3-dimensional maps promise to be very helpful in the understanding of mantle

[1] Laboratoire de Géomagnétisme, I.P.G.P., 4 Place Jussieu – Tour 14, F-75252 PARIS CEDEX 05, U.A. CNRS 922.

convection; and more regional studies evidenced that the behavior of the upper mantle is related to surface tectonics. All these variations in the physical properties of the mantle materials, which were found to be related to the geodynamical and tectonical history of the Earth, might correspond to variations in the thermodynamical state of the mantle.

The electrical conductivity of the Earth's mantle materials is directly related to the value of parameters, such as temperature, partial melting or percent of fluids (SHANKLAND and WAFF, 1977; SHANKLAND and ANDER, 1983): one can then expect to derive reliable information about the thermodynamical state of the upper mantle from electromagnetic studies. Taking the penetration depth of the varying electromagnetic field into account, the determination of the mantle conductivity at depths ranging from 100 to 500 kilometers involves studying variations of frequency smaller than some cycles per day (cpd), and firstly the outstanding daily variation and its harmonics.

With the so-called deep geomagnetic sounding technique, of general use when dealing with this problem (e.g., SCHULTZ, 1985), one studies the relationship between the variations of the vertical component of the geomagnetic field Z and those of the horizontal components, X (geographic North) and Y (geographic East). However the (Z/X) and (Z/Y) ratios do depend on the geometry of the source magnetic field, and deriving the Earth's mantle conductivity from the daily variation data through this method requires a network of observatories whose dimensions are large enough for this geometry to be determined (several thousand kilometers for the daily variation; CAMPBELL and SCHIFFMACHER, 1986). And the method chiefly provides a general average designation of the conductivity profile $\sigma(z)$ (z is depth) in the area covered by the network.

On the other hand, COUNIL *et al.* (1984, 1986a) have shown that, after correcting for the effect of local short scale superficial conductivity heterogeneities, the magnetotelluric (MT) tensor in a single station provides an estimate of the conductivity profile $\sigma(z)$ representative of the crust and mantle in a neighbourhood of the station; for example, in northern France and for the period from 6 h to 24 h, the linear dimensions of this neighbourhood are larger than some hundreds of kilometers (COUNIL *et al.*, 1984).

In the present paper, we will first recall and extend the MT tensor analysis proposed by COUNIL *et al.* (1984, 1986a). Then we will present a study along these lines of the daily variation of the electromagnetic field at six observatories located in North America (Tucson, USA), western and central Europe (Saint Maur, France; Ebro and Toledo, Spain; Nagycenk, Hungary) and Australia (Watheroo). The results will be discussed in terms of lateral variations of conductivity in the upper mantle at a regional scale.

II. The data

The study of the daily variation of the electromagnetic field, according to the following lines (see section III), requires recordings over several years of the variations of both Earth's electric and magnetic field components at the same station. Such data sets are only available for a limited number of observatories, unevenly distributed at the surface of the Earth. A rather extensive description of these data is given in YOSHIMATSU (1957). Let us briefly describe the data sets we selected for this study (see also Table 1).

Table 1

Description of the data used in this study. The azimuths of the lines are reckoned counterclockwisely from the geographic East

	Longitude	Latitude	Line 1 Direction	Line 1 length (km)	Line 2 Direction	Line 2 length (km)	Used data set
Saint Maur	2°22′E	48°42′N	0°	15.000	90°	15.000	1894–1896
Ebro	0°33′E	40°45′N	25°46′	1.420	115°16′	1.280	1950–1955
Toledo	4°02′W	39°53′N	0°	1.585	90°	1.520	1948: 1956–1961
Nagycenk	16°43′E	47°38′N	0°	0.500	90°	0.500	1961–1983
Tucson	110°51′W	32°15′N	89°30′	93.8	19°30′ [1]	56.81 [1]	1932–1942
Watheroo	115°53′E	30°19′S	0°	9.960	90°	3.220	1932–1942

[1] after 1936: 72°2 and 58.4 km

1. Saint Maur (France)

Magnetic variations were continuously recorded at Saint Maur observatory from 1883 to 1901 using a Mascart variometer. Hourly mean values have been published in 'Annales du Bureau Central Météorologique' (MOUREAUX, 1895, 1986, 1897). Moreover, from 1893 to 1895, electric potentials along two 15 km long lines oriented N–S and E–W were recorded. Mean monthly, hourly values of the electric field have been published and analyzed by ROUGERIE (1942). Saint Maur data have already been discussed by COUNIL *et al.* (1984, 1986a).

2. Ebro (Spain)

Magnetic and electric variations were simultaneously recorded at Ebro observatory from 1913 to 1970. The electric potentials were recorded using two approximately 1.2 km long lines oriented N25.3°W and N64.2°E, respectively. BAUER (1922) has given an early discussion of the data. CARDUS and GALDON (1962) have published mean hourly values for the decade 1950–1959 that we will use in the present study. These data have previously been studied by COUNIL *et al.* (1986a).

3. Toledo (Spain)

Magnetic variations have been recorded at Toledo since 1934, and hourly mean values have been regularly published by the 'Instituto Geografico y Cadastral'. Moreover, electric potentials were recorded from 1948 to 1961 along two orthogonal 1.5 km long NS and EW lines (see De Miguel (1951a) for a description of the experiment). The electric hourly mean values have also been published by the same Institute. A first analysis has been made by De Miguel (1951b, 1955).

4. Nagycenk (Hungary)

Magnetic and electric variations have been recorded at Nagycenk observatory since 1961. The electric potentials are measured along two orthogonal 500 meters long NS and EW lines (see Adam *et al.* (1967) for a description of the experiment). The results have been regularly published by the Geodetical and Geophysical Research Institute of the Hungarian Academy of Science. In the present study, we will use the 1961–1983 data set.

5. Tucson (Arizona, USA)

The Tucson observatory was founded in 1909 and has been operating since this date. Earth potentials along two telegraphic lines were measured from 1932 to 1942. The first line was 93.8 km long and oriented N89.5°E; the second line was 56.8 km long and oriented N19.5°E from 1932 to 1936, then 58.4 km long and oriented N17.8°E from 1936 to 1942. Rooney (1949) has published the mean values of the electric potentials in the Proceedings of the Carnegie Institution of Washington. Rooney (1944) made a first interpretation of these data; a detailed analysis has been given by Larsen (1980) and Counil *et al.* (1986b).

6. Watheroo (Australia)

Settled in 1919, the Watheroo magnetic observatory was operated until 1958. Recording of Earth's potentials started in January, 1924 and continued without serious interruption until July, 1947. The potentials were measured along two aerial NS and EW lines. The NS line was 2 miles (3.22 km) long; the EW line was first 2 miles long, then, from 1926, it was extended to 6.2 miles (9.9 km) long. Fleming *et al.* (1947a; 1947b) have published the 1919–1944 magnetic data in the Proceedings of the Carnegie Institution of Washington. Rooney and Gish (1949) have published the electric data for the decade 1932–1942, and given a first analysis. In the present study, we will use the 1932–1942 data set.

III. The Determination of the Mean Stratified Substratum Impedance

1. The static distortion approximation

Consider an Earth model consisting of a stratified substratum topped by an heterogeneous layer of thickness h and write the conductivity:

$$\sigma(P) = \sigma_n(z) + \sigma_a(P) \tag{1}$$

$\sigma_n(z)$ is the so-called normal conductivity, i.e., the conductivity of the neighbouring mean stratified substratum for which the station is representative. The existence and lateral extension L of this substratum are inferred from the data themselves: at all the considered observatories, after removing the perturbations due to the local heterogeneities ($\sigma_a(P)$), it is possible to compute in a unique way a phase value which is the argument of the mean stratified substratum impedance. The very uniqueness of this determination proves the existence of the mean stratified medium (COUNIL *et al.*, 1986b). As for the lateral extension L, it cannot be inferred from the data from a single observatory. In Europe, using long (several hundreds kilometers) telluric lines, then various observatories thousands kilometers apart, L can be shown to be at least on the order of some thousands of kilometers.

The actual electromagnetic field can then be written:

$$\begin{aligned} \mathrm{E} &= \mathrm{E_n} + \mathrm{E_a} \\ \mathrm{B} &= \mathrm{B_n} + \mathrm{B_a} \end{aligned} \tag{2}$$

where ($\mathrm{E_n}$, $\mathrm{B_n}$) is the normal field, which would be observed above the normal substratum ($\sigma_a = 0$; $\sigma = \sigma_n(z)$, $z > 0$; the external sources being unchanged); ($\mathrm{E_a}$, $\mathrm{B_a}$) is the anomalous field, due to the lateral variations of the actual conductivity in the upper layer. Let l be the horizontal characteristic dimension of these conductivity heterogeneities. LE MOUEL and MENVIELLE (1982) have shown that, when the penetration depth δ of the electromagnetic field in the normal substratum is such that $\delta \gg (lh)^{\frac{1}{2}}$, the static distortion of currents approximation holds: the self induction of the distorted currents and their mutual induction with the normal sheet of currents can be neglected; the anomalous field $\mathrm{E_a}$ is then a Coulombian field due to the electric charges on the lateral conductivity contrasts. Then $\mathrm{E_a}$ can be written:

$$\mathrm{E_a} = \mathbf{E_a}\mathrm{E_n} \tag{3}$$

where $\mathbf{E_a}$ is a real tensor which only depends on $\sigma_a(P)$.

Let us further assume a uniform elliptically polarized primary magnetic field of angular frequency ω and write the fundamental MT relationship in the frequency domain and in the (e_1, e_2) basis (e_1 = geographic East; e_2 = geographic North); first we have:

$$\mathrm{E_n} = (1/\mu)\, Z_c \begin{bmatrix} 0 & 1 \\ -1 & 0 \end{bmatrix} \mathrm{B_n} \tag{4}$$

Z_c being the impedance of the normal substratum of conductivity $\sigma_n(z)$.

From (2), (3) and (4) it comes:

$$\mathrm{E} = (1/\mu) Z_c \mathbf{Z}\mathrm{B} = (1/\mu) Z_c \mathbf{D} \begin{bmatrix} 0 & 1 \\ -1 & 0 \end{bmatrix} \mathrm{B} \tag{5a}$$

where **Z** and **D** are real tensors and:

$$\mathbf{D} = \mathbf{I} + \mathbf{E_a} \tag{5b}$$

D is the deviatoric tensor (COUNIL *et al.*, 1984, 1986a; see also LARSEN, 1975).

Let a_0 be any unit real vector. It follows from (5) that it is possible to find a unit vector $\alpha_0(\mathrm{a}_0)$ such that:

$$\mathrm{E.a}_0 = (1/\mu) Z_0(\mathrm{B}.\alpha_0) \tag{6}$$

a_0 and α_0 are called associate directions (COUNIL *et al.*, 1986a). If $\mathrm{E_a}$ is not zero, a_0 and α_0 are not perpendicular. The ratio:

$$\zeta(\mathrm{a}_0) = \mu \mathrm{E.a}_0 / \mathrm{B}.\alpha_0(\mathrm{a}_0) \tag{7}$$

does not depend on the polarization of B (let us recall that we are considering a uniform periodic elliptically polarized primary magnetic field). When the static distortion of currents approximation holds, α_0 is real and ζ is equal to the impedance Z_c of the mean stratified medium within a multiplicative real scalar (see COUNIL *et al.* (1986a) for the demonstration of the reverse proposition). That implies:

$$\mathrm{Arg}\, \zeta = \mathrm{Arg}\, Z_c \tag{8}$$

Let $\mathrm{a_{0m}}$ and $\mathrm{a_{0M}}$ the two real electric directions which make $|\zeta|$ respectively minimum and maximum. These directions are called respectively directions of minimum or maximum current: the real $\mathrm{a_{0m}}$ (respectively $\mathrm{a_{0M}}$) direction is the one along which the intensity of the electric field is minimum (respectively maximum) for a unit inducing magnetic field along the magnetic associate direction α_{0m} (respectively α_{0M}). The ratio $|\zeta_{\mathrm{m}}|/|\zeta_{\mathrm{M}}|$ is an indicator of the intensity of the telluric currents deflected by the local conductive structures.

Following COUNIL *et al.* (1984), let us finally introduce the apparent resistivity,

$$\rho(\mathrm{a}_0) = (\mu/\omega)|\mathrm{E.a}_0/\mathrm{B}.\alpha_0(\mathrm{a}_0)|^2 = (\mu/\omega)|\zeta|^2.$$

$\rho(\mathrm{a}_0)$ is equal, within a real dimensionless factor, to the actual apparent resistivity of the normal substratum, ρ_a; $\rho_m = (\mu/\omega)|\zeta_m|^2$ and $\rho_M = (\mu/\omega)|\zeta_M|^2$ provides respectively with estimates of lower and upper bounds for ρ_a. When ρ_m and ρ_M are on the same order of magnitude (typically $\rho_M/\rho_m < 5$), we adopt $\rho_a = (\rho_m \rho_M)^{\frac{1}{2}}$ as a

Table 2a

Maximum (ρ_M) over minimum (ρ_m) apparent resistivity ratio values at the different observatories. The apparent resistivity estimate ρ_a is deduced from ρ_m and ρ_M values; the uncertainties on these estimates are given in percent (see text). The missing values correspond to the cases where the data did not allow obtainment of reliable estimates.

	m	ρ_M/ρ_m	ρ_a Ω.m	
Saint Maur	1	3.4	16	46%
	2	2.5	10	54%
	3	/	/	
	4	/	/	
Ebro	1	/	12	77%
	2	975	11	75%
	3	903	11	57%
	4	878	16	63%
Toledo	1	8.5	10	80%
	2	/	/	
	3	/	/	
	4	/	/	
Nagycenk	1	3.2	21	55%
	2	5.4	18	71%
	3	2.0	18	60%
	4	/	/	
Tucson	1	4.0	1.4	53%
	2	1.4	1.4	40%
	3	1.8	1.6	30%
	4	1.7	1.7	40%
Watheroo	1	1.2	12	60%
	2	1.2	7	25%
	3	1.2	8	50%
	4	1.2	13	75%

generalization of the choice made in the 2-D case by RANGANAYAKI (1984), 'by analogy with mixtures where the geometric mean gives an accurate estimate of the physical properties (MADDEN, 1976)'; the errors bars, given in Table 2a, are deduced from the uncertainties in ρ_m and ρ_M estimates, and they are not allowed to be smaller than $(\rho_m \rho_M)^{\frac{1}{2}} - \rho_m$. At the Ebro observatory, where ρ_M/ρ_m reaches 1000, we take $\rho_a = \rho_m$ and the errors bars are those on the ρ_m determination.

Then, analysing the MT tensor using couples of electromagnetic associate directions makes it possible:

1) to estimate the intensity of the electric currents distorted by the local conductivity heterogeneities;

2) to compute the argument of the impedance of the mean stratified substratum for which the station is representative;
3) to estimate bounds for the actual apparent resistivity of this substratum.

Remark: The results of the analysis of the MT tensor following the lines indicated above provides an *a posteriori* check of the validity of the static distortion approximation: if the phase of $\zeta(a_0)$ does not depend on a_0, the **D** tensor is real. Indeed, when the static distortion approximation does not hold, the $\mathbf{E_a}$ tensor is complex and the phase is no longer independent of a_0. Then, obtaining equal phases—within the experimental errors—for various real electric directions is a clue that the approximation is valid.

2. *The case of the daily variation*

In the case of a uniform primary inducing field, we have shown how to correct the data for local heterogeneities to obtain the value of the argument of the mean stratified medium impedance Z_c: it is enough to consider the ratio of the electric component $E.a_0$ along a given unit real vector a_0 to the magnetic component $B.\alpha_0$ along the associated unit vector $\alpha_0(a_0)$ (formula (7)).

Reaching the spherical geometry (we are considering large periods) and considering an inducing primary field which has the geometry of an elementary surface harmonic of degree n and order m, formulas (7) and (8) still hold but Z_c is now expected to vary with n. And we will show how considering several harmonics (n, m) permits a high accuracy determination of the associate directions.

Let us now consider specifically the daily solar variation S_R (MAYAUD, 1967). At a given observatory P, the magnetic potential corresponding to the m-th time harmonic can be written:

$$V(P, t) = a \sum_{n=m}^{\infty} (i_n^m + e_n^m) P_n^m(\cos\theta) e^{im(\varphi + \omega t)} \tag{9}$$

a is the Earth's radius, (θ, φ) the colatitude and longitude of P, $P_n^m(\cos\theta)$ the associate Legendre function of degree n and order m, i_n^m and e_n^m complex coefficients with the dimension of a magnetic field and $T = 24$ hours ($\omega = 2\pi/T$). As usual, only real parts of complex quantities such as (9) are considered.

It has been well-known that the solar daily variation presents a day-to-day variability (MAYAUD, 1965a, 1965b). In the following applications we will use monthly averaged daily variations (see section III.3). But, due to the inclination of the Earth's rotation axis, the monthly averaged coefficients i_n^m and e_n^m vary with the month number k; to account for this variation, we will write them $i_n^m(k)$ and $e_n^m(k)$.

Assume now a spherically symmetric Earth (we will come back to this point later). The electric current vector (for the m-th harmonic at P) can be expressed (e.g, LE MOUEL, 1976):

$$E(k, P, t) = i\omega m \sum_n \operatorname{curl} \mathbf{r}\left(\frac{i_n^m(k)}{n} - \frac{e_n^m(k)}{n+1}\right) P_n^m(\cos\theta) e^{im(\varphi + \omega t)} \tag{10}$$

with $\omega = 2\pi/T$, $\mathrm{r} = \mathrm{OP}$, O being the Earth's center.

The impedances $Z^m(k)$ corresponding to the different harmonics m ($m = 1, 2, 3, \ldots$) of the monthly averaged daily variation at the given observatory (θ, φ):

$$Z^m(k) = \mu\left(\frac{E_\varphi(k)}{B_\theta(k)}\right)^m = -\mu\left(\frac{E_\theta(k)}{B_\varphi(k)}\right)^m \tag{11}$$

$$= i\omega\mu a m \frac{\sum_n (n e_n^m(k) - (n+1) i_n^m(k)) \frac{\partial}{\partial\theta} P_n^m(\theta)}{\sum_n n(n+1)(e_n^m(k) + i_n^m(k)) \frac{\partial}{\partial\theta} P_n^m(\theta)}$$

are not expected to vary with k. Indeed, the theoretical expressions of the impedances corresponding to the different modes (n, m) (for a radial Earth) are:

$$Z_n^m(a) = i(2\pi m/T)\, a/(1 + aR_n'(a)/R_n(a)) \tag{12}$$

where $R_n(r)$ is the solution of a second order differential equation the coefficients of which depend on $\sigma(r)$, n and m (e.g., LE MOUEL, 1976). It is well-known that the $(\mu E_\varphi/B_\theta)_n^m$ quantities vary only within a few percent for n varying from 1 to 6 and m from 1 to 6 for reasonable conductivity distributions $\sigma(r)$.

Then, the constancy of the ratios $Z_n^m = (\mu E_\varphi/B_\theta)_n^m$ when k varies results from the elementary arithmetic rule:

$$a_n/b_n = s \,\forall\, n \longrightarrow \Sigma a_n/\Sigma b_n = s.$$

This constancy of Z_m has been experimentally checked by COUNIL *et al.* (1984; 1986a) after disposing of the effects of local heterogeneities using associate directions (see section 3). Figure 1a gives some illustrations of this observation.

Let us now observe the condition when considering only magnetic data, as usually done in the deep geomagnetic sounding technique. The following ratio of the north component to the vertical component, for a given frequency m, is computed:

$$\left(\frac{X}{Z}\right)^m(k) = \frac{\sum_n X_n^m}{\sum_n Z_n^m} = \frac{\sum_n (e_n^m(k) + i_n^m(k)) \frac{\partial}{\partial\theta} P_n^m(\cos\theta)}{\sum_n (n e_n^m(k) - (n+1) i_n^m(k)) P_n^m(\cos\theta)}. \tag{13}$$

This ratio changes with the month number k since the elementary ratios X_n^m/Z_n^m depend on k. This variability is illustrated in Figure 2.

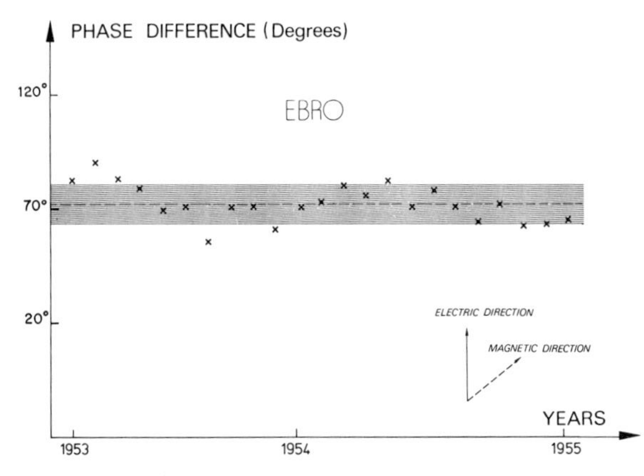

Figure 1a.
Phase of the impedance computed at Ebro using couples of associated directions (North electric component and its associate magnetic direction) for the 12 h harmonic. The mean value is represented by a dashed line and the width of the grey stripe is two times the standard deviation.

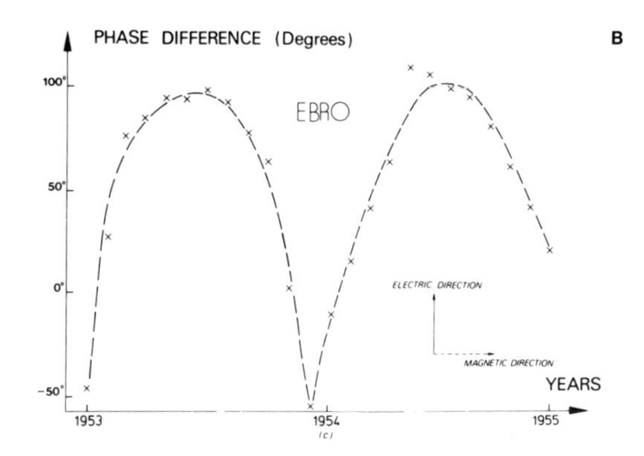

Figure 1b.
Variation of the monthly mean value of the phase difference between the North electric component and the East magnetic component at Ebro for the 12 h harmonic.

Then, estimating the impedance phase for the m cpd frequency from the daily variation data it is impossible using magnetic data alone from a single observatory. On the contrary, magnetotelluric observations allow such an estimate. This limitation in the deep geomagnetic sounding technique originates in the day-to-day variability of the geometry of the sources of the daily variation. For the ring current origin variations, indeed, the stability of the P_0^1 source geometry can allow the estimate of the impedance phase using data from a single observatory (e.g., SCHULTZ, 1985).

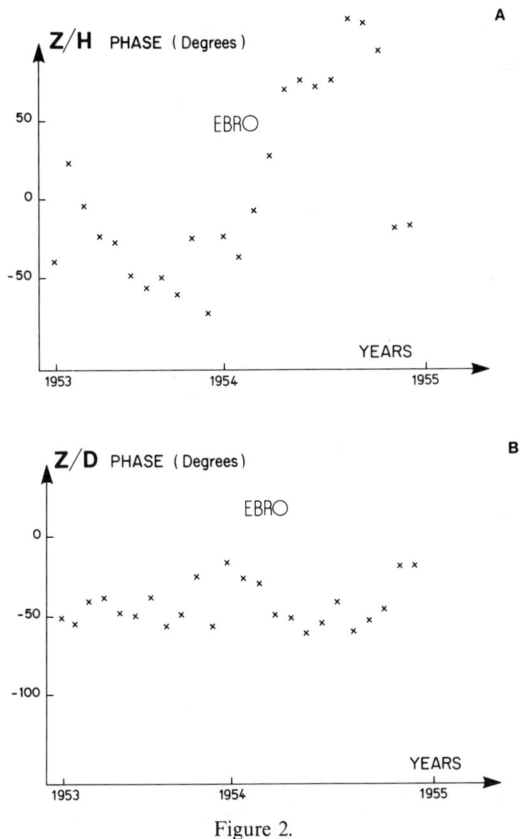

Figure 2.
Variation of the monthly value of the Z/H(A) and Z/D(B) ratios phases computed at Ebro for the 12 h harmonic.

3. *Practical determination of the associate directions*

In this study, we use monthly mean hourly values (i.e., the hourly value labelled i for a given month is the average of all the hourly mean values labelled i of this month) to describe the daily variation of a component of the electromagnetic field. Furthermore, as the following computations rely on the variability of the quiet days solar daily variation throughout the year, we have selected, using the *aa* index (MAYAUD, 1973), the quietest January month among those available for each station, the quietest February month, and so on. For each observatory, we will use such series of 12 electric and magnetic daily variations labelled 1 (January) to 12 (December).

As stated in section III.2, the phase of the elementary impedance Z_n^m, does not depend significantly on the degree n. Then the same holds for the ratio $\zeta^m(a_0)$ i.e., $\zeta(a_0)$ computed for the m-th harmonic of the daily variation using a couple of associate directions. On the contrary, the phase of the ratios $E.a/B.\alpha$ varies with m and then with k, when a and α are not associate directions. Thus, for a given station P, the associate direction $\alpha(a)$ of a given real electric direction a is the direction for which the variation with k of Arg(E.a/B.α) is minimum, namely the direction for which the standard deviation of Arg(E.a/B.α) for the set of selected 12 months will be minimum. Figure 1b illustrates this situation.

IV. The Results

For each observatory, the daily variation of the electromagnetic field has been analyzed using the method described in section III.

We have first checked that the static distortion of currents approximation holds for the daily variation and its harmonics at each of the studied observatories. The phase of the ratio $\zeta(a_0)$ does not depend on a_0, within the error bars. Then, for each observatory, the notion of mean stratified substratum is relevant: in a given neighbourhood, the crust and the upper mantle can be described as a stratified substratum with local heterogeneities of conductivity, if any, such that the static distortion approximation holds (COUNIL *et al.*, 1986b). The lateral extent (characteristic dimension L) of this neighbourhood has been experimentally found to be at least on the order of several hundreds of kilometers for the Saint Maur observatory (COUNIL *et al.*, 1984); in fact, as it will be seen, L is roughly several thousands of kilometers in Europe.

Then we estimated the apparent resistivities ρ_m and ρ_M at each station for the four first harmonics of the daily variation (i.e., for periods $T = 24/\mathrm{m}$ hours, $m = 1$, 2, 3, 4). ρ_m and ρ_M values are found to be significantly different—the one from the other: the ρ_M/ρ_m ratio is generally between 1.5 and 10, but at the Ebro observatory where it reaches 1000 (see Table 2a). That indicates the existence, close to each observatory (but, maybe, Watheroo) of local heterogeneities which deflect (and, at the Ebro observatory, channelize) the telluric currents. The horizontal characteristic dimension l of these heterogeneities is at least on the order of the length of telluric lines, but it is small enough for the self and mutual induction of deflected telluric currents to be neglected (see section III.1). From these ρ_m and ρ_M values, we deduced an estimate of the apparent resistivity ρ_a of the mean stratified substratum for each harmonic of the daily variation (see section III.1). These ρ_a values are given in Table 2a. They are in good agreement, one with another, for the European observatories; those observed at Watheroo are on the same order of magnitude as at European observatories.

We have also computed the argument Φ of the Cagniard magnetotelluric impedance Z_c corresponding to the mean stratified substratum for which each station is representative (formula (7)). Table 2b gives the estimated values of this phase at each station for the first harmonics of the daily variation. Again, the values observed at the European observatories are in good agreement one with another; those observed at Tucson are about 10° lower and those observed at Watheroo are about 10° higher.

Table 2b

Z_c argument for the six studied observatories and the corresponding standard deviations computed as in COUNIL et al. *(1986a). The missing values correspond to the cases where the data did not allow obtainment of reliable estimates.*

m	1		2		3		4	
Saint Maur	71.8	3.0	72.0	19.1	75.9	16.7	/	
Ebro	73.8	8.1	70.6	8.7	65.1	11.9	/	
Toledo	72.2	19.1	/		/		/	
Nagycenk	70.3	5.8	68.7	15.1	77.2	14.7	/	
Tucson	62.9	2.4	60.1	3.1	57.6	2.7	54.1	4.9
Watheroo	83.0	12.3	76.8	7.3	/		/	

Table 2c

Phase of the impedance $Z_n^m(n = m + 1)$ as deduced from planetary analyses of the daily variation.

Planetary models	1	2	3	4
CHAPMAN	71.5	66.5	68.7	/
MATSUSHITA	71.3	66.5	68.8	/

We have compared these values of ρ_a and Φ with the values provided by planetary analyses of the daily variation (e.g., CHAPMAN, 1919; MATSUSHITA, 1967). In these analyses, global data are expressed in terms of $s_n^m = i_n^m/e_n^m$, the ratio of the internal field to the external field for the spherical harmonic of degree n and order m.

The corresponding impedance is:

$$Z_n^m = i(2\pi m/T)\mu a(n - (n + 1)s_n^m)/(n(n + 1)(1 + s_n^m))$$

where a is the radius of the Earth. Table 2b gives the phase values of Z_n^m from CHAPMAN (1919) and MATSUSHITA (1967).

These planetary values are in good agreement with the values we found in this study for European observatories, but significantly different from the Watheroo and Tucson ones (it is probable that European observatories have an excessive weight in worldwide analyses).

V. Discussion

Our results evidence the existence of lateral heterogeneities of conductivity in the upper mantle at a large regional scale. A clearer description of these lateral variations of conductivity can be obtained using the Bostick transform, which relates the apparent resistivity ρ_a and the phase Φ of the impedance Z_c at period T to the conductivity $\sigma(D)$ at depth D through (BOSTICK, 1976):

$$D = (\rho_a T/2\pi\mu)^{\frac{1}{2}}$$
$$\sigma(D) = (1 + K)/(\rho_a(1 - K)) \quad (14)$$
$$K = \Phi/45 - 1$$

where Φ is expressed in degrees.

The correspondence between (ρ_a, Φ) and (D, $\sigma(D)$) is not to be taken as the result of a true inversion. We only use it to compare (D, $\sigma(D)$) couples relative to different places, the computation being performed the same way at the different places.

Table 3 gives the $\sigma(D)$ values deduced from ρ_a and Φ estimates given in Table 2. Figure 3 gives a comparison between $\sigma(D)$ values and conductivity models deduced from global studies of the daily and ring current variations (BANKS, 1969, 1972; PARKER, 1970; SCHULTZ, 1985). Figure 3 shows that:

— our western European values are in good agreement with Parker's model;
— in southwestern North America, the $\sigma(D)$ values are about 5 times greater than the Parker's model conductivity at the corresponding depths (70–120 km);
— in western Australia, the conductivity $\sigma(D)$ is about $1\Omega^{-1}$ m^{-1} for depths ranging from about 200 to 360 kilometers, significantly higher than the Parker's model values at the same depths.

Table 3

D and $\sigma(D)$ values deduced from ρ_a and Φ value of Table 2 through the Bostick *transform. The uncertainties on the determinations are given in percent*

	Saint Maur		Ebro		Toledo		Nagycenk		Tucson		Watheroo	
m	$\sigma(D)$	D	$\sigma(D)$	D	$\sigma(D)$	D	$\sigma(D)$	D	$\sigma(D)$	D	$\sigma(D)$	D
1	0.25	420	0.38	360	0.41	330	0.17	480	1.70	120	0.99	360
	45%	23%	78%	39%	81%	40%	55%	28%	53%	27%	62%	30%
2	0.40	230	0.33	250	/	/	0.18	310	1.40	88	0.83	200
	55%	27%	76%	38%			72%	36%	40%	20%	26%	13%
3	/	/	0.24	200	/	/	0.34	260	1.10	76	/	/
			58%	29%			61%	30%	30%	15%		
4	/	/	/	/	/	/	/	/	0.89	68	/	/
									40%	20%		

Subsequently significant differences in the upper mantle conductivity between Europe, southwestern North America and western Australia are evidenced.

The variation with temperature and pressure of the mantle materials conductivity can be expressed as:

$$\sigma = \sigma_0 \exp(-(E + P\Delta V)/kT) \quad (15)$$

where σ_0, E and ΔV are respectively a conductivity, an activation energy, and an activation volume which are characteristic of the material and the conduction process; T is absolute temperature, P is pressure and k is the Boltzmann's constant. Laboratory experiments have provided estimates of the conductivity of olivine under a variety of thermodynamical conditions, (DUBA *et al.*, 1974; SHANKLAND, 1975;

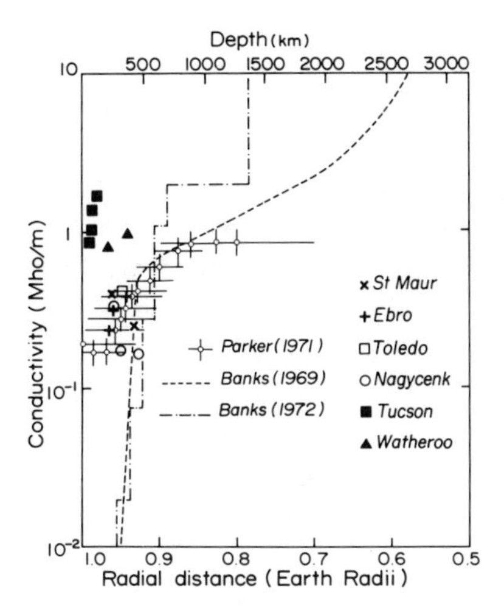

Figure 3.
Comparison between two upper mantle conductivity models and $\sigma(D)$ values (see text and Table 3 for further explanation).

SHANKLAND and WATT, 1977) as well as for some other minerals and rocks (see, for instance, OLHOEFT (1981) or PARKHOMENKO (1982)). But the upper mantle is not only made of dry solid materials. There is a liquid phase, made either of melt, or H_2O rich fluids, or both. The actual conductivity of the upper mantle thus depends on both conductivities of the solid and liquid phases, as well as on the percentage of liquids.

Thereafter, as an approximation of the actual conductivity of the upper mantle, let us adopt the effective conductivity

$$\sigma^* = (\sigma_1 + (\sigma_1 - \sigma_s)(1 - 2f/3))/(1 + (f/3)(\sigma_s/\sigma_1 - 1)) \tag{16}$$

where σ_1 and σ_s are respectively the conductivities of the liquid and solid phases, and f is the percentage of liquid. This relation may be simplified when $\sigma_s \ll \sigma_1$, as it can generally be assumed in the upper mantle (where the conductivity σ_s of the solid phase is expected to be less than $10^{-2}\ \Omega^{-1}\ \mathrm{m}^{-1}$ (DUBA *et al.*, 1974; SHANKLAND and WAFF, 1977)). In these conditions (16) leads to:

$$f = 3\sigma^*/(2\sigma_1 + \sigma^*) \tag{17}$$

Accordingly, the determination of f knowing $\sigma^* = \sigma(D)$ needs knowledge of the conductivity of the liquid phase σ_1, which depends on the composition of this phase. Following TARITS (1986), we will first consider successively the two limit cases of quasi-anhydrous (less than 0.1% H_2O rich fluids content (TARITS, 1986)) upper mantle with partial melting on the one hand, and of an unmelted hydrous upper mantle on the other hand. Thereafter we will come to what could be the behavior of the Earth upper mantle.

The quasi-anhydrous case

Consider first the case of the upper mantle conductivity at depths of about 100 kilometers. Figure 4 gives the variation of temperature and partial melting as a function of conductivity calculated by SHANKLAND and WAFF (1977) using the σ_0, E and ΔV parameters values given in Table 4 for the solid phase and the melt basalt. Our conductivity estimates under southwestern North America led to a melt fraction larger than 15% and to a temperature larger than 1450 °C at these depths. The Parker's model conductivity led to a melt fraction about 5% and a temperature between 1350 and 1400 °C. Thus a significant degree of partial melting is necessary to account for our conductivity estimates, as well as those of Parker, in the case of a quasi-anhydrous upper mantle. This result might indicate that a significant degree of partial melting might occur in the upper mantle at depths of about 100 kilometers.

Consider now the depths of about 300 kilometers and take again the estimates of

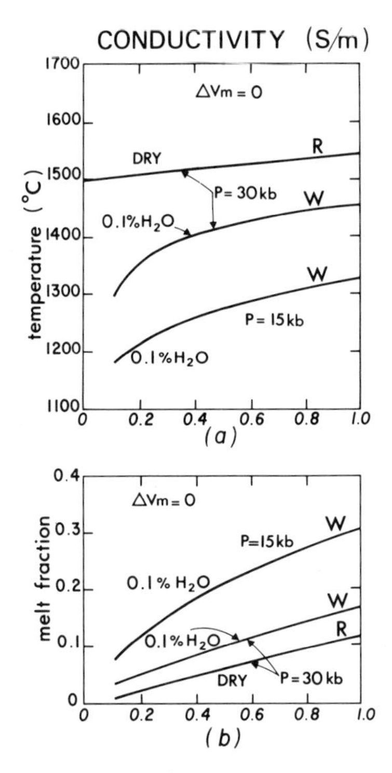

Figure 4.
Calculated temperatures (a) and partial melt fractions (b) at 15 and 30 kbar pressure as a function of effective conductivity. Uncertainties in conductivity determinations cause greater uncertainty for the calculated melt fraction than for the calculated temperature. Curve *W* is from WILLIE (1971); curve *R* from RINGWOOD (1975) (after SHANKLAND and WAFF, 1977).

Table 4

Electrical conduction parameters for olivine and basalt melt up to 30 kbar as selected by SHANKLAND and WAFF (1977).

	σ_0 Ω^{-1} m^{-1}	E eV	ΔV eV/kbar
Solid phase			
Low temperature	22.3	1.37	-3.0 10^{-3}
High temperature	4.56 10^7	3.435	6.23 10^{-3}
Basalt melt	1.82 10^4	1.147	0

electrical conduction parameters σ_0, E and ΔV given in Table 4. Following (15) and (17), the effective conductivity can be written:

$$\sigma^* = (2f/(3 - f))\sigma_0 \exp(-E/kT)$$

where f is the melt fraction.

Consistent with the mean continental geotherm, the temperature at a depth of 300 kilometers is about 1400 °C with a vertical gradient of 0.4 °C/km. Significant departures from this mean geotherm may occur, in relation, for instance, with convective motions in the upper mantle.

For the purpose of estimating an upper bound of the difference in temperature ΔT between western Australia and western Europe, let us assume further that the percentage of partial melting does not vary with temperature. This obviously provides with an upper bound for the difference in temperature: an increase in the percentage of partial melting would probably correspond to an increase in temperature and the actual ΔT is them smaller than the one corresponding to our limit case. It follows that:

$$\log(\sigma_2/\sigma_1) > E(T_2 - T_1)/kT_1T_2$$

$$\Delta T = T_2 - T_1 < kT_1T_2 \log(\sigma_2/\sigma_1)/E$$

where (σ_1, T_1) and (σ_2, T_2) are the conductivity and the temperature under western Europe and western Australia respectively at the considered depth. Accordingly, a difference in temperature smaller than 100 °C between the European upper mantle and the Australian one $(T_2 > T_1)$ at depths of about 300 kilometers would make clear the observed variation in conductivity.

The unmelted hydrous case

According to Equation (17), the knowledge of σ^* allows the determination of the percentage of fluid f, provided σ_1 is known.

At the present time, laboratory measurements of the conductivity of electrolytes have been made only at rather low temperatures ($T < 600$ °C), low pressure ($P < 4$ kbar) and low molar concentration ($M < 0.001$) (FRANTZ and MARSHALL, 1982). TARITS (1986) extrapolated these results up to pressure approaching 30 kbars, temperature about 1400 °C and molar concentration close to 0.1 M corresponding to the upper mantle at depths of nearly 100 kilometers (SHANKLAND and ANDER, 1983). He found values of σ_1 ranging from 5 to 15 Ω^{-1} m^{-1}. But it is impossible to derive accurate estimates of electrolytes conductivity at the thermodynamical conditions prevailing at depths verging on 300 kilometers (T about 1400 °C and P about 100 kbar) through an extrapolation of the available laboratory measurements.

Consider first the upper mantle conductivity estimates at depths of roughly 100 kilometers obtained in section IV. With values of σ_1 ranging from 5 to 15 Ω^{-1} m^{-1}, Equation (17) leads to percentages of fluid from 2 to 6% for $\sigma^* = 0.2$ Ω^{-1} m^{-1} as provided by Parker's model and from 10 to 30% for $\sigma^* = 1$ as provided by the present study under southwestern North America. In such a case, the conductivity estimates did not provide any information about temperature.

Consider now the depths of generally 300 kilometers and assume that the difference in conductivity between western Europe and western Australia is only due to a difference in the percentage of fluids.

Let us again assume that σ_1 is on the order of 10 Ω^{-1} m^{-1} for $P = 100$ kbars. Then, formula (17) leads immediately to:

$$f = 3\sigma^*/(2\sigma_1)$$

and the difference in conductivity observed between western Australia and western Europe might be accounted for by a percentage of connected fluids four to five times greater under western Australia than under western Europe.

The actual mantle (intermediary case)

Neither the quasi-anhydrous case, nor the unmelted hydrous case correspond to the actual upper mantle. But they provide bounds for temperature partial melting and percentage of fluids.

a. The upper mantle at depths of nearly 100 km

The conductivity corresponding to the Parker's model can be accounted for by the solution suggested by TARITS (1986) when studying the upper mantle under the Pacific Ocean (1% of either fluid and temperature about 1150 °C). In the case of southwestern North America, the anomalously high conductivity is probably associated with a significant anomaly in the percentage of H_2O rich fluids (the other limit case gives unreasonable values of T and partial melting). This regional increase of H_2O rich fluids in the upper mantle might be related to the melting of a subducted oceanic lithosphere under southwestern North America (ENGEBRETSON *et al.*, 1984).

In this latter case, the knowledge of the upper mantle conductivity does not provide any further constraint on temperature.

Table 5

Estimated upper mantle state at depths of about 100 kilometers as deduced from our conductivity estimates in the two limit cases of a quasi-anhydrous and of an unmelted hydrous upper mantle (see text for further explanation).

	Quasi-anhydrous limit		Unmelted hydrous limit	
	Partial melting %	T °C	H_2O rich fluid %	T °C
Southwestern North America	>15	>1450	10–30	/
Parker's model	5	1350–1400	2–6	/

b. The upper mantle at depths about 300 kilometers

Our actual knowledge about the upper mantle seems to rule out large variations in the percentage of fluids. On the contrary, variations on the order of 100 °C in the upper mantle temperature at depths around 300 km might easily be attributable to the result of convective motions. Thus we are led to interpret the observed variation in conductivity as the effect of variation of the upper mantle temperature.

VI. Conclusion

The results we present in this paper illustrate the efficiency of the analysis of the magnetotelluric tensor using associate directions. For each of the six studied observatories, we have been able to:

— evidence the existence, close to each station, of local heterogeneities;

— monitor that the static distortion of currents approximation applies for the daily variation and its harmonics, and remove the effects of the local heterogeneities;

— estimate at each station and for frequencies from 1 to 4 cpd both the argument of the impedance and the apparent resistivity ρ_a of the mean stratified substratum in the neighbourhood of the station. The characteristic dimension L of this stratified substratum is experimentally found to be on the order of several thousands of kilometers in western Europe.

Using the Bostick transform, we have then transformed our ρ_a and Φ estimates into couples $(D, \sigma(D))$, $\sigma(D)$ being the conductivity at depth D. We have thus evidenced the existence of significant conductivity variations at a regional scale (characteristic dimension $\lambda \gg L$) in the upper mantle up to depths of roughly 400 kilometers. A comparison between our conductivity estimates and planetary models leads us to consider the Parker's model as a more acceptable global conductivity model for the uppermost hundreds of kilometers of the Earth *vis-a-vis* the Bank's models. Nevertheless the very existence of regional heterogeneities of conductivity in the upper mantle might induce the questioning of planetary models based upon the spherical approximation to provide an accurate description of the upper mantle conductivity.

Interpreting our conductivity estimates in terms of temperature and percentage of fluids in the upper mantle, we suggest that the high conductivity values at depths of about 100 kilometers under southwestern North America are related to a significant increase in percentage of fluids, which might be related to the presence of a subducted lithosphere. On the other hand, the difference in the upper mantle conductivity between Europe and Australia at depths of about 300 kilometers is interpreted as the effect of an increase of the upper mantle temperature.

Finally, let us compare our results with those obtained by WOODHOUSE and DZIEWONSKI (1984) when mapping 3-dimensional seismological structures of the Earth's upper mantle. The high upper mantle conductivity values found at Tucson and Watheroo correspond to low seismic wave velocities while the European lower conductivities correspond to high seismic wave velocities (we refer the reader to our Figure 4 and to the plate 2 in WOODHOUSE and DZIEWONSKI (1984)). Although an interesting observation, such a correlation, based upon scant points, is not worth discussing further. However, it might confirm that more long period MT studies could contribute efficiently to the 3-dimensional mapping of the mantle. A last occurrence worth noticing is that, whereas delays in travel times of the seismic waves are on the order of 10%, variations in the magnetotelluric impedance can reach 50%.

Acknowledgments

The authors are deeply indebted to C. Jaupart and J. P. Poirier for fruitful scientific discussions. Contribution I.P.G.P. no. 955

REFERENCES

ADAM, A., VERO, J. and WALLNER, A. (1967), *Tellurische und erdmagnetische Messungen im Observatorium bei Nagycenk*. Observatoriumsberichte des Geophysikalischen Forschungslaboratoriums der Ungarischen Akademie der Wissenschaften vom Jahre 1966, 129–141.

BANKS, R. J. (1969), *Geomagnetic variations and the electrical conductivity of the upper mantle*. Geophys. J. R. Astr. Soc. *17*, 457–487.

BANKS, R. J. (1972), *The overall conductivity distribution of the Earth*. J. Geomagn. Geoelectr. *24*, 337–351.

BAUER, L. A. (1922), *Some results of recent Earth current observations and relations with solar activity*. Terr. Mag. and Atm. Elect. *27*, 1–34.

BOSTICK, F. X., Jr. (1976), *A simple and almost exact method of magnetotelluric analysis* (Abstract). Workshop on Electrical Method in Geothermal Exploration, Snowbird, Utah.

CAMPBELL, W. H. and SCHIFFMACHER, E. R. (1986), *A comparison of upper mantle subcontinental electrical conductivity for North America, Europe and Asia.* J. Geophys. *59*, 56–61.

CARA, M. (1978), *Regional variations of higher Rayleigh mode velocities: A spatial filtering method.* Geophys. J. R. Astr. Soc. *54*, 439–460.

CARDUS, J. O. and GALDON, E. (1962), *Observaciones magneticas y telluricas en el observatorio del Ebro durante el decenio 1950–1959.* Publicaciones del observatorio del Ebro no° 21.

CHAPMAN, S. (1919), *The solar and lunar diurnal variation of the Earth magnetism.* Phil. Trans. R. Soc. *218*, 95–125.

COUNIL, J. L., LE MOUEL, J. L. and MENVIELLE, M. (1984), *A study of the diurnal variation of the electromagnetic field in northern France using ancient recordings.* Geophys. J. R. Astr. Soc. *78*, 831–845.

COUNIL, J. L., LE MOUEL, J. L. and MENVIELLE, M. (1986a), *Associate and conjugate direction concepts in magnetotellurics.* Ann. Geophys. *4B*, 115–130.

COUNIL, J. L., MENVIELLE, M. and LE MOUEL, J. L. (1986b), *Study of the daily variation at Tucson (Arizona; USA): Experimental evidence of static distortion of telluric currents* (in preparation).

DE MIGUEL, L. (1951a), *Observatorio central geofisico de Toledo: corrientes telluricas ano 1948.* Instituto Geografico y Cadastral, Madrid.

DE MIGUEL, L. (1951b), *Observatorio central geofisico de Toledo: corrientes telluricas ano 1949.* Instituto Geografico y Cadastral, Madrid.

DE MIGUEL, L. (1955), *Observatorio Central geofisico de Toledo: corrientes telluricas ano 1952.* Instituto Geografico y Cadastral, Madrid.

DUBA, A., HEARD, H. C. and SHOCK, R. (1974), *Electrical conductivity of olivine at high pressure and under controlled oxygen fugacity.* J. Geophys. Res. *79*, 1667–1673.

ENGEBRETSON, D. C., COX, A. and THOMPSON, G. A. (1984), *Correlation of plate motion with continental tectonics: Laramide to basin and range.* Tectonics *3*, 115–120.

FLEMING, J. A., JOHNSTON, H. F., FORBUSH, S. E., MCNISH, A. G. and SCOTT, W. E. (1947a), *Magnetic results from Watheroo Observatory, Western Australia, 1919–1935.* Researches of the Department of Terrestrial Magnetism, Carnegie Inst. Wash. Pub. *175*, vol. VII-A.

FLEMING, J. A., JOHNSTON, H. F., MCNISH, A. G., PARKINSON, W. C., FORBUSH, S. E., GREEN, J. W. and SCOTT, W. E. (1947b), *Magnetic results from Watheroo Observatory, Western Australia, 1936–1944.* Researches of the Department of Terrestrial Magnetism, Carnegie Inst. Wash. Pub. *175*, vol. VII-B.

FRANTZ, J. D. and MARSHALL, W. L. (1982), *Electrical conductances and ionization constants of calcium chloride and magnesium chloride in aqueous solutions at temperatures to 600 °C and pressures to 4000 bars*, Am. J. Sci. *282*, 1666–1693.

LARSEN, J. C. (1975), *Low frequency (0.1–6.0 cpd) electromagnetic study of the deep mantle electrical conductivity beneath the Hawaiian islands.* Geophys. J. R. Astr. Soc. *43*, 17–46.

LARSEN, J. (1980), *Electromagnetic response functions from interrupted and noisy data.* J. Geomagn. Geoelectr. *32*, Suppl. I., 89–103.

LE MOUEL, J. L. (1976), *L'induction dans le Globe.* Traité de Géophysique Interne, (eds J. Coulomb and G. Jobert) tome 2, Masson, Paris.

LE MOUEL, J. L. and MENVIELLE, M. (1982), *Geomagnetic variation anomalies and deflection of telluric currents.* Geophys. J. R. Astr. Soc. *68*, 575–587.

MADDEN, T. R. (1976), *Random networks and mixing laws.* Geophysics *41*, 1104–1125.

MATSUSHITA, S. (1967), *Solar quiet and lunar daily variation fields*, In *Physics of Geomagnetic Phenomena*, Vol. 1, (Academic Press, New York).

MAYAUD, P. N. (1965a), *Analyse morphologique de la variabilité jour-à-jour de la variation journalière régulière* S_R *du champ magnétigue terrestre, 1. Le système de courants* C_P *(régions polaires et subpolaires).* Ann. Geophys. *21*, 369–401.

MAYAUD, P. N. (1965b), *Analyse morphologique de la variabilité jour-à-jour de la variation journalière régulière* S_R *du champ magnétique terrestre. 2. Le système de courants* C_M *(régions non polaires).* Ann. Geophys. *21*, 514–545.

MAYAUD, P. N. (1967), *Atlas des Indices.* K. IAGA Bull. *21*, 113 pp (IUGG Publ. Office, Paris).

MAYAUD, P. N. (1973), *A hundred year series of geomagnetic data, 1868–1967; indices* aa, *storm sudden* commencements. IAGA Bull. *33*, 252 pp. (IUGG Publ. Office, Paris).

MONTAGNER, J. P. (1986), *Three dimension structure of the Indian Ocean inferred from long period surface wave.* Geophys. Res. Lett. *13*, 315–318.

MONTAGNER, J. P. and JOBERT, N. (1983), *Variation with age of the deep structure of the Pacific Ocean inferred from very long period Rayleigh wave dispersion.* Geophys. Res. Lett. *10*, 273–276.

MORSE, S. A. (1980), *Basalts and Phase Diagrams*, (Springer-Verlag) 493 pp.

MOUREAUX, T. (1895), *Observations magnétiques faites au Parc Saint-Maur durant l'année 1893.* Annales du Bureau Central Météorologique, Paris.

MOUREAUX, T. (1896), *Observations magnétiques faites au Parc Saint-Maur durant l'année 1894.* Annales du Bureau Central Météorologique, Paris.

MOUREAUX, T. (1897), *Observations magnétiques faites au Parc Saint-Maur durant l'année 1895.* Annales du Bureau Central Météorologique, Paris.

NATAF, H. C., NAKANISHI, I. and ANDERSON, D. L. (1984), *Anisotrophy of shear velocity inhomogeneities in the upper mantle.* Geophys. Res. Lett. *11*, 109–112.

OLHOEFT, G. R. (1981), *Electrical properties of rocks,* In *Physical Properties of Rocks and Minerals,* (eds Y. S. Touloukian, W. R. Judd, and R. F. Roy, (McGraw Hill, New York 548 pp.

PARKER, R. L. (1970), *The inverse problem of electrical conductivity in the mantle.* Geophys. J. R. Astr. Soc. *22*, 121–138.

PARKHOMENKO, E. I. (1982), *Electrical resistivity of minerals and rocks at high temperature and pressure.* Rev. Geophys. Space Phys. *20*, 193–218.

RANGANAYAKI, R. P. (1984), *An interpretive analysis of magnetotelluric data.* Geophysics *49*, 1730–1748.

RINGWOOD, A. E. (1975), *Composition and Petrology of the Earth's Mantle*, (McGraw-Hill, New York).

ROONEY, W. J. (1944), *Summary of Earth current records from Tucson, Arizona, for a complete sunspot-cycle.* Terr. Mag. and Atm. Elect. *49*, 147–157.

ROONEY, W. J. (1949), *Earth current results at Tucson magnetic observatory 1932–1942.* Researches of the Department of Terrestrial Magnetism, Carnegie Inst. Wash. Pub. *175*, vol. IX.

ROONEY, W. J. and GISH, O. H. (1949), *Earth current results from Watheroo Observatory, Western Australia, 1932–1942.* Researches of the Department of Terrestrial Magnetism, Carnegie Inst. Wash. Pub. *175*, vol. XVI.

ROMANOWICZ, B. A. (1979), *Seismic structure of the upper mantle beneath the United States by three-dimensional inversion of body wave arrival times.* Geophys. J. R. Astr. Soc. *57*, 479–506.

ROMANOWICZ, B. A. (1980), *A study of large-scale lateral variations of P-velocity in the upper mantle beneath western Europe.* Geophys. J. R. Astr. Soc. *63*, 217–232.

ROUGERIE, P. (1940), *Contribution à l'étude des courants telluriques.* Thése de spécialité, Paris.

SCHULTZ, A. (1985), *On the electrical heterogeneity of the Earth's interior: A global study of mid-mantle conductivity.* Ph.D. Thesis, University of Washington.

SCLATER, J. G., JAUPART, C. and GALSON, D. (1980), *The heat flow through oceanic and continental crust and the heat loss of the Earth.* Rev. Geophys. Space Phys. *18*, 269–311.

SHANKLAND, T. J. (1975), *Electrical conduction in rocks and minerals: Parameters for interpretation.* Phys. Earth Planet. Interiors *10*, 209–219.

SHANKLAND, T. J. and ANDER, M. E. (1983), *Electrical conductivity, temperatures and fluids in the lower crust.* J. Geophys. Res. *88*, 9475–9484.

SHANKLAND, T. J. and WAFF, H. S. (1977), *Partial melting and electrical conductivity anomalies in the upper mantle.* J. Geophys. Res. *82*, 5409–5417.

TARITS, P. (1986), *Conductivity and fluids in the oceanic upper mantle.* Phys. Earth Planet. Inter. *42*, 215–226.

WOODHOUSE, J. H. and DZIEWONSKI, A. M. (1984), *Mapping the upper mantle: Three-dimensional modelling of Earth structure by inversion of seismic waveforms.* J. Geophys. Res. *89*, 5953–5986.

WYLLIE, P. J. (1971), *The Dynamic Earth: Textbook in* Geosciences (John Wiley, New York).

YOSHIMATSU, T. (1957), *Universal Earth currents and their local characteristics.* Memo. Kakioka Mag. Obs. Suppl. *I*, 1–75.

(Received 1st July, 1986, revised/accepted 16th September, 1986)

PAGEOPH, Vol. 125, Nos. 2/3 (1987) 0033-4553/87/030341-27$1.50+0.20/0

Substitute Conductors for Electromagnetic Response Estimates

ULRICH SCHMUCKER[1]

Abstract—Various concepts exist to define substitute conductors for empirical response estimates at singular frequencies: Chapman's shell-core model, the Cagniard-Tikhonov apparent resistivity, the Niblett-Bostick and Molochnov transformation, the $\rho^* - z^*$ transformation. They are all interrelated and assign comparable resistivities to the substitute conductor at a given frequency. Applications to synthetic response data of plane and spherical conductors show under which conditions these substitutions come closest to the model and which influence of source dimensions and Earth's sphericity can be expected. $\rho^* - z^*$ transformed global response data for *S* and *Dst* variations demonstrate how substitute conductors may serve as useful guides in inverse procedures.

Key words: Electromagnetic induction, magnetotelluric and geomagnetic sounding, electric conductivity of crust and mantle.

1. *Introduction*

Substitute conductors essential purpose is to reproduce the electromagnetic response of the conducting Earth to external source fields at one singular frequency. This response can be (i) the *Q*-ratio of internal to external fields, (ii) the scalar *Z*-impedance between orthogonal telluric and magnetic horizontal fields, (iii) BANKS' *W*-ratio of vertical to horizontal magnetic fields which measures the depth of penetration *C* against lateral source dimensions. All these responses are interrelated and are to be understood as transfer functions between complex FOURIER amplitudes of surface field components.

If amplitude and phase of the response are to be interpreted simultaneously, the substitute conductor must have two adjustable parameters; one of them usually the resistivity of a uniform half-space or sphere. If the phase is ignored, this will be the only parameter to be determined. In either case, different substitutions may have to be adopted for different frequencies. They translate empirical response estimates for a sequence of frequencies with their error limits into sets of model parameters, which as a function of frequency, characterize the true resistivity within the respective depth range of penetration.

After reviewing the basic properties of response functions in Section 2, various concepts to define substitute conductors are presented in Section 3. In Section 4

[1] Institut für Geophysik, Herzberger Landstr. 180, 3400 Göttingen, Federal Republic of Germany.

limitations are considered for conclusions drawn from them, while Section 5 suggests their use as guidelines for inverse procedures.

2. *Response Functions*

They refer by definition to time and space harmonic electromagnetic fields. For flat Earth models, occupying the lower half-space $z \geq O$ of Cartesian (x, y, z) coordinates, the FOURIER amplitudes of the field components are assumed to have in horizontal planes the time-space factor $\exp\{i(\omega t + k_x x + k_y y)\}$. This introduces a horizontal wavenumber vector $\mathbf{k} = (k_x, k_y)$ to describe the sourcefield geometry with $|\mathbf{k}|^{-1}$ as scale length of source dimensions. For spherical Earth models of radius R, the spherical harmonic $P_n^m(\cos\theta)\exp(im\lambda)$ will express the dependence of the FOURIER amplitudes on colatitude θ and longitude λ on surfaces $r =$ const. in spherical (r, θ, λ) coordinates. Source dimensions are now given by R/n or, more properly, by $R/\sqrt{n(n+1)}$.

In either case, a laterally uniform Earth is assumed of magnetic permeability $\mu = 1$ throughout. This implies that all responses are directional invariant, i.e., they depend only on the absolute value $|k|$ for a flat Earth and on the degree n of the spherical function for a spherical Earth. Finally, it is assumed that the induction is by tangential electric sources with no vertical or radial electric fields appearing anywhere. Consequently the theory of induction in plane conductors with $\sigma = \sigma(z)$ or in spherical conductors with $\sigma = \sigma(r)$ establishes the following linear relations (see, SCHMUCKER, 1970; WEIDELT, 1972; or ROKITYANSKI, 1982).

Let E and B denote electric telluric and magnetic FOURIER amplitudes in the above sense on surfaces $z = O$ or $r = R$ and let affices (i) and (e) denote their parts from internal and external sources, respectively, with respect to these surfaces. The linear relations which reflect the response of the conducting matter below a flat surface and thus define the response functions $Q = Q(\omega, |k|)$ etc. are

$$B_{x,y}^{(i)} = QB_{x,y}^{(e)}, \qquad B_z^{(i)} = -QB_z^{(e)}$$

$$E_x = +ZB_y, \qquad E_y = -ZB_x \tag{1}$$

$$B_z = iWB_x|k|/k_x, \qquad B_z = iWB_y|k|/k_y$$

or

$$B_z = C\{ik_xB_x + ik_yB_y\}$$

with $W = |k|C$.

The corresponding relations for the field on a spherical surface, defining the responses $Q_n(\omega)$ etc. are

$$B^{(i)}_{\theta,\lambda} = Q_n B^{(e)}_{\theta,\lambda}, \qquad B^{(i)}_r = -\frac{n+1}{n} Q_n B^{(e)}_r$$

$$E_\theta = -Z_n B_\lambda, \qquad E_\lambda = +Z_n B_\theta \tag{2}$$

$$B_r = W_n \frac{P^m_n}{dP^m_n/d\theta} B_\theta, \qquad B_r = W_n \frac{\sin\theta}{\text{im}} B_\lambda \; (m \neq 0)$$

or

$$B_r = -\left\{ dB_\theta/d\theta + \cot\theta \, B_\theta + \frac{\text{im}}{\sin\theta} B_\lambda \right\} C_n/R$$

with $W_n = (n+1)nC_n/R$. In deriving the last relation, note that $B_r \sim P^m_n$ and that

$$d^2P^m_n/d\theta^2 + \cot\theta \, dP^m_n/d\theta + \{(n+1)n - m^2/\sin^2\theta\}P^m_n = O.$$

Because electric and magnetic fields are connected also by rot $E = -i\omega B$ and because both fields are continuous at the surface, response functions are interrelated (cf., SCHMUCKER, 1970; BANKS, 1972):

$$Z_{(n)} = i\omega C_{(n)}, \qquad Q = \frac{1 - |k|C}{1 + |k|C} = \frac{1 - W}{1 + W}$$

$$Q_n = \frac{n}{n+1} \frac{1 - (n+1)C_n/R}{1 + nC_n/R} = \frac{n - W_n}{n + 1 + W_n}. \tag{3}$$

Hence, only one of them has to be considered to find substitute conductors. In the following this will be the C-response because of its direct bearing on the penetration depth.

As an example, and for later reference, the C-response of a uniform half-space and sphere is considered, given in SI units by

$$C(\omega,|k|) = (i\omega\mu_o\sigma + k^2)^{-1/2} = (2i/p^2 + k^2)^{-1/2}$$

$$C_n(\omega) = Rj_n(i\alpha)/\{i\alpha j_{n-1}(i\alpha) - nj_n(i\alpha)\} \tag{4}$$

with $p = \sqrt{2\rho/\omega\mu_o}$ as 'skindepth' and $\alpha = (1+i)R/p$; j_n denotes modified spherical BESSEL functions of the first kind and order n.

From these expressions the generally valid asymptotic properties of the C-response are readily inferred. Observe that ij_{n-1} is approximately equal to j_n for large arguments $|\alpha|$ and that j_n becomes $(i\alpha)^n/\{1{\cdot}3{\cdot}\ldots(2n+1)\}$ for small arguments $|\alpha|$. Allow now, at a sufficiently high frequency, the skindepth to be small against source dimensions as introduced above. Then with $|k|p \ll 1$ and $np/R \ll 1$ the response of plane and spherical conductors approaches the same limiting value

$$C_o(\omega) = \frac{1}{2} p(\omega)(1 - i) \tag{5}$$

which is independent of source dimensions. It may be regarded as 'zero-wavenumber' response, but definitely not in the sense that $|k|$ or its spherical equivalent n/R are assumed to be zero, i.e., there will be finite W and Q responses for $C \to C_o$ because—for plane conductors—$dC/dk \to O$ for $|k| \to O$. If to the contrary, $|k| \cdot p$ and $n\ p/R$ are large against unity, a frequency and conductivity independent C-response of 'no induction' is approached: $|k|^{-1}$ for Cartesian coordinates and $R/(n+1)$ for spherical coordinates.

Generalized to any laterally uniform Earth model, the zero-wavenumber approximations

$$C(\omega,|k|), C_n(\omega) \to C_o(\omega) \qquad Z(\omega,|k|), Z_n(\omega) \to Z_o(\omega)$$

$$Q_o(\omega,|k|) \to 1 - 2|k|C_o(\omega) \tag{6}$$

$$Q_n(\omega) \to \frac{n}{n+1}\{1 - (2n+1)C_o(\omega)/R\}$$

apply, when $|C_o| \cdot |k|$ and $n \cdot |C_o|/R_E$ are small against unity; $C_o(\omega)$ denotes now the respective zero-wavenumber response of the model. If these expressions are large against unity, the limiting values for 'no induction' from above apply.

In addition, the C-response of simplified 2-layer models will be needed. Let h be the thickness of a top layer or outer shell above a uniform half-space or sphere of radius $q \cdot R$, $q = 1 - h/R$ with $h \ll R$. Let $p_l = \sqrt{2\rho_l/\omega\mu_o}$ be the skindepth values for the top layer ($l = 1$) and for the uniform substratum ($l = 2$) of resistivity ρ_l. The general formula of the C-response for plane 2-layer models is

$$C(\omega,|k|) = \frac{1}{K_1}\frac{K_1 + K_2\tanh(K_1 h)}{K_2 + K_1\tanh(K_1 h)}$$

with

$$K_l^2 = 2i/p_l^2 + k^2 \text{ (e.g., SCHMUCKER, 1970; eq. (5.44)).}$$

Suppose now that the top layer is by comparison a poor conductor, i.e., $p_1 \gg p_2$, and that, at the considered frequency, p_1 is also large against h, but small enough to give $|k|p_1 \ll 1$ and of course also $|k|p_2 \ll 1$. Then with the zero-wavenumber approximations $K_l = (1+i)/p_l$, and with $\tanh(K_1 h) \approx K_1 h$ the C-response from above reduces to

$$C_o(\omega) = h + \frac{p_2(\omega)}{2}(1 - i). \tag{7}$$

The C-response of spherical 2-layer models becomes identical, when $np_l/R \ll 1$ and $h \ll p_1$. Note that in these 'h-type models' the phase of the impedance, $\varphi(\omega) = \arg\{i\omega C_o\}$, lies between 45 and 90 degrees.

Assume for a complementary model that the top layer (or shell) is a thin conductor of conductance $\tau = h/\rho_1$, again with $h/p_1 \ll 1$, but $p_1 \ll p_2$, i.e., the top layer is by comparison a good conductor. Setting as before kp_l and np_l/R small against unity, the C-response from above has the approximation

$$C_o(\omega) = \frac{p_2(\omega)/(1 + i)}{1 + i\omega\mu_o\tau p_2(\omega)/(1 + i)} \tag{8}$$

which again is also valid for spherical shell-core models. Note that for this 'τ-type model' the phase of the impedance $\varphi(\omega)$ lies between zero and 45 degrees.

If the uniform substratum is either an extremely good conductor with $p_2 \ll h$ or an extremely poor conductor with $\omega\mu_o\tau p_2 \gg 1$, the C-response for h and τ-models is only determined by their respective parameter h or τ:

$$C_o(\omega) = h, \qquad C_o(\omega) = (i\omega\mu_o\tau)^{-1}. \tag{9}$$

3. *Substitute Conductor Concepts*

Substitute conductors refer exclusively to the zero-wavenumber responses $Q_o(\omega)$, $Z_o(\omega)$ and $C_o(\omega)$ of eq. (6). Their concepts are now listed in chronological order:

3.1 CHAPMAN'S Uniform-Core Model

It was introduced to interpret the Q_n responses of a spherically symmetric Earth to solar and lunar daily variations (CHAPMAN and BARTELS, 1942; Section 22.2). The substitute model is an h-type model, consisting of a nonconducting shell of thickness h above a uniform 'core' of radius $q \cdot R$ and resistivity ρ_c. For each time-harmonic of S and L variations Chapman defines a parameter $\beta = qR/p_c(\omega) = qR\sqrt{\omega\mu_o/2\rho_c}$ and observes that the phase of the Q_n-response depends for large values of β almost exclusively on this parameter.

This observation is readily verified from eq. (6) by substituting C_n from eq. (7). The condition $\beta \gg 1$ implies that the zero-wave-number can be used provided that q is near unity:

$$Q_n(\omega) = \frac{n}{n+1}\left\{1 - (2n+1)\frac{h + p_c/2}{R} + (2n+1)\frac{p_c}{2R}i\right\}.$$

Hence for $np_c/R \ll 1$ and $2nh/R \ll 1$, the phase of Q_n is

$$\psi_n(\omega) \approx (2n + 1)p_c/2R \tag{10}$$

which is Chapman's eq. (74) for $\beta \gg 1$ and $\mu = 1$. It defines the resistivity ρ_c of the 'core' in terms of phase ψ_n. Similarly, the modulus for sufficiently small phases determines in

$$|Q_n| \approx \frac{n}{n+1}\{1 - (2n+1)(h + p_c/2)/R\}, \tag{11}$$

the thickness of the shell.

An example: Chapman finds for the second time harmonic of S $|Q_n(\omega)| = 1/2.2$ and $\psi_n(\omega) = 18.8$ degrees, $\omega = 4\pi$ *cpd*. The spherical harmonic used for the *m-th* S-harmonic is in Chapman's analysis of degree $n = m + 1$. Thus, with $n = 3$ the following model parameters evolve: $h = 60$ km, $p_c = 597$ km or $\rho_c = 32.6$ Ωm—not far from modern estimates. Chapman quotes on the basis of all S and L harmonics $h = 250$ km and $\rho_c = 27.8$ Ωm.

3.2 *The Uniform Substitute Conductor of CAGNIARD-TIKHONOV*

This substitution is for magnetotelluric impedance estimates Z, assuming that it is sufficiently close to its zero-wavenumber value Z_o (CAGNIARD, 1953). Since no source geometry enters then into the determination, estimates will be available for a wide range in frequency and the phase information may be thought to be contained in the frequency dependence of $|Z|$. Therefore in this substitution the phase is ignored and, with $|Z| = \omega|C| = \omega p/\sqrt{2}$ from eq. (5), the true Earth is replaced for frequency ω by a uniform conductor of 'apparent resistivity'

$$\rho_a(\omega) = \frac{\mu_o}{\omega}|Z|^2 = \omega\mu_o|C|^2 \tag{12}$$

at zero depth.

BERDICHEVSKY and DMITRIEV give the following interpretation of $\rho_a(\omega)$: Obtain the squared telluric surface amplitude $|E|^2$, $E = E_x$ or E_y, by integration over its derivatives, with respect to depth z:

$$|E(O)|^2 = -\int_o^\infty \{\partial|E|^2/\partial z\}\, dz = -\int_o^\infty \{2|E|\partial|E|/\partial z\}\, dz.$$

Substitute E by $\pm\mu_o^{-1}\rho\partial B/\partial z$ (from rot $\mathbf{B} = \mu_o\, E/\rho$ for a laterally uniform field) and $\partial E/\partial z$ by $\pm i\omega B$ (from rot b.f. $= -i\omega B$) with B standing for B_x or B_y. This gives with $|E|^2/|B|^2$ for $|Z|^2$ at $z = 0$

$$\rho_a = -\int_o^\infty \rho\{\partial|B(z)/B(O)|^2/\partial z\}\, dz$$

and, with corresponding substitutions in $\partial|B|^2/\partial z$, for the apparent conductivity

$$\sigma_a = \rho_a^{-1} = -\int_o^\infty \sigma\{\partial|E(z)/E(O)|^2/\partial z\}\, dz.$$

The apparent resistivity or conductivity thus sample the true Earth resistivity or conductivity according to the attenuation which the squared magnetic or electric field undergo as functions of depth.

3.3 The NIBLETT-BOSTICK Transformation

It assigns the substitute resistivity ρ_{NB} to the depth $|C|$, taking frequency derivatives of $|C|$ into account. No specific model is used and this transformation does not attempt to find a substitute conductor which reproduces the observed response. Instead NIBLETT (1960) defines for the depth range down to $z = |C(\omega)|$ the conductance

$$\tau_{NB}(\omega) = \int_{o}^{|C(\omega)|} \{\rho_{NB}(z)\}^{-1}\, dz$$

in terms of the substitute resistivity to be found. Regarding this depth range as 'thin sheet' and assuming that the remaining half-space $z \geq |C|$ is an extremely poor conductor, eq. (9) applies and connects τ_{NB} with $|C|$:

$$\tau_{NB}(\omega) = \{\omega\mu_o|C(\omega)|\}^{-1}.$$

Differentiating both relations with respect to ω leads in

$$\{\rho_{NB}(|C|)\}^{-1} d|C|/d\omega = -\frac{d|C|/d\omega + |C|/\omega}{\omega\mu_o|C|^2}$$

to an expression which defines ρ_{NB} at depth $|C(\omega)|$. Rewriting it for $\rho_a = \omega\mu_o|C|^2$ and $d\rho_a/d\omega$ gives

$$\rho_{NB}\{|C(\omega)|\} = \rho_a(\omega)\frac{1 + m(\omega)}{1 - m(\omega)} \tag{13}$$

with $m = -\omega/\rho_a \cdot d\rho_a/d\omega$. WEIDELT's 3rd inequality (WEIDELT, 1972; eq. (2.32)) ensures that for consistent response estimates $|m|$ does not exceed unity.

The same inequality implies a monotonic increase of $|C|$ with decreasing frequency, i.e., the transformation should give a depth profile $\rho_{NB}(|C|)$ which with decreasing frequency continues monotonically downwards.

Because $d|C|/d\omega$ is not an observable quantity and error estimates are difficult to assign to it, WEIDELT's formula

$$\varphi \approx \frac{\pi}{4}(1 - m)$$

(WEIDELT, 1972; eq. (2.28)) presents the possibility to approximate $m(\omega)$ by the phase $\varphi(\omega)$ and its well defined error. With $\Phi = \varphi - \pi/4$ as phase difference against the phase of uniform conductors, the approximated transformation becomes

$$\rho_{NB} = \rho_a \frac{1 - \frac{4}{\pi}\Phi}{1 + \frac{4}{\pi}\Phi} \tag{13a}$$

or, for small phase differences,

$$\rho_{NB} \approx \rho_a(1 - \frac{8}{\pi}\Phi). \tag{13b}$$

3.4 The MOLOCHNOV Transformation

It defines with similar arguments a substitute resistivity ρ_M for depth $|C|$, but uses a different modeling concept, yielding

$$\rho_M = \rho_a(1 + m)^2 = \rho_{NB}(1 - m^2) \tag{14}$$

(MOLOCHNOV, 1968). Obviously, both transformations lead to similar results for $m \ll 1$. Because ρ_M will always be smaller than ρ_{NB} (for $m \neq 0$), this substitution reproduces the true resistivity better than ρ_{NB}, when the resistivity decreases downward ($m < O$), but less well where the resistivity increases with depth ($m > O$).

3.5 The $\rho^* - z^*$ Transformation

It is based on plane 2-layer models with two adjustable parameters: an h-model according to eq. (7) for response estimates with a phase $\varphi = \arg\{Z\}$ above 45 degrees, a τ-model according to eq. (8) when this phase is below 45 degrees. In the original concept (SCHMUCKER, 1970; KUCKES, 1973) h-models were used in both cases. The supplementary τ-model was added later (SCHMUCKER, 1971; Sec. 3.2).

Let ρ^* denote in either model the resistivity of the underlying uniform half-space with skindepth $p^* = \sqrt{2\rho^*/\omega\mu_o}$. Then for $\varphi \geq \pi/4$, thickness h^* of the resistive top layer and skindepth p^* of the uniform half-space follow readily from eq. (7) as

$$\begin{aligned} h^*(\omega) &= \mathrm{Re}\{C(\omega)\} + \mathrm{Im}\{C(\omega)\} \\ p^*(\omega) &= -2\,\mathrm{Im}\{C(\omega)\} \end{aligned} \tag{15}$$

which gives the half-space resistivity as

$$\rho^*(\omega) = 2\omega\mu_o(\mathrm{Im}\{C(\omega)\})^2 = 2\rho_a(\omega)\cos^2\varphi(\omega) \tag{16}$$

with $\arg\{C\} = \varphi - \pi/2$ and ρ_a from eq. (12).

If $\varphi \leq \pi/4$, the model parameter of the τ-model are found with similar expressions from the admittance $A = Z^{-1}$. Replacing C in eq. (8) by $A = (i\omega C)^{-1}$ leads with τ^* as top layer conductance and

$$A = \mu_o\tau^* + \frac{1 - i}{\omega p^*} \tag{17}$$

to

$$\mu_o\tau^*(\omega) = \mathrm{Re}\{A(\omega)\} + \mathrm{Im}\{A(\omega)\}$$
$$p^*(\omega) = -(\omega\ \mathrm{Im}\{A(\omega)\})^{-1} \tag{18}$$

and, with $\arg\{A\} = -\varphi$, to the complementary definition

$$\rho^*(\omega) = \frac{1}{2}\frac{\mu_o}{\omega}/\mathrm{Im}\{A(\omega)\}^2 = \frac{1}{2}\rho_a(\omega)/\sin^2\varphi(\omega). \tag{19}$$

Thus, by the use of phase information, a new apparent half-space resistivity $\rho^*(\omega)$ is found, which is freed from the influence of surface layers, providing at the same time in $h^*(\omega)$ or $\tau^*(\omega)$ estimates of thickness or conductance of these layers.

By joining ρ^*-values from a sequence of frequencies, a depth profile $\rho^* - z^*$ can be constructed by assigning $\rho^*(\omega)$ as a substitute resistivity to the frequency depending depth $z^*(\omega) = \mathrm{Re}\{C(\omega)\}$ of a perfect substitute conductor, eq. (9). This conductor would produce the in-phase part of the internal magnetic surface field and z^* is therefore an indicator of the depth from which thc in-phase induction currents flow at the respective frequency.

WEIDELT's 2nd inequality (WEIDELT, 1972; eq. (2.31)] ensures that z^* increases monotonically with decreasing frequency. Furthermore, if $\hat{\imath}_u(z)$ denotes the density of the in-phase currents at a given frequency, the integral $\int_0 \hat{\imath}_u(z - z)\, dz$ is zero for any laterally uniform half-space (WEIDELT 1972; Sec. 2g). This establishes the depth z^* as 'center depth of the in-phase currents' in the sense of a center of mass, even though there will be weak currents with a phase of 180 degrees and negative densities $\hat{\imath}_u$ at some greater depth. The choice of z^* as depth estimate for ρ^* comes from models with an exponentially decreasing resistivity as discussed in Sec. 4.5.

3.6 Comparison

This section concludes with a comparison of the various substitutions. Chapman's procedure is identical with the $\rho^* - z^*$ transformation, if the phase φ is close to 90 degrees and $\mathrm{Im}\{C\}$ about equal to $|C|\ (\frac{\pi}{2} - \varphi)$.

There exists also a simple relationship between the NIBLETT-BOSTICK and the $\rho^* - z^*$ transformation, when the phase φ is not too far from 45 degrees. Expressing ρ^* from eqs. (15) and (19) in terms of the phase difference $\Phi = \varphi - \pi/4$ leads to

$$\rho^* = \rho_a \cdot \begin{cases}(1 - \sin 2\Phi), & \Phi \geq O \\ (1 + \sin 2\Phi)^{-1}, & \Phi \leq O\end{cases}. \tag{20}$$

Hence, for small values of Φ, ρ^* approaches $\rho_a(1 - 2\Phi)$ whereas from eq. (13b) $\rho_{NB} \to \rho_a(1-8/\pi\cdot\Phi)$, implying a 25% larger deviation of ρ_{NB} from ρ_a. But since ρ_{NB} is assigned to a larger depth $|C|$ than ρ^*, both transformations will not be too different, if the resistivity changes monotonically with depth.

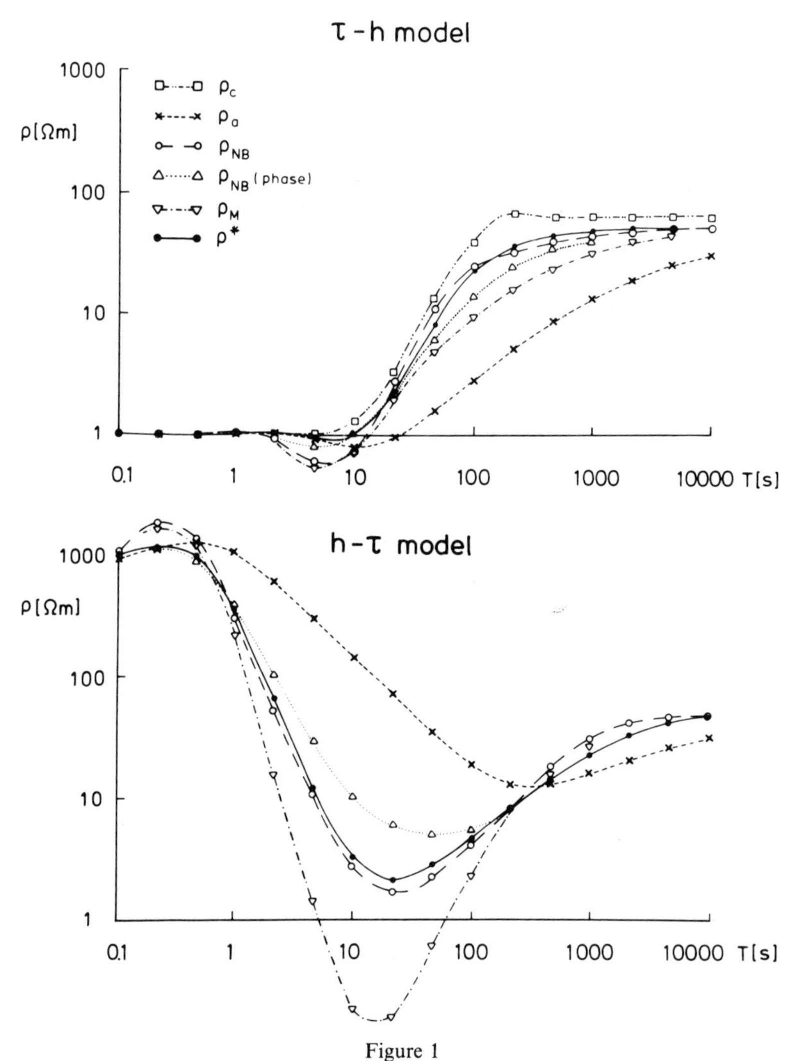

Figure 1

Comparison of various concepts to define substitute resistivities for the zero-wavenumber response. The chosen model response is for the two 3-layer models of Figure 4 over five decades in period T.—ρ_c: CHAPMANs 'core' resistivity; ρ_a: CAGNIARD-TIKHONOV apparent resistivity; ρ_{NB}: NIBLETT-BOSTICK resistivity, also as phase-approximation eq. (13a); ρ_M: MOLOCHNOV resistivity; ρ^*: h, τ-model resistivity.

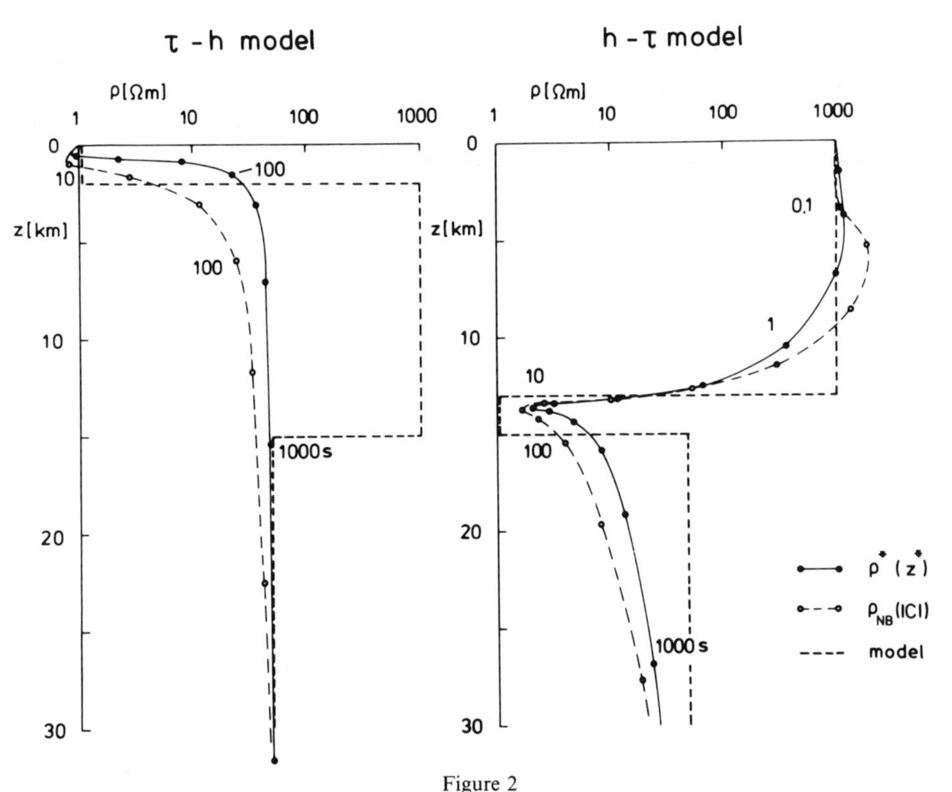

Figure 2

NIBLETT-BOSTICK transformation and $\rho^* - z^*$ transformation of the zero-wavenumber response of the same models as in Figure 1. Curve parameter is the period T. The transformations both reproduce the right-hand model, but not the intermediate high resistivity layer in the left-hand model. Note in either case the shielding which conductors exert upon underlying poor conductors, evidenced by large deviations of ρ_{NB} and ρ^* from the true resistivity.

To illustrate these relationships, substitutions are carried out with response data of two models (Figure 1 and 2). They are combined h- τ-models, in which a highly resistive top layer is underlain by a thin conductive sheet above a uniform half-space or *vice versa* with the thin sheet on top. The combined responses follow from eqs. (7) and (8) as

$$C_o(\omega) = h + \frac{p/(1 + i)}{1 + i\omega\mu_o\tau p/(1 + i)} \tag{21a}$$

for the $h - \tau$ model and as

$$C_o(\omega) = \frac{h + p/(1 + i)}{1 + i\omega\mu_o\tau\{h + p/(1 + i)\}} \tag{21b}$$

for the $\tau - h$ model. The intercomparison uses, however, not these approximations, but the exact formula for 3-layer models with finite values assigned to all model parameters. Substitutions requiring $d|C|/d\omega$ have been carried out with analytically derived frequency derivatives. For comparison, phases Φ also have been used in their place. To include Chapman's 'core' resistivity ρ_c, the approximation for $\mathrm{Im}\{C\}$ from above was employed.

Figure 1 shows the substitute resistivities from all definitions as functions of period $T = 2\pi/\omega$. They all reveal the essentials of the respective model, even though those with phase information more distinctively and within a narrower range in period. However note that the poorly conducting intermediate layer in the $\tau - h$ model is hardly resolved. In Figure 2 the same responses are converted into $\rho_{NB}(|C|)$ and $\rho^*(z^*)$ plots. Both transformations lead to comparable depth profiles in which the intermediate good conductor of the $h - \tau$ model appears at about the correct depth, while the intermediate poor conductor of the $\tau - h$ model is not visible. This inherent and irrevocable limitation in the resolving power of electromagnetic response data will bc considered in greater detail in the following section.

4. Application to Response Functions of Models

The transformation of response estimates into substitute conductors provides insight into the class of models which is to be studied for final interpretation. Limitations are twofold: from the limited resolving power of response functions as such and from finite source dimension. They are considered now on the basis of models of variable complexity, resembling real Earth conditions. The substitution to be used is the $\rho^* - z^*$ transformation from Sec. 3.5.

4.1 2-Layer Models

These are plane Earth models with great resistivity contrasts in the sense that, at sufficiently low frequency, the top layer acts either as nonconductor ('h-model') or as thin-sheet conductor ('τ-model'). The chosen top layer resistivities of 1000 and 1 Ωm, respectively, are representative values for crystalline rocks and well-conducting sediments. With a thickness of 15 and 2 km the top layers may resemble the upper crust or a sedimentary basin above a moderately conducting substructure, here given a resistivity of 50 Ωm.

Figure 3 shows the $\rho^* - z^*$ transformed model responses over five decades in period $T = 2\pi/\omega$. Responses over three decades, here from 1 to 1000s, are sufficient to derive all model parameters directly from a visual display of the $h^*(T)$, $\tau^*(T)$ and $\rho^*(T)$ curves. The response at longer periods alone would determine only thickness or conductance of the top layer and the half-space resistivity. Nearly correct values are then obtainable from the response at just one singular frequency.

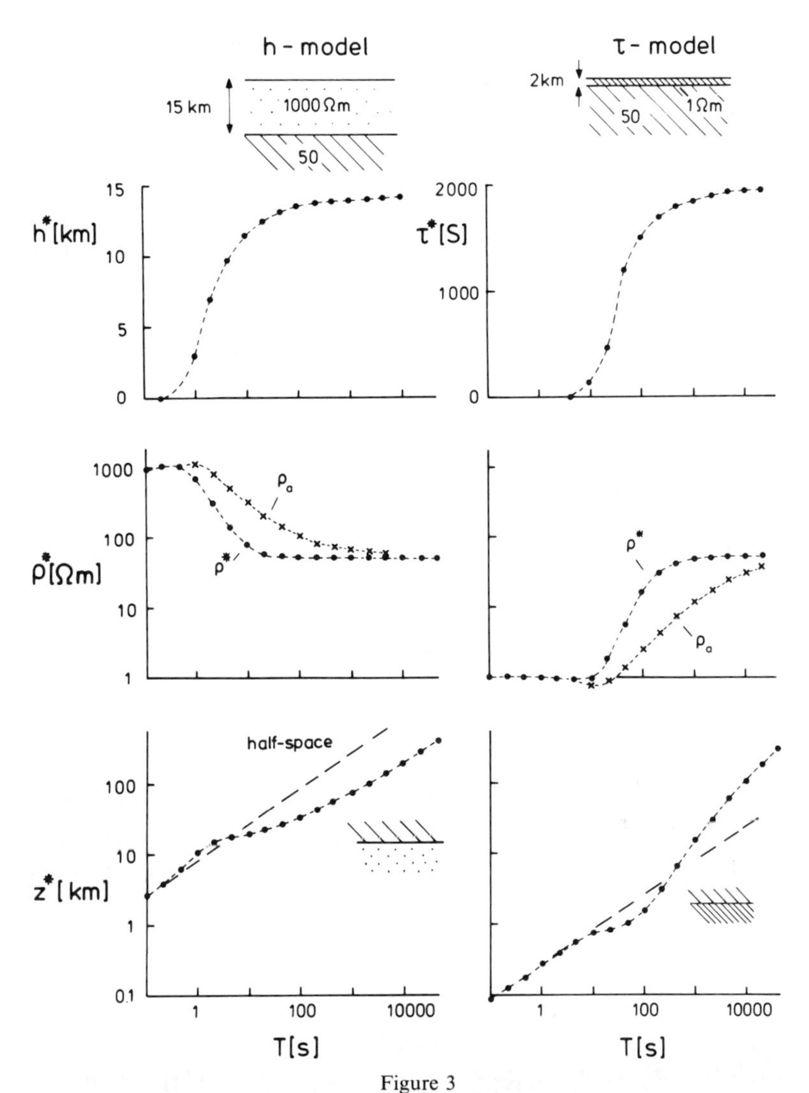

Figure 3
$\rho^* - z^*$ transformation of the response of plane 2-layer models over five decades in period T. From top to bottom: the model, thickness or conductance estimates for the top layer, estimated resistivity ρ^* for the half-space (for comparison also the apparent resistivity ρ_a), depth of the perfect substitute conductor.

The display of the depth $z^*(T)$ of the perfect substitute conductor illustrates how the center depth of the in-phase induced currents moves downward with increasing period. Where this curve levels off in relation to the indicated uniform half-space slope $z^* = p/2 \sim \sqrt{T}$, the bottom or top of a good conductor is reached, slowing down the descent of the currents to greater depth. Where the $z^*(T)$ curve is steeper, the currents pass more quickly through a depth range of high resistivity.

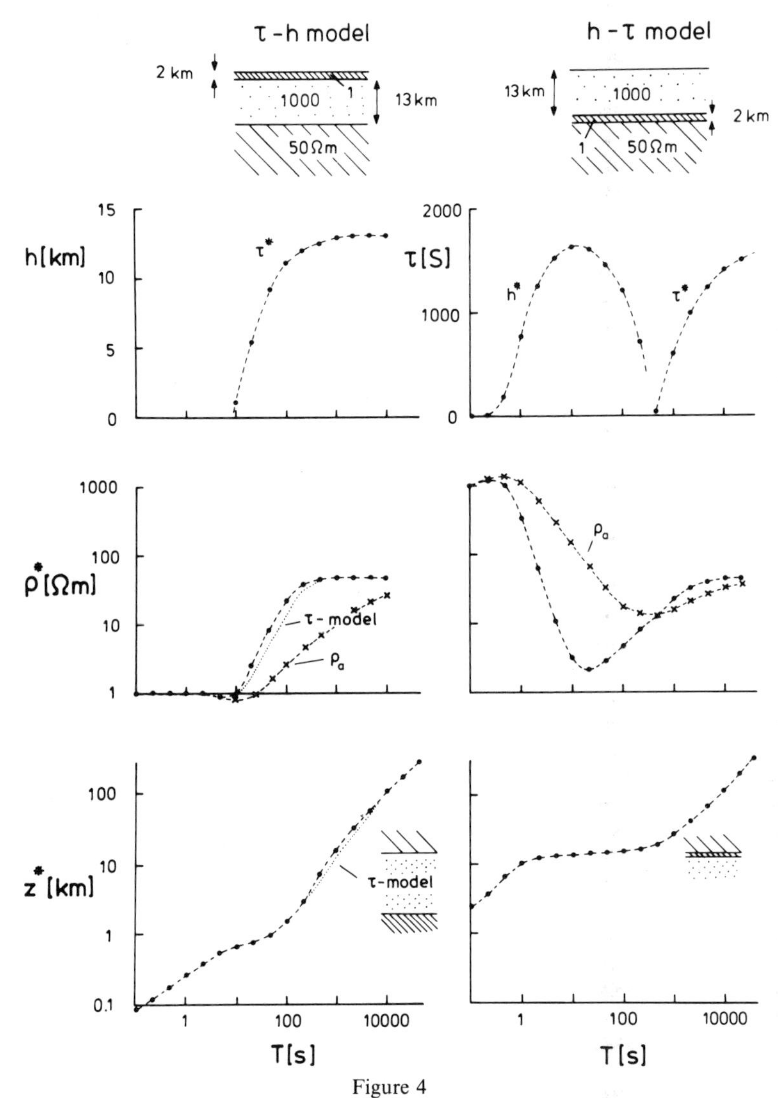

Figure 4
$\rho^* - z^*$ transformation of the response of plane 3-layer models. See Figure 3 for further explanations.—While both top layers are well resolved, in the model to the right, the second layer of high resistivity in the model to the left is not recognizable in the transformed response curves.

4.2 3-Layer Models

They are combined *h*- and τ-models, i.e., the resistivity contrasts are again sufficiently large so that the two top layers become at long periods effective nonconductors or thin-sheet conductors. The $\tau - h$ model to the left in Figure 4 represents the typical case that well-conducting sediments overlie highly resistive basement rocks;

the $h - \tau$ model to the right the case of a thin mid-crustal conductor below such basement rocks in the absence of sediments. The following conclusions are readily verified from eqs. (21).

In the $\tau - h$ model the resolving power of the response is now limited due to the shielding which the good conductor at the top exerts on the poor conductor below. In fact, this second layer is totally invisible. At long periods τ^* merges into the correct conductance of the first layer. But at shorter periods, when the phase is barely above 45 degrees, there is no indication from h^* values for the mere existence of a second layer, nor does this evidence come from ρ^* at any period. Where the center depth of induced currents passes through the depth range of the poor conductor, the slope of the $z^*(T)$ curve is indeed slightly increased, but again the difference against the τ-model without such a conductor is minute.

A reversed sequence of the top layers restores in the $h - \tau$ model the resolving power of the response. There are now sections with h^* at short periods ($\varphi > \pi/4$) and with τ^* at long period ($\varphi < \pi/4$) which give unambiguous evidence for the existence of two top layers, providing almost correct values for their thickness and conductance. The same can be inferred from ρ^* and z^*, which level off quite visibly at the correct depth of the thin conductor. In summary, conductance and thickness of the top layers, as well as the half-space resistivity, are well determined model parameters. Responses at very short periods would establish also a finite top layer resistivity, while thickness and resistivity of the second layer remain uncertain except for the fact that it acts like a thin-sheet and has a known conductance.

4.3 *Multi-layer Models*

If such models are constructed from a sequence of h and τ-conductors, PARKER's model (1980) evolve which can reproduce any given set of response data. But it becomes increasingly difficult to identify individual layer parameters just from a visual inspection of h^*/τ^*, ρ^* or z^*-curves. It may be even difficult to estimate only the right number of layers, as it was demonstrated for the case of $\tau - h$ models, and only inverse procedures can yield at least an approximate answer (cf., Sec. 5).

4.4 *Continuous Models*

They provide an alternative way for a direct interpretation of response data, if not more than two or three adjustable parameters are involved. Firstly, models with a monotonic increase or decrease of resistivity with depth will be studied, i.e., exponential or power law models. They have been instrumental in the original formulation of the $\rho^* - z^*$ transformation (SCHMUCKER, 1970; KUCKES 1973). LAHIRI and PRICE (1939) used such analytic models to interpret the global Q-response of S-and Dst variations (cf. Figure 7).

Consider then a model with an exponentially decreasing resistivity: $\rho(z) = \rho_o$

$\exp(-2\lambda z), \lambda > O$. Let $E(z)$ denote the complex FOURIER amplitude of the tangential electric field (E_x or $-E_y$) at depth z. Then, for a zero-wavenumber approximation, the C-response has to be found from the solution of

$$d^2E/dz^2 = \alpha^2\lambda^2 e^{2\lambda z}E$$

with

$$\alpha = \sqrt{i\omega\mu_o/\rho_o}/\lambda = (1 + i)/(p\lambda)$$

and $p = \sqrt{2\rho_o/\omega\mu_o}$ as skindepth at zero depth. The substitution $u = \exp(\lambda z)$ converts this equation into

$$u^2 d^2E/du^2 + u dE/du - \alpha^2 u^2 E = O,$$

which has as a solution modified Bessel functions of order zero. Those of the second kind have the right behaviour for $u, z \to \infty$:

$$E(u) = A \cdot K_o(\alpha u) \text{ with } E(\infty) = O.$$

The tangential magnetic field B (for B_y or B_x) follows from $\partial E/\partial z = -\alpha\lambda u A K_1(\alpha u) = -i\omega B$ and gives at $z = O$ ($u = 1$) the desired response:

$$C_o(\omega) = i\omega\frac{E(1)}{B(1)} = \frac{1}{\alpha\lambda}K_o(\alpha)/K_1(\alpha). \tag{22}$$

In order to see how well the $\rho^* - z^*$ transformed response resembles the model, two limiting cases are considered. At sufficiently high frequency and shallow depth of penetrations $p\lambda$ will be small and $|\alpha|$ large against unity. The appropriate approximations for Bessel functions yield

$$K_o(\alpha)/K_1(\alpha) = 1 - 1/2\alpha + 3/8\alpha^2 + O\{\alpha^{-3}\}$$

and thereby

$$C_o = \frac{p}{2}\left\{1 - i + \frac{1}{2}\lambda p i - \frac{3}{16}(\lambda p)^2(1 + i)\right\}. \tag{23}$$

Since the phase $\varphi = \pi/2 - \arg\{C_o\}$ is above 45 degrees, eqs. (15) are to be used, yielding the substitute resistivity

$$\rho^*(z^*) = \rho_o\{1 - \lambda p + \frac{5}{8}(\lambda p)^2\}$$

for depth $z^* = p/2 \cdot \{1 - 3/16\,(\lambda p)^2\}$ to second order in (λp). The model resistivity at this depth is, again to second order,

$$\rho(z^*) = \rho_o{}^{-2\lambda z^*} \approx \rho_o\left\{1 - \lambda p + \frac{1}{2}(\lambda p)^2\right\},$$

i.e., the $\rho^* - z^*$ transformation reproduces the shallow part of the model quite correctly with a small difference in the second order term.

At low frequencies and great depth of penetration $p\lambda$ will be large against unity and $|\alpha|$ small. The appropriate approximations are now $K_o = -ln(\alpha)$ and $K_1 = \alpha^{-1}$, which when inserted into eq. (22) give

$$C_o = -ln(\alpha)/\lambda = ln\left(\frac{p\lambda}{1+i}\right)/\lambda \tag{24}$$

and therefrom the substitute resistivity eq. (16)

$$\rho^*(z^*) = \omega\mu_o\,\pi^2/8\cdot\lambda^{-2}$$

for depth

$$z^* = ln\{p\lambda\sqrt{2}\}/\lambda.$$

A comparison with the model value

$$\rho(z^*) = 2\rho_o/(\lambda p)^2 = \omega\mu_o/\lambda^2$$

reveals again a remarkable fit, except for a constant factor of $\pi^2/8 = 1.23$, independent of depth and model parameter λ or p. Since in the intermediate frequency range also no larger deviations occur, as demonstrated in Figure 9, the substitution provides a close approximation to the true resistivity at all depth.

To first order in λp, the model parameter can be derived—as in the case of 2 layer models—directly from the response at one singular frequency:

$$p = 2\mathrm{Re}\{C_o\},\ \lambda p = 4(\mathrm{Re}\{C_o\} + \mathrm{Im}\{C_o\}) \tag{23a}$$

at high frequencies from eq. (23) and

$$\lambda^{-1} = -4/\pi\cdot\mathrm{Im}\{C_o\},\ \lambda p = \sqrt{2}\exp(\lambda\,\mathrm{Re}\{C_o\}) \tag{24a}$$

at low frequencies from eq. (24).

KUCKES (1973) came with power law models $\rho(z) \sim z^{-m}$, $m > O$, to similar conclusions and, as Figure 6 shows, the Lahiri and Price power law model for a spherical Earth is also well represented by $\rho^* - z^*$ transformed S and Dst responses, calculated for this model. If an overlying effective nonconductor or thin-sheet conductor shall be included, simply add its thickness to the C-response of eq. (22) or its conductance, multiplied with μ_o, to the admittance $A = 1/i\omega C$.

Models with an exponentially increasing resistivity $\rho(z) = \rho_o\exp(2\lambda z)$, $\lambda > O$, are treated in similar ways. By substitution of $u(z) = \exp(-\lambda z)$ the solution is now in terms of modified Bessel functions of the first kind: $E(u) = AI_o(\alpha u)$ with α from eq. (22) and $E(0) = A$, yielding from $i\omega B(u) = \alpha\lambda I_1$ as C-response

$$C_o(\omega) = \frac{1}{\alpha\lambda}I_o(\alpha)/I_1(\alpha). \tag{25}$$

Note that the electric field has a finite value at $z = \infty$, reflecting the problematic zero-wavenumber approximation, when σ is or approaches zero at infinite depth.

For simplicity, it may be assumed that source dimensions are infinite.

Using again the approximations for large arguments $|\alpha|$, the high-frequency response becomes

$$C_o(\omega) = \frac{p}{2}\{1 - i - \frac{1}{2}\lambda p i - \frac{3}{16}(\lambda p)^2(1 + i)\}, \tag{26}$$

i.e., the phase is below 45 degrees and the substitute resistivity ρ^* has to be derived from the admittance according to eq. (19). This gives

$$\rho^*(z^*) = \rho_o\left\{1 + \lambda p + \frac{5}{8}(\lambda p)^2\right\}$$

for depth $z^* = p/2\ \{1 - 3/16\ (\lambda p)^2\}$, which again agrees well to second order in λp with the correct resistivity

$$\rho(z^*) \approx \rho_o\left\{1 + \lambda p + \frac{1}{2}(\lambda p)^2\right\}$$

at that depth.

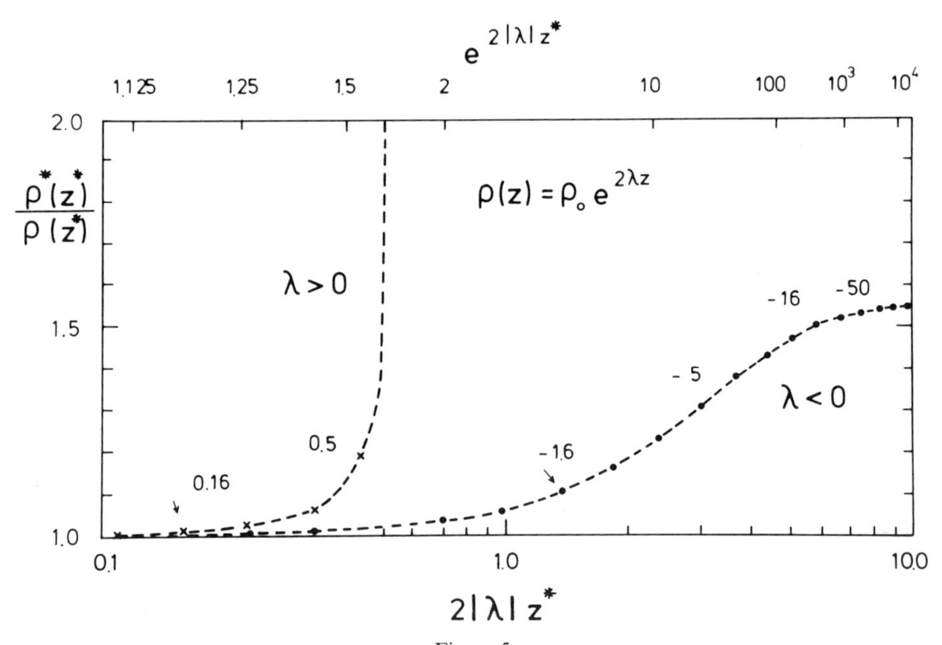

Figure 5

$\rho^*(z^*)$-resistivity in relation to the true resistivity $\rho(z^*)$ as function of the exponent in exponential models. On top the resistivity at depth z^* in relation to the surface resistivity ρ_o. Curve parameter is λp with p as skindepth for ρ_o, i.e., the period and depth of penetration increase along the curves from the left to the right. Since the response, multiplied with λ, depends only on λp (cf., eqs. (22) and (25)), the shape of the curves is the same for all exponential models with a shifted curve parameter λ.—While ρ^* never deviates more than by a factor of 1.5 from the true resistivity at any depth, when the resistivity is exponentially decreasing (λ negative), such agreement exists only at short periods with $\lambda p \leq 0.5$, when the resistivity is exponentially increasing (λ positive).

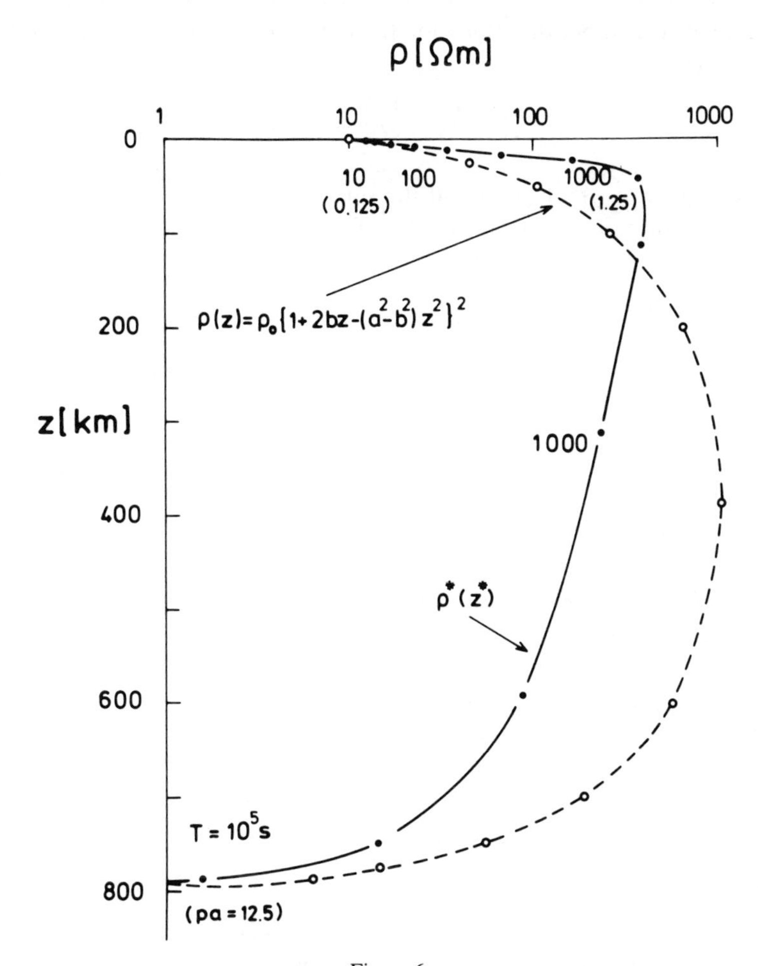

Figure 6

$\rho^* - z^*$ transformation of the response for a continuous model with maximal resistivity at 388 km, zero resistivity at 800 km and $\rho_o = 10\ \Omega$m at zero depth ($a^{-1} = 40$ km, $b = 0.95 \cdot a$). The transformed response between 100 000s and 10s reproduces the uppermost and lowermost portion of the model near the ultimate depth of 800 km, when $p \cdot a = \sqrt{2\rho_o/\omega\mu_o} \cdot a$ is either small or large against unity. In the intermediate depth range ρ^* is at first above the model until a cross-over at $pa \sim 2.8$, and then below.

For small arguments the approximations $I_o(\alpha) = 1 + (\alpha/2)^2$ and $I_1(\alpha) = \alpha/2 \cdot \{1 + \frac{1}{2}(\alpha/2)^2\}$ lead to the, not quite realistic low frequency response

$$C_o(\omega) = -ip^2\lambda + 1/4\lambda \tag{27}$$

and from the admittance to the substitute resistivity

$$\rho^*(z^*) = 16\rho_o \cdot (p\lambda)^6$$

at constant depth $z^* = (4\lambda)^{-1}$. This is in total disagreement with the model, similar to KERTZ's (1978) conclusion in a corresponding 2-layer case with a nonconducting substratum. Due to shielding by shallow low resistivities, the increasingly poor conductor at depth is not resolvable. The fields simply do not penetrate deeply enough (Figure 5).

Finally, power series models are considered which have maximum resistivity at some intermediate depth. Appropriate fourth order polynomials of the form

$$\rho(z) = \rho_o \{1 + 2bz - (a^2 - b^2)z^2\}^2 \quad a > b > O, \qquad z \leq (a - b)^{-1},$$

have an analytically derivable response: From WEIDELT (1972, eqs. (2.16)/(2.17))

$$C_o(\omega) = p/\{\sqrt{(pa)^2 + 2i} - pb\} \tag{28}$$

with $p = \sqrt{2\rho_o/\omega\mu_o}$ as zero depth skindepth. The model starts off with a surface value ρ_o, reaches at depth $b/(a^2 - b^2)$ a peak value $\rho_o/\{1 - (b/a)^2\}^2$ and approaches zero resistivity at depth $z = (a - b)^{-1}$. As it is readily verified from the high and low frequency approximations of eq. (28), when $p \cdot a$ is either small or large against unity, there is again a good agreement between the substitute conductor $\rho^*(z^*)$ and the model at shallow and great depth near $(a - b)^{-1}$. But, as Figure 6 shows, ρ^* does not reach the peak resistivity at intermediate depth, demonstrating once again the basic limitations of substitute conductors, when shielding by low resistivities near the surface is involved.

4.5 Source Effects

They are of second order in depth of penetration versus source dimensions. For uniform Earth models this is verified by approximating their responses in eq. (4) to second order in (kp) and (np/R), respectively:

$$\begin{aligned} C(\omega,k) &= C_o(\omega)\cdot\{1 + (kp/2)^2 i\} \\ C_n(\omega) &= C_o(\omega)\cdot\{1 + n(n + 1)(p/2R)^2 i\} \end{aligned} \tag{29}$$

In the case of the spherical conductor second order approximations in α^{-1},

$$j_n(i\alpha) = i^n e^\alpha/2\alpha \left[1 - \frac{n(n + 1)}{2\alpha} + \frac{n(n + 1)\cdot\{n(n + 1) - 2\}}{(2\alpha)^2 2} \right],$$

for spherical Bessel functions with large arguments$|\alpha|$ were used.

This result can be generalized to any Earth model, but excluded are plane models which approach or have infinite resistivity at infinite depth. PRICE (1962) based his assessment of source effects mainly on models of this type, where this effect is unavoidable. But since realistic Earth models always conclude with a good conductor at some ultimate depth, source effects will not exceed those from uniform conductors.

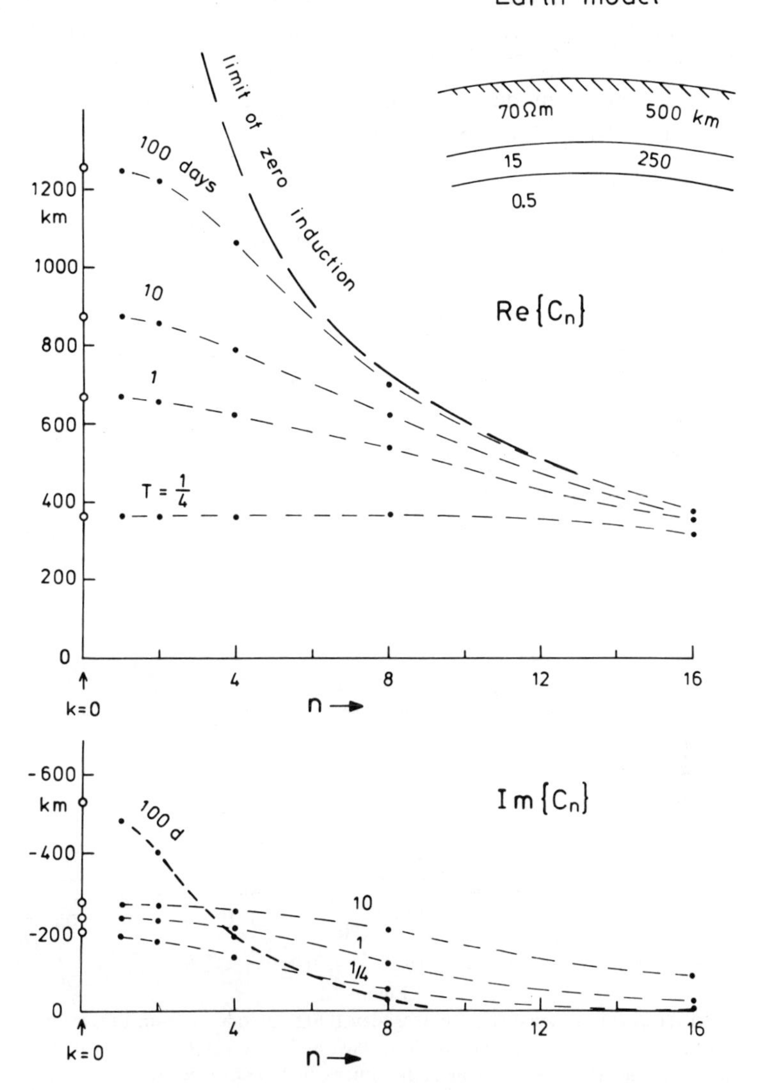

Figure 7
Source effect on the C-response of the 3-layer spherical Earth model A from Figure 11. The real and imaginary parts of the calculated response function $C_n(\omega)$ is shown for increasing degrees of spherical harmonics of the source until the limit of 'no-induction' $C_n = R/(n + 1)$ is reached, $R = 6371$ km. Curve parameter is the period T in days.—The source dependence of the real part of C_n is up to $n = 16$ very small for variations with T smaller than 6 hours, but clearly visible in the imaginary part and thus in the phase of the response. For periods of one and ten days source dimensions affect also the real part, but for n less than about 4 the response is far from its limiting value of 'no-induction', i.e., mainly determined by the internal resistivity distribution. For $T = 100$ days this limit is practically reached, when n exceeds 4.

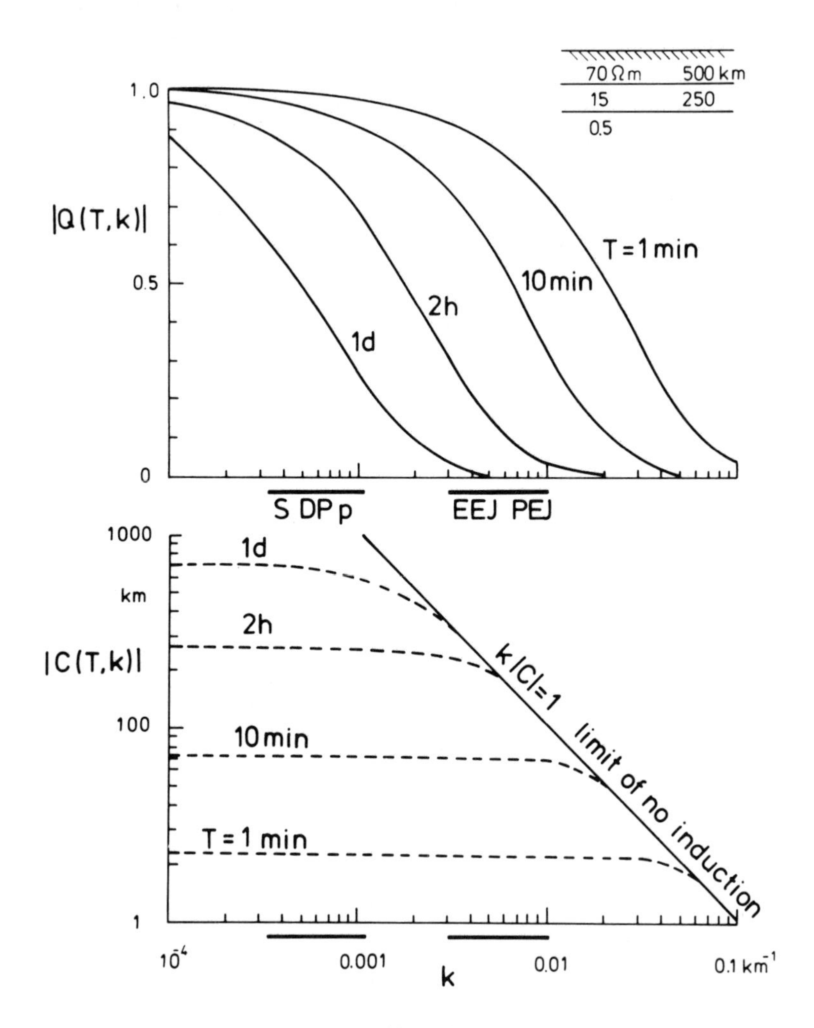

Figure 8

Source effect on the Q and C-response of plane 3-layer Earth model A from Figure 11, shown for the moduli as function of horizontal wavenumber k for fast and slow variations. Bars indicate the spectral range in k for slow and fast variations in mid-latitudes (left) and in jet regions (right). The ultimate wavenumber for ground observations should be 0.01 km^{-1} corresponding to source dimensions of 100 km which is the height of the nearest source region in the ionospheric E-layer.—For the chosen Earth model, source effects on the C-response will be neglible for pulsations everywhere, but note that their internal part is visibly below $Q = 1$ in jet regions, i.e., their depth of penetration will become here comparable to source dimensions without marked effect on C, however. Slow variations in jet regions, on the other hand, are close to the limit of 'no-induction' $Q = O$ and $C = k_j^{-1}$ their use for electromagnetic investigations is therefore critically dependent on the exact knowledge of the spatial source configuration. In mid-latitudes only very slow variations have a slightly source-dependent C-response, but note again the clearly reduced Q-response against unity, which allows response estimates also from magnetic vertical variations.

This is demonstrated in Figure 7 for the C-response of the spherical Earth model from Sec. 5, in Figure 8 for the Q- and C-response of an equivalent plane model. It appears that the zero-wave-number response is a valid approximation throughout except for slow DP and S variations in jet regions. In the period range of diurnal variations jet fields are close to the limit of 'no induction', i.e., they are without any significant internal part, where the chosen Earth model applies.

If source dimensions are known and independent of frequency, WEIDELT's transformation formulas (1972; Sec. 3) can be used to convert response estimates $C(\omega,k)$ or $C_n(\omega)$ into zero-wavenumber responses $C_o(\omega)$. However this is not possible for diurnal variations, for instance, because each time harmonic requires spherical harmonics of different degrees n. But, as Figure 9 shows, source effects and effects

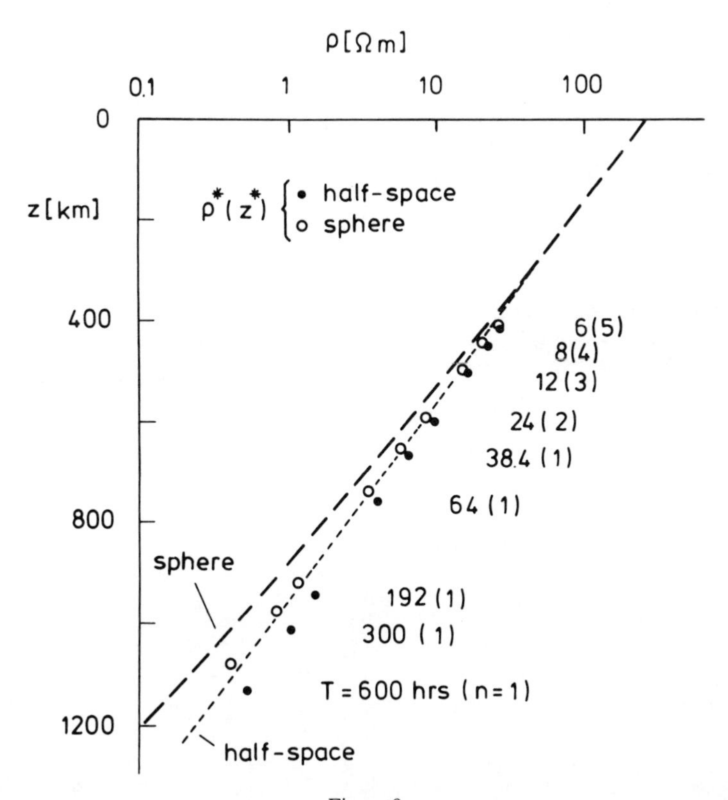

Figure 9

$\rho^* - z^*$ transformation of the model response, when the resistivity decreases in the Earth's mantle exponentially with depth. To demonstrate, how source dimensions and sphericity influence the response, two models are compared: (i) a spherical model $\rho(z) = \rho_o\{1 - z/R\}^m$ and a source field of variable degrees n of spherical harmonics, as indicated; (ii) a plane model $\rho(z) = \exp\{-2\lambda R\}$ and a quasi-uniform source field, $k = 0$. With $\rho_o = 250\ \Omega\text{m}$, $m = 37$, $R = 6371$ km and $2\lambda = m/R$ the models represent model 'd' by LAHIRI and PRICE (1939) to interpret global response estimates for S and Dst variations. See Table 1 for such estimates at the selected periods from 6 to 600 hours.

from the Earth's sphericity are small even for these deeply penetrating variations of rather complicated source field structure.

5. *Application to Response Estimates from Observations*

They refer to discrete frequencies and are from a harmonic analysis of regular variations or from a spectral analysis of irregular variations. In either case, averages of squared, spectral amplitudes are involved which assigns *rms* errors δ to response function estimates from a sequence of frequencies or frequency bands. They apply as relative errors to the modulus or as absolute errors to the phase, i.e., in the case of the C-response either an error $\delta|C|$ is assigned to $|C|$ or an error δ in angular measure to the phase. Since all definitions of substitute and apparent resistivities involve the squared modulus, their relative errors are 2δ, assuming the phase to be correct. Similarly, the relative errors of h^*, τ^* and z^* are δ.

The selected sample data set consists of response estimates for S and *Dst* variations at nine frequencies, ranging in period T from 6 hours for the 4-th *Sq*-harmonic to 600 hours for the *Dst*-continuum near the spectral peak from the 27-day reoccurrence tendency of magnetic storms. They have been derived from 18 months of observation at 13 European observatories, which are not situated on or near the coast, and have been taken from a recent collection of such data (SCHMUCKER, (1985); Tables 6 and 7, Sec. 4.2.2, column 'SCH').

Table 1 lists the response estimates and their transformation into substitute conductor resistivities. All phases are clearly above 45 degrees and, as Figure 10 shows, a simple h-type model will apply. The $h^*(T)$ curve merges into a constant depth of 750 km which would be the top of the underlying good mantle conductor.

Table 1

Sq *and* Dst *response estimates for continental Europe*

T hrs	C km		δ	ρ_a Ωm	φ degrees	h^* km	ρ^* Ωm
6	365	$-215i$	0.12	66 ± 16	60 ± 7	150 ± 20	34 ± 8
8	405	$-295i$	0.07	69 ± 10	54 ± 4	110 ± 10	48 ± 7
12	565	$-320i$	0.04	77 ± 6	60 ± 3	245 ± 10	37 ± 3
24	750	$-155i$	0.05	54 ± 5	78 ± 4	600 ± 30	4.4 ± 0.4
38.4	690	$-150i$	0.06	28 ± 3	78 ± 5	540 ± 30	2.6 ± 0.3
64	780	$-160i$	0.05	22 ± 2	78 ± 4	620 ± 30	1.8 ± 0.2
192	860	$-120i$	0.05	8.6 ± 0.9	82 ± 4	740 ± 40	0.3 ± 0.03
300	900	$-200i$	0.09	6.2 ± 1.1	77 ± 7	700 ± 60	0.6 ± 0.1
600	1020	$-290i$	0.12	4.1 ± 1.0	74 ± 9	730 ± 90	0.6 ± 0.2

The $\rho^*(T)$ assigns to it a half-space resistivity of 0.6 Ωm, while the overlying top layer appears to have a resistivity in excess of 50 Ωm. In the $z^*(T)$ curve an incon-

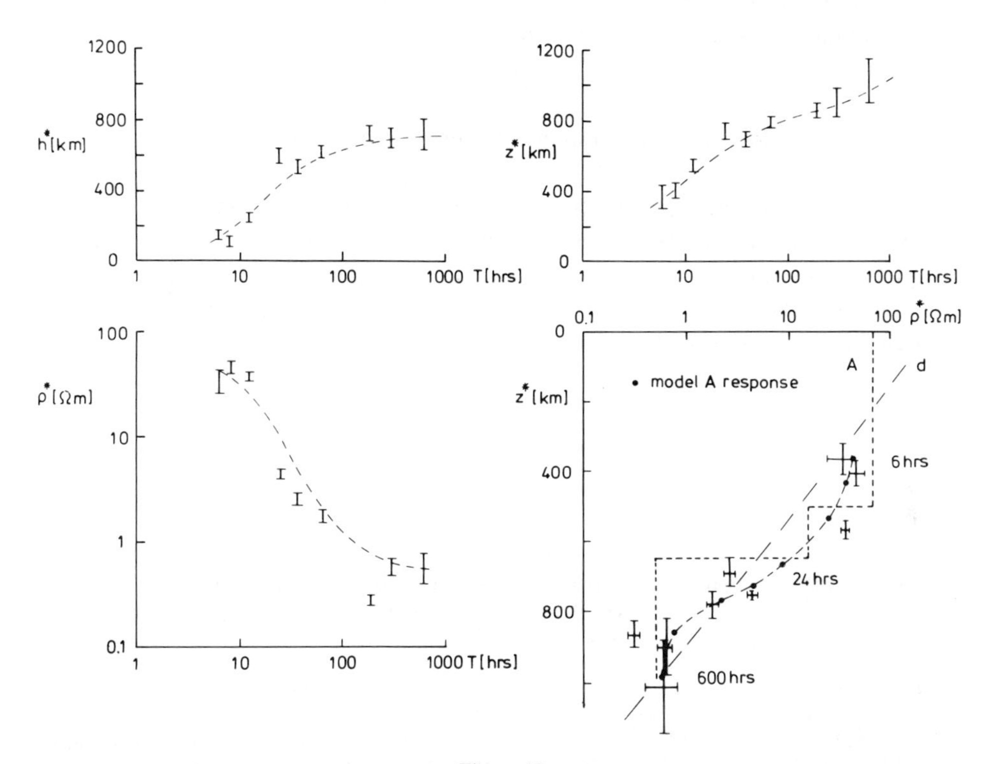

Figure 10

$\rho^* - z^*$ transformation of nine response estimates for S and Dst variations, derived from vertical to horizontal magnetic variations at twelve continental European observatories (Table 1). The estimates range in period from $T = 6$ to 24 hours in S and from 38.4 to 600 hours (= 25 days) in Dst.—The transition depth from high to low mantle resistivity is well established by a nearly constant h^*-values of 700 km at long periods, also by the reduced slope of $z^*(T)$ around that depth. From $\rho^*(T)$ follows at short periods a mean upper mantle resistivity above 50 Ωm and from ρ^* at long periods a mean lower mantle resistivity below 1 Ωm. The $\rho^* - z^*$ plot combines these depth and resistivity estimates into a smoothed image of the mantle resistivity distribution below the European continent.—Empirical estimates are compared with calculated $\rho^* - z^*$ values for the 3-layer model A from Figure 11. For comparison the LAHIRI-PRICE model 'd' (without surface conductor) is shown. But note that its $\rho^* - z^*$ would have followed closely the model line (cf. Figure 9), while the empirical values are in better agreement with the response of the step model A.

sistency is noticeable, when at the transition from Sq to Dst the depth of the perfect substitute conductor decreases slightly with increasing frequency, but the reduced slope between 600 and 800 km depth clearly indicates, where the transition to low mantle resistivities occurs.

The interpretation by layered models allows three or four layers at the most (Figure 11). If a continuous Earth model with an exponential decrease in resistivity $\rho_o \exp(-2\lambda z)$ is adopted, the response at the longest period (600 hrs) gives from eq.

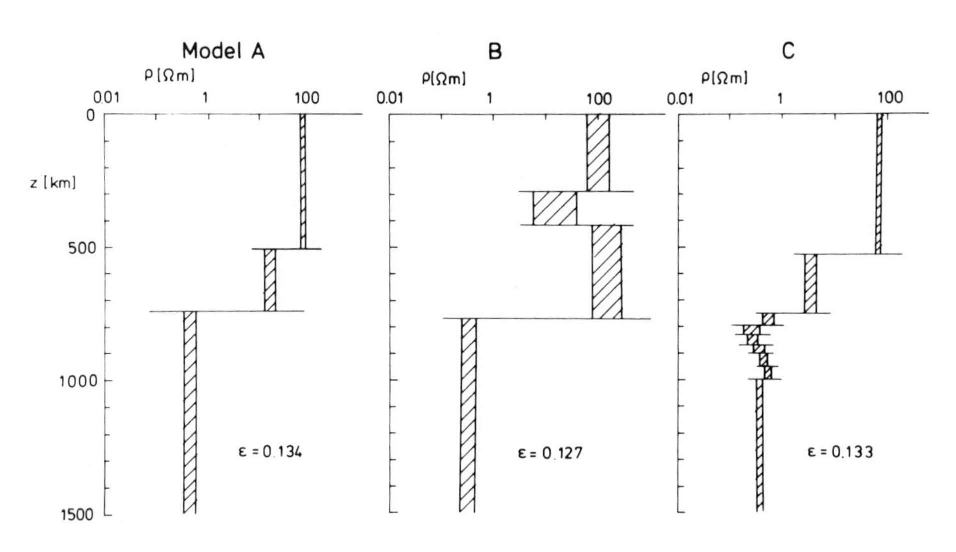

Figure 11
Least squares model interpretation of the response data in Table 1 by layered spheres. For details cf. SCHMUCKER (1985)—Model A: The best fitting 3-layer model with the constraint that layer thickness times square root of layer conductivity is an adjustable constant to minimize the quoted *rms* misfit ε between observed and calculated logarithmic responses, i.e., model A explains the data within 13.4% which is roughly also the *rms* error of the data. Models B and C with additional layers (model C from a generalized inversion) do not improve the fit, but are thought to bring out possible details which seem to be supported by certain characteristics of the data: A reduced resistivity in the uppermost mantle and a thin conducting zone in the transition region between the high and low resistivity region of the mantle, starting at 750 km depth.

(24a) the following parameter: $\lambda = 2.71 \cdot 10^{-3}$ km^{-1}, $p = 8270$ km, $\rho_o = 125$ Ωm. Converting it into a spherical power law model $\rho_o(1 - z/R)^m$, then from the equivalence of both models for $z \ll R$ the exponent $m = 2\lambda R = 34.5$ evolves which is close to $m = 37$ in the LAHIRI and PRICE model 'd'.

Since $\lambda p = 22.4$ is well above unity, the low frequency approximation is justified. For verification by the high frequency approximation $\lambda p \ll 1$, responses at less than one hour period would be needed, which are not available.

Finally, the $\rho^* - z^*$ plot in Figure 10 will be examined again to see, whether the data might yield further resolvable properties beyond an interpretation by two or three uniform layers. There are indeed indications for such properties, the first from the slightly reduced ρ^* value of the 4-th *Sq* harmonic, the second from the *Dst* response at 192 hours or 8 days, which has the smallest imaginary part of all responses. The first indication is for an upward reduction of resistivity, when moving into the depth range of less than 400 km, the second indication is for a thin zone of extremely low resistivity at 800 km depth, where the transition takes places from high to low mantle resistivity.

But it will be noted that both indications are from response estimates at only one frequency, i.e., attempts to use more complicated models, as shown also in Figure 11, will not improve the overall fit between observed and calculated responses to any extent. On the other hand, such models are not only compatible with the data within error limits, but they are also a motive for further investigations.

This concludes the final discussion regarding how substitute conductors, as a direct expression of the collected data, guide the search for models which fully exploit their information content. But regardless of the specific inverse procedure used for the final interpretation, it should not produce models with properties which are not visible already in the data themselves.

References

BANKS, R. J. (1972), *The Overall Conductivity Distribution of the Earth.* J. Geomagn. Geoelectr. *24*, 337–351.

CAGNIARD, L. (1953), *Basic Theory of the Magnetotelluric Method of Geophysical Prospecting.* Geophysics *18*, 605–635.

CHAPMAN, S. and BARTELS, J., *Geomagnetism.* (Oxford Clarendon Press 1940) 1049 pp.

KERTZ, W. (1978), *Das $\rho^*(z^*)$-Verfahren bei mit der Tiefe zunehmendem Widerstand.* Protokoll Kolloq. Erdmagnetische Tiefensondierung, Neustadt/Weinstraße, Nieders. Landesamt f. Bodenforschung Hannover, 77–80.

KUCKES, A. F. (1973), *Relations Between Electrical Conductivity of a Mantle and Fluctuating Magnetic Fields.* Geophys. J. R. Astr. Soc. *32*, 119–131.

LAHIRI, B. N. and PRICE A. T. (1939), *Electromagnetic Induction in Non-uniform Conductors and the Determination of the Conductivity of the Earth from Terrestrial Magnetic Variations.* Phil. Trans. Roy. Soc. London *A 237*, 509–540.

MOLOCHNOV, G. V. (1968), *Magnetotelluric Sounding Interpretation Using the Effective Depth of Penetration of Electromagnetic Fields*, in Russian. Izv. Akad. Nauk SSSR Fiz. Semli 9, 88–94.

NIBLETT, E. R. and SAYN-WITTGENSTEIN, C. (1960), *Variation of Electrical Conductivity with Depth by the Magnetotelluric Method.* Geophys. *25*, 998–1008.

PARKER, R. L. (1980), *The Inverse Problem of Electromagnetic Induction: Existence and Construction of Solutions Based on Incomplete Data. J. Geophys. Res. 85*, 4421–4428.

PRICE, A. T. (1962), *The Theory of Magnetotelluric Methods when the Source Field is Considered.* J. Geophys. Res. *67*, 1907–1918.

ROKITYANSKY, I. I., *Geoelectromagnetic Investigation of the Earth's Crust and Mantle* (Springer-Verlag, Berlin–Heidelberg–New York 1982) 381 pp.

SCHMUCKER, U. (1970), *Anomalies of Geomagnetic Variations in the Southwestern United States.* Scripps Institution of Oceanography Bulletin *13*, Univ. of California Press, 165 pp.

SCHMUCKER, U. (1971), *Neue Rechenmethoden zur Tiefensondierung.* Protokoll Kolloq. Erdmagnetische Tiefensondierung Rothenberge, Inst. für Geophysik Göttingen, 1–39.

SCHMUCKER, U., *Magnetic and electric fields due to electromagnetic induction by external sources, Electrical Properties of the Earth's Interior*, Landolt-Börnstein (Numerical Data and Functional Relationships in Science and Technology) New Series Group V Vol. 2b (Springer-Verlag 1985).

WEIDELT, P. (1972), *The Inverse Problem of Geomagnetic Induction.* Z. f. Geophysik *38*, 257–289.

(Received 20th January, 1987, accepted 21st January, 1987)

PAGEOPH, Vol. 125, Nos. 2/3 (1987) 0033-4553/87/030369-24$1.50+0.20/0

The Planetary Scale Distribution of Telluric Currents and the Effect of the Equatorial Electrojet: An Investigation by Canonical GDS

G. P. GREGORI,[1] L. J. LANZEROTTI,[2] B. ALESSANDRINI,[3] G. DEFRANCESCHI,[1] and R. CIPOLLONE[1]

Abstract—The planetary scale distribution of electrical currents in the Earth is still largely unknown. The role of the oceans for long period (hours to days) inducing electromagnetic fields $\mathbf{B}_e$ of external origin has been investigated by several authors, while the role of telluric current channelling, from the planetary viewpoint, is still far from a satisfactory understanding. Canonical geomagnetic depth sounding (GDS) analysis can yield locally a direction parallel to the strike of a telluric current density flowing in a region around the recording site and which also has the property of being the most relevant source for the internal origin field $\mathbf{B}_i$ observed at the given site at the given frequency. The use of such local information from 64 geomagnetic observatories is discussed here in a study to infer evidences of (a) the role of the polarization properties of $\mathbf{B}_e$ and (b) the role of the telluric current channelling within conductivity anomalies relevant to the planetary scale circuitry. The results show clear evidence of the influence of the equatorial electrojet on the polarization of $\mathbf{B}_e$ in a latitudinal band between $\pm(15^\circ - 20^\circ)$ latitude. There is also evidence that the $\mathbf{B}_e$ associated with the equatorial electrojet produces telluric currents which flow at a much shallower depth than the skin depth to be expected in the case of a plane Earth. This implies that the Parkinson planes in these regions reflect the conductivity structure underground more than the polarization of $\mathbf{B}_e$ due to the equatorial electrojet. Further, it clearly appears that some regular planetary scale pattern of telluric currents plays a more significant role than current channelling within some conductivity anomalies of fixed strike close to some geomagnetic observatories. Finally, the number of observatories used in this study appears to be insufficient to deduce any information concerning a seasonal evolution of the telluric current pattern on a planetary scale.

Key words: Telluric currents, equatorial electrojet, GDS, skim depth, induction, oceans.

1. Introduction

The planetary scale distribution of telluric currents has been investigated by several authors and by means of different approaches (e.g., FAINBERG and ZINGER, 1981; BEAMISH *et al.*, 1983; FAINBERG, 1983; ROBERTS, 1984; TAKEDA, 1985; and references therein). A point of considerable concern has been the assessment of the actual role of the oceans in induction phenomena when long period variation fields are considered. In fact, in such cases, the skin depth must prospectively enlarge

[1] I.F.A., C.N.R., p. L. Sturzo 31, 00144 Roma (Italia).
[2] AT & T Bell Laboratories, Murray Hill, NJ 07974 (USA).
[3] I.N.G., via di Villa Ricotti 42, 00161 Roma (Italia).

substantially compared to the average depth of the ocean; e.g., a magnetic field wave with period $T = 24$ hours can be expected to have a skin depth $S \sim 600$–800 km; see, e.g. HUTTON (1976). In this respect, several authors have attempted to estimate, as a function of frequency, the percentage of telluric currents flowing in the ocean with respect to the total telluric currents, providing the following results:

—BEAMISH *et al.* (1980a, b) find the percentage to be ~ 10–30% depending on the order and degree of the spherical harmonic term being considered;

—FAINBERG and SANIN (1981) provide an estimate of the oceans' role in terms of a few tens percent when computing the apparent specific resistance, and a few tens degrees when computing the phase of the impedance (for the *Sq* field); they find, however, only scant percent originating from induction by the *D*st field (48 hours $< T <$ 72 hours);

—ROBERTS (1984) applies a combined technique of SHA (Spherical Harmonic Analysis) and complex demodulation, in the period range 2 days $< T <$ 200 days, and finds that the oceans still play a relevant role. This is partially a contradiction with the long period results of FAINBERG and SANIN (1981).

—TAKEDA (1985) estimates 30% for the *Sq* field-produced oceanic telluric currents with respect to the total telluric currents.

In addition to the role of large area structures like the oceans, considerable attention has been paid to the phenomenon of channelling (JONES, 1983), by which local geologic and tectonic structures can eventually affect even the planetary scale pattern of telluric currents. Mid-oceanic ridges are a planetary scale network of conductivity anomalies which prospectively play some role in telluric current channelling. Direct search, however, for magnetic effects that might be associated with any telluric currents flowing in a magma chamber within a mid-oceanic ridge have produced no conclusive evidence (LAW and GREENHOUSE, 1981; FILLOUX, 1981; LAW, 1983). FILLOUX (1981) hypothesizes some suitable return currents in the ocean water in order to explain the non-observation of telluric current effects.

In general, results have been reported in the literature of strong telluric currents flowing within specific elongated conductivity anomalies either on a local or regional scale. Extension to a planetary scale has normally been the source of considerable debate (JONES, 1983, and references therein).

Canonical GDS is a specific technique for Geomagnetic Depth Sounding (GDS) or, equivalently, for dealing with the separation of external and internal origin geomagnetic fields. The separated fields can be used to the extent possible for investigating either external sources or underground conductivity structures, respectively. A full theoretical and mathematical account is given elsewhere (GREGORI and LANZEROTTI, 1986). It is sufficient for the purposes of this paper to provide just a few simple, physically intuitive, arguments.

For some suitable frequency range of the external field $\mathbf{B}_e$ (for example, for T ranging from a few minutes to ~ 2 hours) it is well known that the Parkinson plane is observed in an analysis of ground-based geomagnetic variations. This reflects the fact that the incident source field penetrates sufficiently deep to find a 'good' 'ultimate' conductor. Since the skin depth is negligible compared to the source distance of $\mathbf{B}_e$, the recording magnetometer assumedly lies almost 'just above' the ultimate conductor. Therefore, the observed total field at the instrument will be $\mathbf{B} = \mathrm{B}_e + \mathbf{B}_i$ (where $\mathbf{B}_i$ is the internal origin field) and it must be tangent to the ultimate conductor; the Parkinson plane is such a tangent plane (refer to GREGORI and LANZEROTTI, 1980, for a more extensive discussion).

Three issues must be addressed: (i) the polarization properties of **B** in the Parkinson plane within the period range ~few minutes to ~2 hours; (ii) the extension of the same argument to T shorter than a few minutes; i.e., in the geomagnetic pulsation range; and (iii) the case of $T \geqslant 2$ hours.

Regarding the first issue, assume that $\mathbf{B}_e$ has not fixed polarization, changing in time more or less randomly. The observed **B** will lie in a Parkinson plane with an apparently circular polarization when averaged over a sufficiently long time interval (provided that the underground conductivity structure distribution is isotropic in all azimuthal directions). Obviously, from symmetry, an isotropic $\mathbf{B}_e$ and an isotropic flat-layered conductivity structure cannot provide any preferred azimuthal direction.

In contrast, consider the case when the induced telluric current density **j** which originates the $\mathbf{B}_i$ field (at the given site and at the given period T) has some relevant preferred direction. Physically, such a situation can be indicative either of the actual **j** at the recording site, or of **j** at some other site, even distant. In this case, whenever $\mathbf{B}_e$ is isotropic, **B** will be elliptically polarized, with the minor axis of the polarization ellipse perpendicular to the relevant **j** (as defined above). That is, the major axis of the polarization ellipse will be parallel to the strike of the relevant **j**. Thus, there is the possibility of a direct local measurement of the strike of the relevant telluric current, at the given site and for the given period T. This can reflect either the strike of an actual conducting anomaly (e.g., possibly a mid-oceanic ridge) or the strike of the planetary scale **j** at the given recording site. This problem is one of the principal topics of concern in this paper. The direction of the major axis of the polarization ellipse in the Parkinson plane is called, in canonical GDS, the 'second canonical direction' (scd), which is determined except for a multiple of π radians.

In general, the standard use of ground-based measurements of geomagnetic pulsations for magnetospheric investigations is most often based on the assumption that the measured polarization characteristics are only a function of the magnetospheric source $\mathbf{B}_e$ and not of the $\mathbf{B}_i$ from induced telluric currents. Obviously, this is in direct contradiction with the assumption used by investigators studying Parkinson plane phenomenon. That is, for a period T less than a few minutes it is customary to assume that the $\mathbf{B}_i$ effect can be safely and completely neglected, while for T longer than a few minutes the polarization characteristics of $\mathbf{B}_e$ are generally considered as being totally unimportant (generally the $\mathbf{B}_e$ are analyzed over a 'suitable' time interval). Both assumptions are based on such long-standing experimental practice by so many investigators that the assumptions must be correct in their respective application ranges. The problem, in fact, is how the opposite conclusions can be matched to each other in their common period range. This important open question will be addressed elsewhere.

The case of the period of the external variation field $\gtrsim 2$ hours has implications for the subsequent discussions. Consider Figure 1a, which depicts the unphysical case of an infinitely conducting Earth, with the result that telluric currents flow immediately beneath any recording magnetometer; the Parkinson plane in all cases will be coincident with the plane tangent to the Earth at the recording site. Figure 1b refers to the 'real' Earth, where the skin depth of penetration of the source field is 'small'; that is, the average distance l beneath the Earth's surface of the induced electric current is much smaller than the distance L to the source field (the ionosphere). This is the case when the Parkinson plane is experimentally observed. For the *Sq* field, however, which has components with $T \leqslant 24$ hours, with maximum skin depth $S \sim 600 - 800$ km, $l \geqslant L$. In such a case the intuitive argument leading to the Parkinson plane is not valid. In this case the polarization ellipse of the measured field $\mathbf{B}$ expectantly reflects more the polarization properties of $\mathbf{B}_e$ than the effects of $\mathbf{B}_i$. This is increasingly the case as l enlarges compared to L. Of direct concern to the case of the equatorial electrojet is the situation where $\mathbf{B}_e$ is strongly polarized (in fact, in this case almost linearly polarized). While the $\mathbf{B}_i$ effect will reduce the sharpness of the polarization, $\mathbf{B}$ will still reflect the external polarization field.

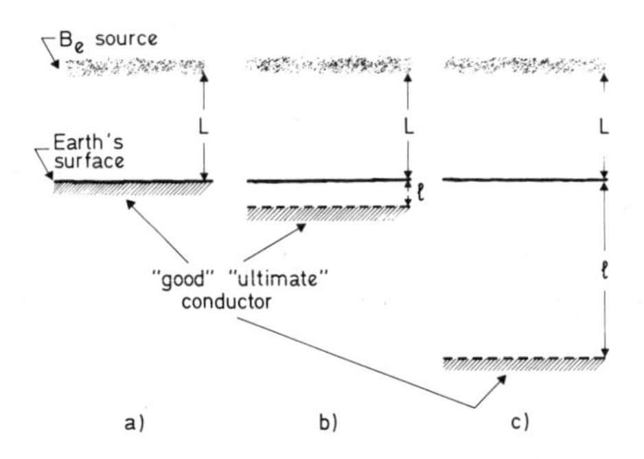

Figure 1

Three ideal situations: (a) An infinitely conducting Earth with the 'ultimate' conductor coinciding with its surface. The Parkinson plane is tangent to the Earth's surface. (b) A 'real' Earth has the 'ultimate' conductor at some depth l. Whenever $l \ll L$ (as for $T \lesssim 2$ hours), the Parkinson plane is actually observed, and it can be conceived as being parallel to the tangent plane to the 'ultimate' conductor beneath the recording site. (c) Whenever the skin depth $S \sim l \gtrsim L$, as can be expected (in principle) for long T, the polarization of the geomagnetic field observed at the Earth's surface reflects basically almost only the polarization properties of the external field $\mathbf{B}_e$.

Thus, whenever $l \ll L$ (as in Figure 1b) and whenever $\mathbf{B}_e$ is not strongly polarized, the polarization characteristics of $\mathbf{B}_e$ prospectively are unimportant. On the contrary, when $l \lesssim L$ and/or when $\mathbf{B}_e$ is strongly polarized, the polarization characteristics of $\mathbf{B}_e$ play a relevant role in the polarization characteristics of the detected **B**.

The equatorial electrojet is the main feature of the $\mathbf{B}_e$ source in an equatorial band centered approximately around the dip equator (e.g. FAMBITAKOYE, 1976; FORBES, 1981). It is important, for the sake of clarity, to distinguish between the 'planetary component' S_R^P of the electrojet; which is a general increase in the observed total geomagnetic field which occurs around the equator, in a reasonably wide latitudinal band, and which can be regarded as some kind of analytical continuation of the geomagnetic field extrapolated from the middle latitudes, and the 'real electrojet' S_R^E, which is a very sharp enhancement in field superimposed over the planetary component S_R^P and which spans a smaller latitudinal band. In this paper, the equatorial electrojet is taken to be S_R^P because, in fact, the effects considered here relate to observing the total **B**.

For the purposes of this paper, the equatorial electrojet has to be considered as a current flowing along the dip equator within some given latitudinal band at an altitude of ~ 100 km. It can be said to appear during local day time. Since the Fourier expansion of a periodic function $F(t)$ of time t, defined for $0 < t < 2\pi$, is

$$F(t) = \frac{1}{2} + \frac{2}{\pi}(\cos t - \frac{1}{3}(\cos 3t + \frac{1}{5}(\cos 5t - \ldots) \tag{1}$$

where $F \equiv 0$ for $0 < t < \pi/2$ and $(3/2)\pi < t < 2\pi$, and $F \equiv 1$ for $\pi/2 < t < (3/2)\pi$, the equatorial electrojet expectedly contributes predominantly waves of periods $T =$ 24, 8, . . ., $24/(2n + 1)$, . . ., hours. Therefore, the principal case of concern is that of Figure 1c (with or without the complication of the polarization characteristics of $\mathbf{B}_e$).

However, there is also a different possibility. According to the estimates by the previously cited authors of the importance of the oceans' role on the Sq induction, the skin depth is minimally important for the definition of the length l in Figures 1b, c. Therefore, *a priori*, it can be expected that even in the case of Sq, or of the equatorial electrojet, perhaps Figure 1b may be the operative illustration. The answer to this dilemma is of direct concern in this paper.

The equatorial electrojet region is usually taken to have a latitudinal extension of $\sim \pm 6°$ (geomagnetic) and $\sim \pm 18°$ (geographic) (Forbes, 1981). GREGORI *et al.* (1982), in a study of the same data as analyzed here, report that effects the electrojet can apparently be observed in the latitudinal band $-20° \lesssim \lambda \lesssim +20°$ (geomagnetic dipolar).

2. *The Data*

PARKINSON (1977) reported an Sq analysis from 64 geomagnetic observatories (see Table 1 and Figure 2) for the time interval July 1957 to December 1958. He analyzed the five international quiet days at each available observatory for each

Table 1

Station log

Prog. no.	Ref. no.	Acronym	Dip lat. (degrees)	Number of available months	Notes
1	6	TIK	65.48	4	(1)
2	12	COL	64.78	3	(1)
3	16	SRE	58.60	10	
4	18	DOB	59.70	18	
5	21	NUR	58.20	18	
6	22	LER	58.50	1	
7	26	SIT	60.40	18	
8	27	SVD	52.28	16	(1)
9	35	IRK	56.50	16	
10	37	NGK	50.00	3	
11	38	VAL	51.14	18	(1)
12	44	VIC	54.90	16	
13	50	TYH	44.50	15	
14	55	MMB	37.40	2	
15	56	AGN	60.80	18	
16	151	VLA	40.00	3	
17	60	TFS	37.03	18	(1)
18	61	TKT	41.50	18	
19	139	ISK	35.53	6	(1)
20	63	TOL	35.64	18	(1)
21	65	BEO	52.20	7	
22	69	FRD	54.40	18	
23	140	SMG	35.48	17	(1)
24	70	ALM	33.00	15	
25	71	KAK	30.10	18	
26	73	SSO	26.83	6	(1)
27	75	TUC	40.30	18	
28	77	HEN	24.90	17	
29	79	HON	22.00	18	
30	141	TEO	28.71	18	(1)
31	80	ABG	12.90	18	(2)
32	81	SJG	32.10	18	
33	82	MBO	8.90	18	(3)
34	83	MUT	7.20	17	(3)
35	84	GUA	6.50	18	(3)
36	87	AAE	−0.50	11	(3)
37	88	TRD	−0.30	13	(3)
38	89	IBA	−3.10	14	(3)
39	90	KOR	−0.04	18	(3)
40	91	PAB	17.60	12	
41	93	BNG	−7.10	16	(3)
42	94	FAN	5.30	15	(3)
43	142	MFP	−7.10	2	(1) (2)
44	95	JAR	1.10	14	(3)
45	96	TTB	8.90	16	(3)
46	98	HNA	−10.90	14	(2)
47	138	KUY	−17.60	11	(2)
48	103	LUA	−24.00	11	(2)

49	105	PMG	−17.30	7	(2)
50	107	HUA	1.00	14	(3)
51	108	API	−16.30	18	(2)
52	110	TAH	−15.80	8	(2)
53	111	TAN	−30.15	18	(1)
54	112	VSS	−11.70	18	(2)
55	143	IML	−20.85	5	(1)
56	113	WAT	−46.40	18	
57	114	HER	−46.80	18	
58	115	TOO	−51.20	18	
59	116	AML	−51.30	18	
60	117	TWA	−22.40	16	
61	118	PAF	−49.70	11	
62	119	MCQ	−64.60	18	(1)
63	120	AIA	−38.10	17	
64	128	HLY	−47.20	11	

Notes

(1)—The dip latitude has been evaluated by making the average between the coordinates evaluated by GUSTAFSSON (1969) and by TSYGANENKO (1979, 1981). All other dip latitudes have been taken from Appendix 2 of MATSUSHITA and CAMPBELL (1967).

(2)—Stations classified like 'EE stations' according to Figure 4 (for T = 24 hours), but not included among the 12 closest stations to the dip equator as for the evidence from Figure 11.

(3)—'EE stations' satisfying to the requirements of Figure 4 (for T = 24 hours), and also to the requirements of the 12 stations in Figure 11 which are closest to the dip equator.

month, reporting the average fields and the first four diurnal harmonics (that is, T = 24, 12, 8 and 6 hour periods, respectively). This paper is based on these reported Fourier amplitudes and phases.

3. The Analysis

For each observatory, and for each month, the polarization plane (the 'Parkinson plane') and the polarization ellipse of the total **B** lying in it has been evaluated for each of the four periods 24, 12, 8, 6 hours. The major axis of each polarization ellipse is then projected on the (local) horizontal plane, providing the azimuthal direction p_3 (modulus π radians), the values of which are the basis for the discussions in this paper.

The amplitude p_5 of the major semi-axis of the polarization ellipse is a measure of the spatial component of **B** which can be shown as of purely external origin (for the case of Figure 1b and approximately isotropic $\mathbf{B}_e$). In the case of either Figure 1c, or of Figure 1b but with $\mathbf{B}_e$ strongly (or almost linearly) polarized, the amplitude p_5 is a component of $\mathbf{B}_e$ (slightly reduced in amplitude because of the telluric currents). In all these cases, the amplitude p_5 will be a measure of the equatorial electrojet effect, as discussed in detail in GREGORI *et al.* (1982).

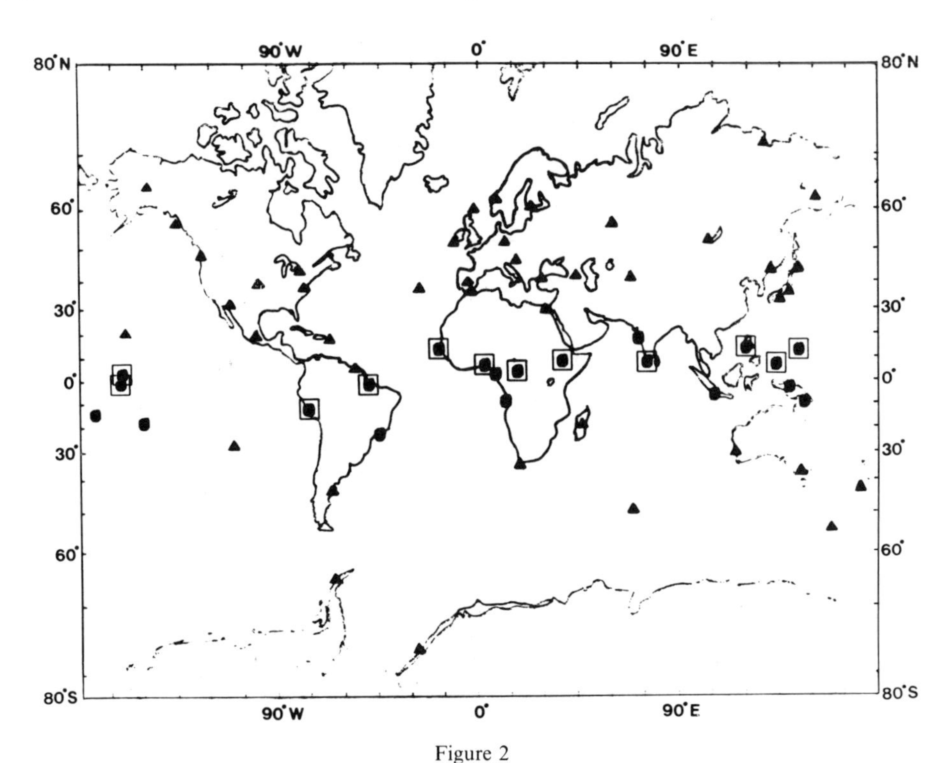

Figure 2
Map of the 64 geomagnetic observatories used in this canonical GDS analysis. The 21 observatories which are affected by the equatorial electrojet ('EE stations') are indicated by small circles. The 12 EE stations which are less than 9° from the dip equator are further marked with an open square (for reference to Figure 11).

The following is a discussion of the azimuthal directions p_3 evaluated at each observatory, monthly for each of the periods $T = 24$, 12, 8 and 6 hours.

4. *The Seasonal Dependence of the SCD at each Observatory*

Figure 3 plots the values of p_3 as a function of month for $T = 24$ hours for 16 of the stations studied by Parkinson. These are representative of the results for all stations and all periods.

A continuous fit line has been drawn through the points of each plot. The line for each panel has been determined in an approximate way as follows. Since each p_3 value is determined apart $\pm 180°$, a formal fit has been made in terms of a Fourier series with fundamental period one year, with a suitable number of higher harmonics, depending upon the total number of available data-months.

An examination of all p_3 results reveals that only a limited number of observa-

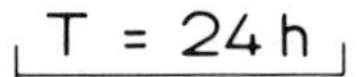

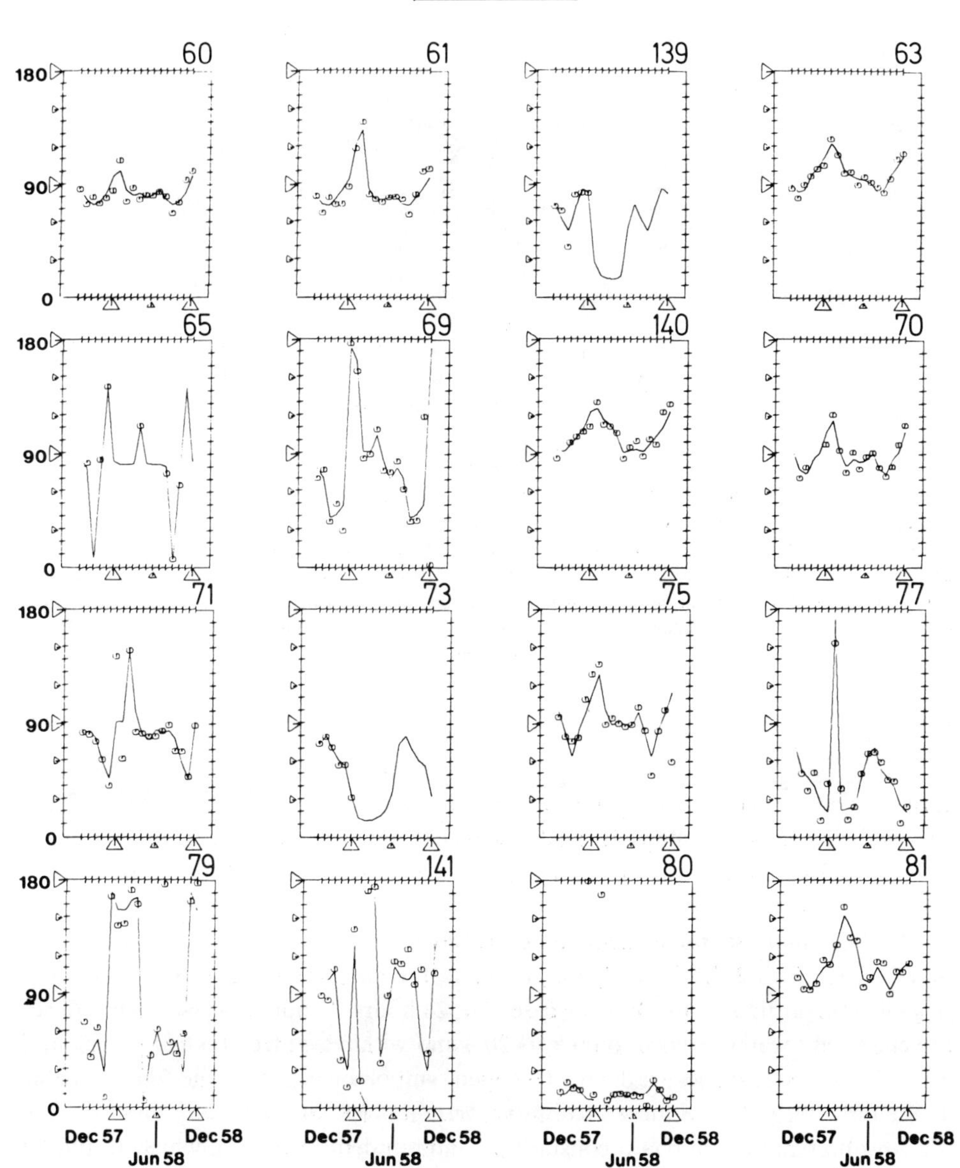

Figure 3
Diagram of p_3 *vs* month for sixteen observatories for $T = 24$ hours. The number over the upper left hand corner of each plot is the reference number of the observatory (see Table 1). The number of points depends on the number of months with available data (see Table 2). The continuous line is the result of an approximate numerical fitting procedure (see text).

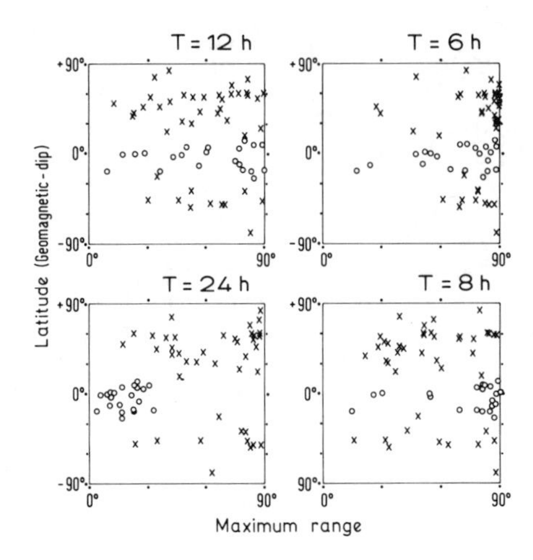

Figure 4
Maximum range or 'swing' w_T (T = 24, 12, 8 or 6 hours) of each observatory *vs* its dip latitude. The 'swing' w_T is defined as the maximum azimuthal difference of p_3 for one year. Since p_3 is defined apart $\pm 180°$, w_T is always $0° < w_T < 90°$. An interesting feature is the clustering in the 24 hour plot of the 21 observatories closer to the dip equator; this is the criterion chosen for defining the 21 'EE stations', which are marked by an open circle on all plots. The remaining non-EE stations are marked by small crosses. Notice also the tendency for clustering, with $w_T \sim 80°$–$90°$, of most EE stations in the $T = 8$ hours plot.

tories have a small seasonal dependence for p_3, while the largest part of the observatories show a somewhat regular seasonal trend for the interval of data available. The observatories with small seasonal variations are mostly found for $T = 24$ hours.

The maximum seasonal excursion of p_3 has been evaluated for each observatory and for each period T. Figure 4 is a plot of such maximum excursions as a function of geomagnetic dip latitude. For the case $T = 24$ hours, 21 out of the 64 observatories are clustered together near 0° dip ($\pm \sim 20°$) and within a range of $\sim 30°$ (horizontal axis). These observatories are denoted by open symbols on Figures 2 and 4 and in Table 1. We tentatively denote these 21 observatories the 'EE stations'.

No observatory exhibits a constant p_3, independent of month, which would be expected in the case, for example, of strong channelling in a large, elongated conductivity anomaly. In this case, while the intensity of a channelled current could have a seasonal dependence, the strike, parallel to p_3, should remain constant. That is, apart from the 21 EE stations, at all the remaining observatories the seasonal excursion of the planetary scale distribution appears, in general, to be definitely more important than possible channelling effects. See also Section 9, below.

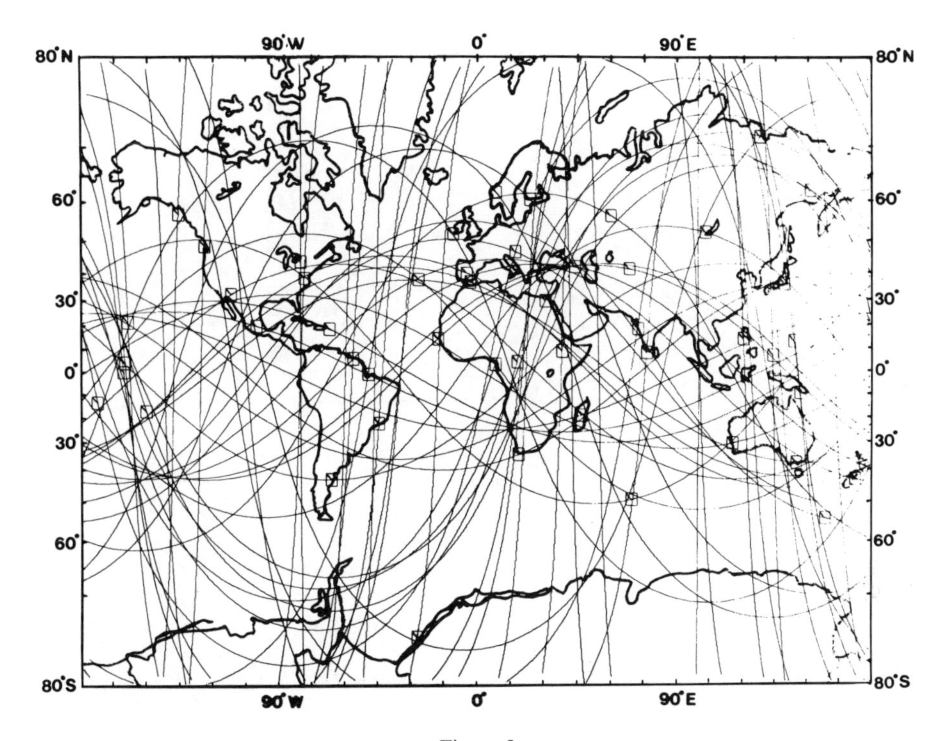

Figure 5

Example of a **j**-GMG for each observatory (for each given month and period T a great circle is drawn on the Earth with strike equal to p_3). All such great circles drawn for all the observatories (at the same month and T; in this example December 1958, $T = 24$ hours) are then drawn either on a Mercator projection map (as in this example), or on a polar stereographic projection. (In the present analysis such maps have been drawn for 18 months, for the 4 values of T, and in 2 projections; i.e., altogether, $18 \times 4 \times 2 = 144$ maps).

5. *The j-GMG's*

At a given observatory point P on the Earth's surface, the great circle through P with strike p_3 can be drawn. Drawing all such great circles for all the available observatories at each given time instant (monthly value) results in a plot we call the TCd-GMG plot (or the **j**-GMG plot, where TCd and GMG designate *T*elluric *C*urrent *d*irection, and *G*eo*M*agneto*G*raphy, respectively). One **j**-GMG can be drawn for each month and for each period studied. For the data analyzed here, such **j**-GMG's have been drawn for all 18 months and for all 4 periods, both in Mercator equatorial projection and in polar stereographic projection, for a total of 144 plots. Such plots give, in effect, the seasonal variations of the distribution of the telluric currents on a planetary scale for the 18 months of data. Figure 5 is one example of the 144 plots, for the month of December 1958, $T = 24$ hours.

In order to discuss the possible existence of a visual trend in the data, at least for

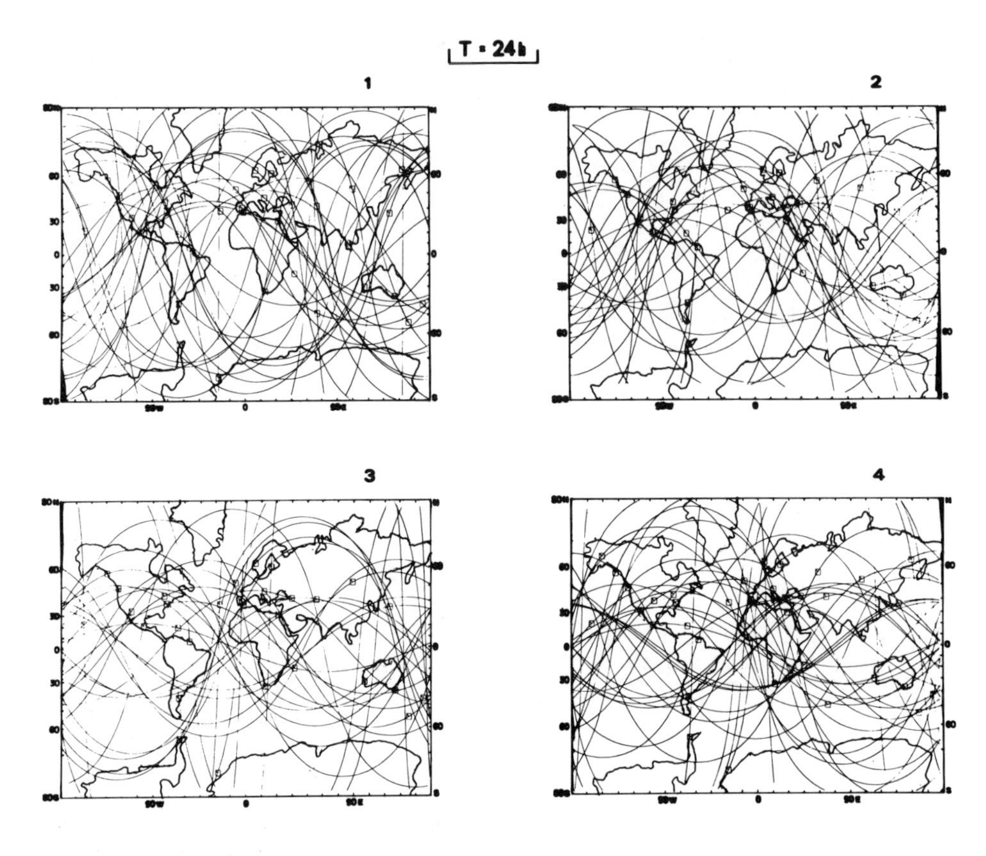

Figure 6
Three month averages of *j*-GMG for $T = 24$ hours, being: 1) months D = Nov-Dec-Jan, 2) months M = Feb-Mar-Apr, 3) months J = May-Jun-Jul, 4) months S = Aug-Sep-Oct. Similar maps have been drawn for $T = 12$, 8 and 6 hours. Only the 43 non-EE stations are included here (for clarity purposes).

some stations, 3-month averages of the p_3's have been considered for each station for the D-months (Nov., Dec., Jan.), M-months (Feb., Mar., Apr.), J-months (May, Jun., Jul.), and S-months (Aug., Sep., Oct.). Whenever in consideration of a given observatory and a given period T, two identical months were available, one for 1957 and the same for 1958, each month was weighted by 1/2. By means of such 3-month average values of p_3, 8 plots (for each T) of **j**-GMG's have been drawn and are shown in Figures 6 and 7. Figure 6 contains only the 43 non-EE stations and Figure 7 contains only the 21 EE stations.

Two conclusions can be drawn:

—The EE stations (Figure 7) have p_3 aligned approximately North-South; the implications are discussed in Section 7.

—It is difficult to recognize a regular seasonal trend from the non-EE stations (Figure

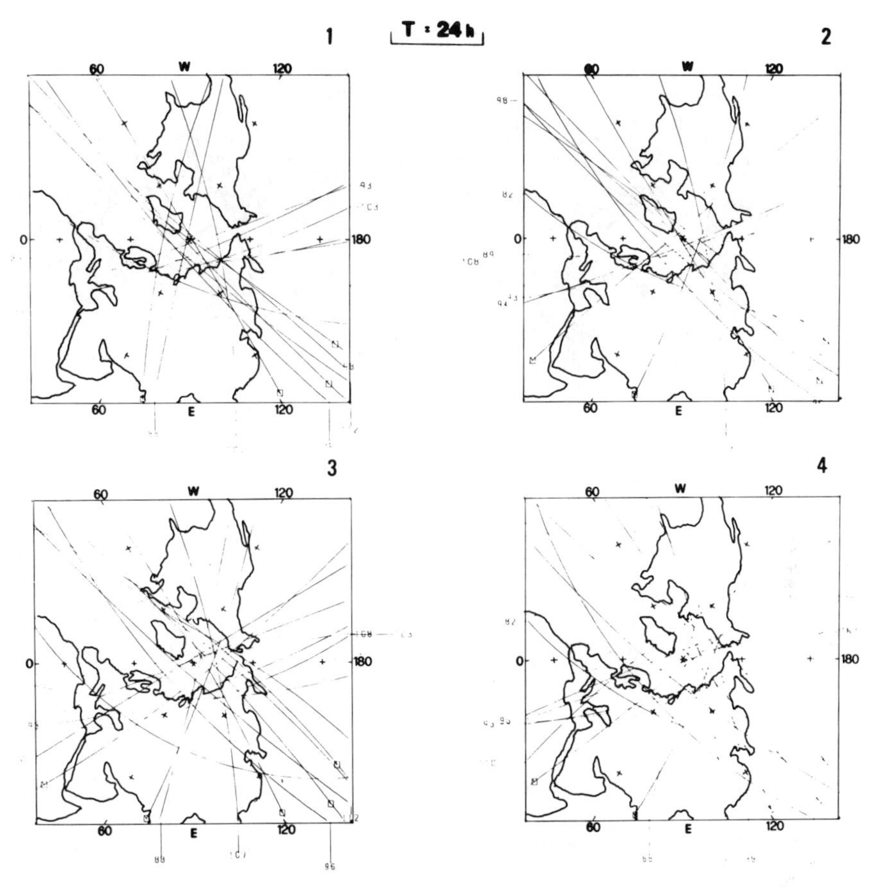

Figure 7
The same as Figure 6, but in polar stereographic projection, plotting only the 3-month average great circle for the 21 EE stations. Notice that such great circles all pass close to the pole; i.e., p_3, is almost N-S for all these observatories. Similar plots (not here shown) have also been drawn for $T = 12$, 8, and 6 hours. The reference number of the observatory of each great circle is shown outside each plot.

6) in this representation. However, see the following Section for results from different representations.

6. *The Seasonal Repetitiveness of the j-GMG's*

In order to inspect quantitatively how regularly the **j**-GMG's are changing as a function of month, the differences $p_{3,12} = p_3(t) - p_3(t + 12 \text{ months})$ and $p_{3,6} = p_3(t) - p_3(t + 6 \text{ months})$ have been evaluated for each observatory and for each period T (where the time t is expressed in integer multiples of a month). The results are shown in Figure 8 (for $p_{3,12}$) and Figure 9 (for $p_{3,6}$). The regularity of the seasonal

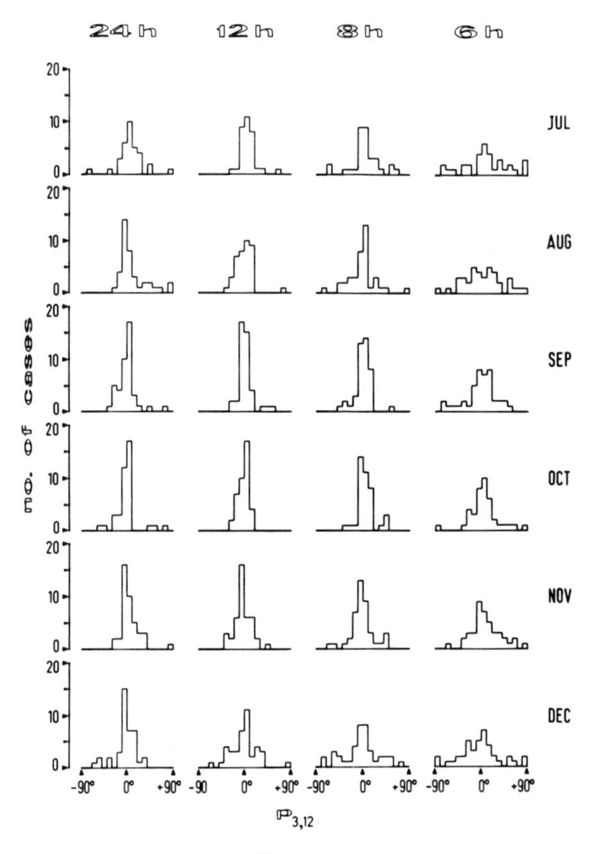

Figure 8
$p_{3,12}$ values; i.e., the p_3 value of a given month minus the p_3 value of 12 months later: (a) $T = 24$ hours; (b) $T = 12$; (c) $T = 8$ hours; (d) $T = 6$ hours. The repetitiveness from one year to the next is evident, even though it is worse for $T = 6$ hours, possibly because of a decrease in the signal to noise ratio in the basic data.

dependence of the **j**-GMG's is beyond any possible doubt from the results of Figure 8. An effect is seen as well in the 6 month difference plots for most observatories.

The equinoctial months appear to have smaller differences, however (as might be expected from symmetry reasons). However, the smallest $p_{3,6}$ values are obtained for the month pairs (Apr. '58–Oct. '57) and (Oct. '58–Apr. '58), smaller than the pair (Sept.–Mar.). No explanation is apparent.

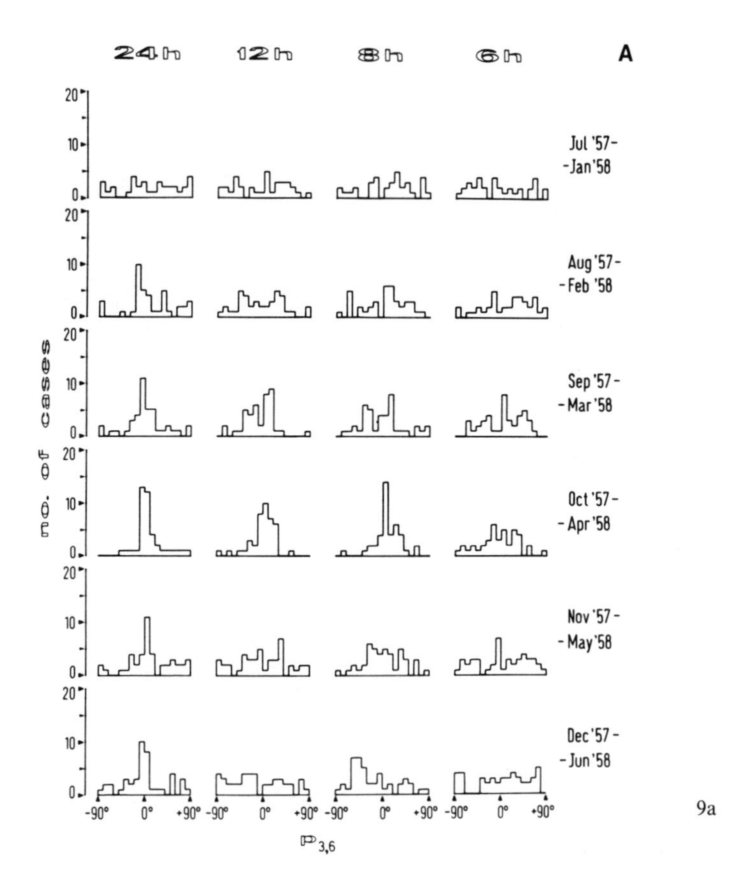

9a

7. *The Role of the Equatorial Electrojet*

The histograms in Figure 10 are compiled from the 21 EE stations by summing the 3-month results (grouped as above). The North-South alignment of p_3 appears very clearly, with some small seasonal trend as well. The half-width of the histogram, however, is basically independent of season.

Histograms of p_3 for 20 of the 21 EE stations are shown in Figure 11, where the seasons are distinguished but the sums have been made over period (the station Moca is excluded because of insufficient data availability; Table 1). The EE stations are ordered in Figure 11 from top to bottom left column and top to bottom right column according to increasing distance from the dip equator. The histogram values are more peaked for the lower dip latitude stations: primarily the left column observatories plus the two observatories at the top of the right column (these sites are all located at less than 9° from the dip equator). These 12 EE stations are

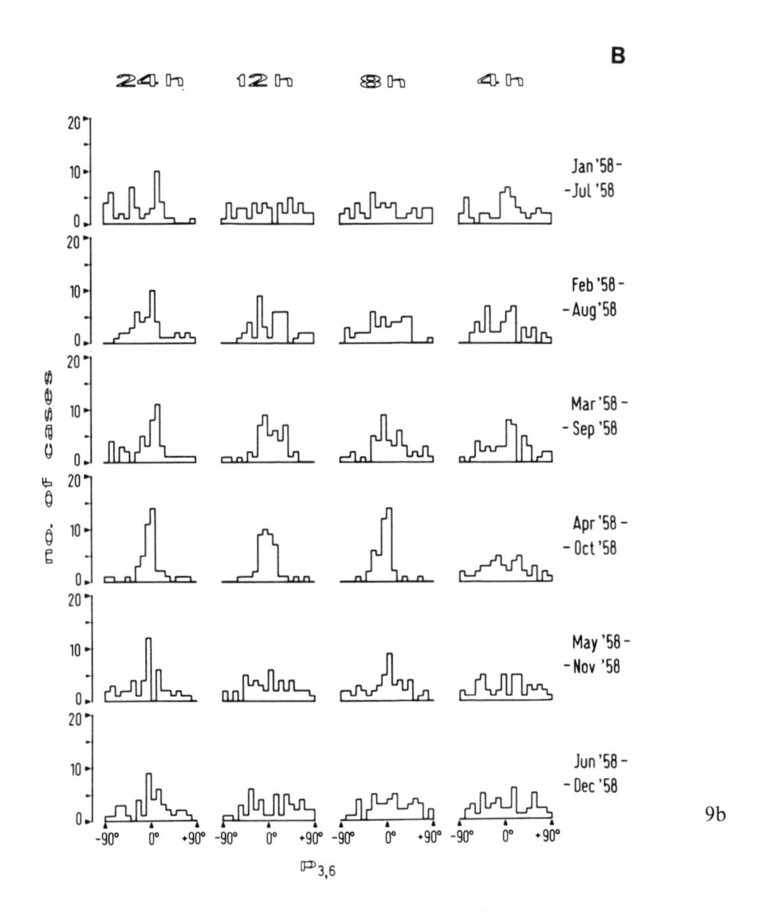

Figure 9
$p_{3,6}$ values; i.e., the p_3 value of a given month minus the p_3 value of 6 months later: (a) July 1957 through Dec. 1957; (b) Jan. 1958 through June 1958.

correspondingly marked by a different symbol in Figure 2 and also indicated in Table 1.

8. *The Role of Conductivity Anomalies*

From the foregoing discussions it appears clear (i) that the seasonal trend in the p_3 is, in general, more important than the influence of local conductivity anomalies; (ii) that the equatorial electrojet plays an important role in the results, and (iii) that 64 stations are insufficient for providing a reasonably complete picture of the planetary scale pattern of telluric currents. Nevertheless, a search can be made to investigate whether a local conductivity anomaly at some of the stations can affect,

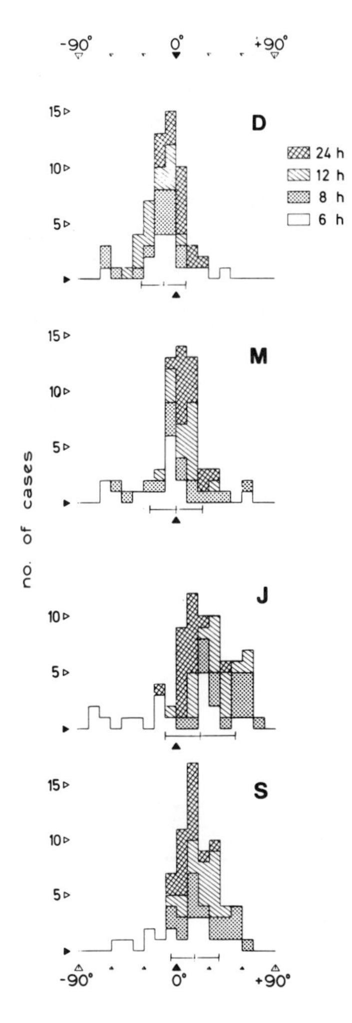

Figure 10
Number-of-cases histograms of three-month-average p_3 for the 21 EE stations. The line under each histogram indicates the individual barycenters and rms deviations. Same month designations as for Figure 6.

by some amount, the results from these stations.

In general, any given feature in local magnetic records depends upon $\mathbf{B}_e$ whenever it correlates with the Earth's orbital motion, while it will depend on local features whenever it correlates with local geomorphological features. By such an argument the Parkinson plane, for example, is recognized to be either of tectonic or of oceanic origin (BULLARD, 1970).

The p_3 were examined to determine if they were related to the nearest mid-ocean ridge. The closest point P_j has been located on the network of the mid-oceanic

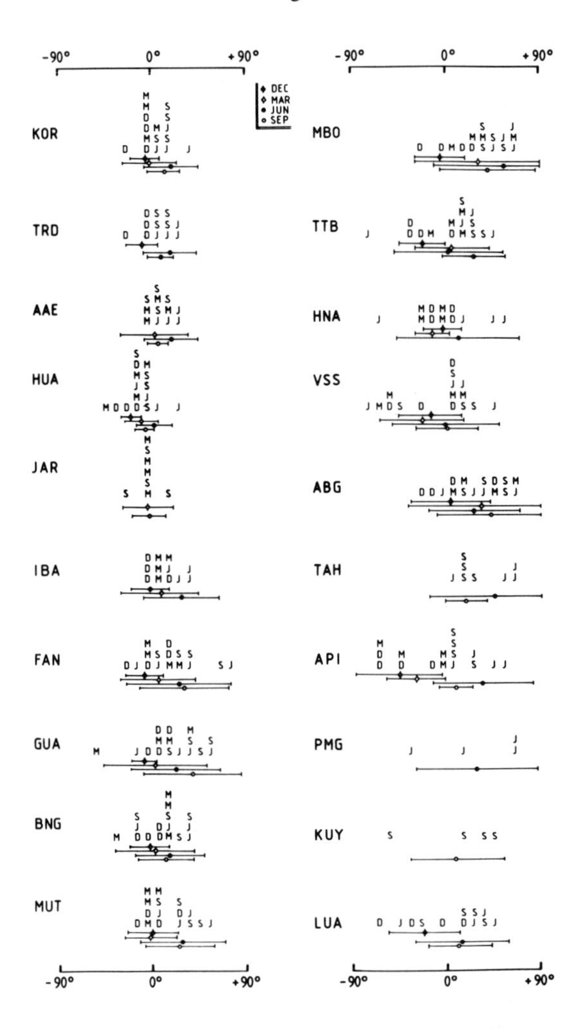

Figure 11
Number-of-cases histograms of three-month-average p_3 for each EE station (excluding Moca because of insufficient data) summing over all periods T. Each histogram is designated by the seasonal designation letter D, M, J, or S. The bars under each histogram give the barycenters and rms deviations for each histogram, distinguishing different seasons. The EE stations are ordered in increasing order of absolute value of dip latitude (as in Table 1), left column first, from top to bottom.

ridges for each observatory O_j ($j = 1, 2 \ldots, 64$). Then, the strike through each O_j has been drawn which is parallel to the tangent to the mid-oceanic ridge at each respective P_j. The difference between each such strike and the corresponding p_3, plotted as a function of the distance $\overline{O_jP_j}$, shows much scatter and no discernable trend (not shown here). The same results were found when only the 43 non-EE stations were considered.

Table 2

EE station with $w_8 < 70°$

EE station				Period							
Progr. no.	Refer. no.	Acronym	*	24 h		12 h		8 h		6 h	
				w_{24}	$\bar{p}_3$	w_{12}	$\bar{p}_3$	w_8	$\bar{p}_3$	w_6	$\bar{p}_3$
43	142	MFP	2	3.92	20	9.28	170	14.82	0	17.18	175
37	88	TRD	13	5.81	0	16.65	10	26.33	5	47.61	0
50	107	HUA	14	10.36	170	28.23	160	30.50	170	85.56	90
38	89	IBA	14	10.76	15	43.08	10	54.69	30	58.05	5
36	87	AAE	11	21.50	0	47.58	10	54.48	15	79.78	175
52	110	TAH	8	32.58	0	78.77	80	69.49	75	65.15	60

* Number of months available.
All w_T and $\bar{p}_3$ values are in degrees. The $\bar{p}_3$ value is a simple estimate, rounded off to the next 5°, by inspection of Figure 3 (and analogous ones) ($\bar{p}_3$ is some average value for p_3 around which the monthly p_3 is oscillating). The azimuth p_3 is measured anticlockwise from the South direction.

Table 3

Non EE stations with either w_T (T = 24, or 12, or 8, or 6 h) < 30°

Non-EE stations				Period							
Progr. no.	Refer. no.	Acronym	*	24 h		12 h		8 h		6 h	
				w_{24}	$\bar{p}_3$	w_{12}	$\bar{p}_3$	w_8	$\bar{p}_3$	w_6	$\bar{p}_3$
10	37	NGK	3	17.29	125	12.73	105	33.68	90	89.79	90
					110		95		80		110
14	55	MMB	2	80.31	—	22.38	—	21.58	—	87.63	—
					50		65		50		110
16	151	VLA	3	42.81	80	63.40	90	76.52	90	29.70	70
					80		55		100		70
17	60	TFS	18	22.73	90	81.22	90	55.25	65	87.88	65–70
					90		70		65		65–70
23	140	SMG	17	44.34	105	31.38	100	29.66	90	68.97	80
					105		100		90		80
26	73	SSO	6	42.64	40	27.57	50	28.69	60	27.49	90
					55		60		65		70
27	75	TUC	18	46.20	90	22.77	80	50.62	75	81.20	85
					90		80		75		85
57	114	HER	18	34.61	75	29.94	85	16.06	80	71.26	110
					75		85		80		95
61	118	PAF	11	23.42	85	62.56	70	59.19	115	82.76	90
					75		90		70		90

* Number of months available.
All w_T and $\bar{p}_3$ values are in degrees. The $\bar{p}_3$ values have been estimated either by making reference to the interpolating line in Figure 3 (upper estimate) or by considering only the single points of each monthly p_3 (lower estimate). For MMB, since only 2 months are available, no interpolating line has been drawn. See also the captions of Table 3.

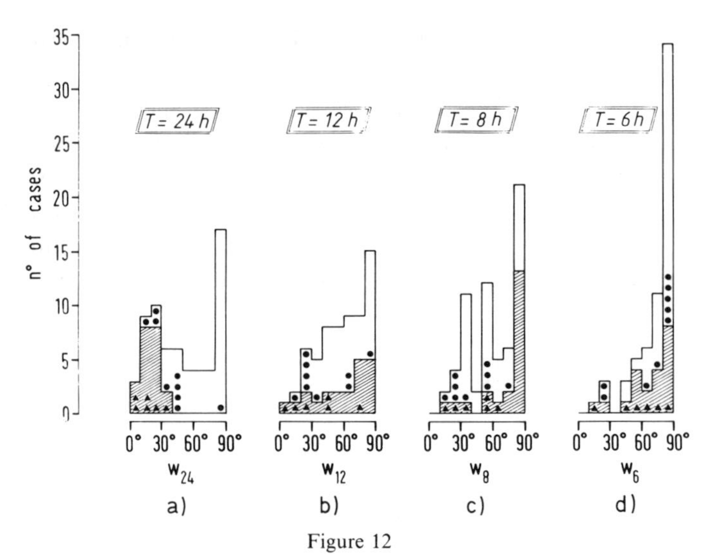

Figure 12

Number-of-cases histograms of 'swings' (w_T) for different T: (a) T = 24 hours; (b) T = 12 hours; (c) T = 8 hours; (d) T = 6 hours. The 'swing' w_T is the abscissa of Figure 4, and represents the maximum observed seasonal azimuthal variation of p_3 at the period T. Notice that w_T is always $0° < w_T < 90°$ by definition. The 21 EE stations are hatched. The 6 EE stations with $w_8 < 70°$ and the 9 non-EE stations with at least one $w_T < 30°$ (T = 24, 12, 8 or 6 hours) are also marked with solid dots, and are listed in Tables 2 and 3, respectively.

In a second attempt, the maximum angle of azimuthal spread (or 'swing', w_T) of the p_3 at each station was examined. Such swings are the abscissas in Figure 4. Notice that because of the indeterminacy by $\pm 180°$ of p_3, w_T is always $0° < w_T < 90°$. Therefore, a 90° swing is indicative of a very strong seasonal dependence. Figure 12 gives histograms of the frequency of occurrence of the w_T's for 63 stations (the station LER is omitted because it has data only for one month). The EE stations are cross-hatched.

Because of the peaks occurring close to 90°, the seasonal effect appears to be the general rule for all O_j's. The seasonal effect appears stronger for the shorter periods. Nine non-EE stations with at least one $w_T < 30°$ (T = 24, 12, 8, 6) and six EE stations with $w_8 < 70°$ are denoted by solid dots in Figure 12 and are listed in Tables 2 and 3 respectively. These sites possibly are candidates for stations affected by channelling within some local conductivity anomaly.

However, one of the identified stations MFP has data only for two months (Table 2) and four out of the five remaining identified stations are very close to the dip equator. The last identified EE station, TAH, is at a reasonably high dip latitude, but it is also the one station among the 6 in Table 2 with the largest w_T's. The average value $\bar{p}_3$ of p_3 (around which the swing occurs) appears reasonably stable as a function of T (except for station HUA for T = 6 hours), and is mostly E-W aligned (except for TAH). Even in these cases it is likely that the equatorial electrojet could be considerably more important than any eventual local conductivity anomalies.

Call (T_1, T_2) the regression plot with w_{T_1} as abscissa and w_{T_2} as ordinate. Figure 13 shows that w_{24} inclines to be correlated with w_{12}, w_8, or w_6 and w_{12} is correlated with w_8. However, the 6 stations of Table 2 have $w_{12} \approx w_8$, while for all other EE stations $w_8 \approx 80°–90°$. The (12, 6) and (8, 6) plots show wide scatter, possibly a consequence of the errors in the data for $T = 6$. In summary, no final conclusion appears possible for the eventual role of local conductivity anomalies at the EE stations.

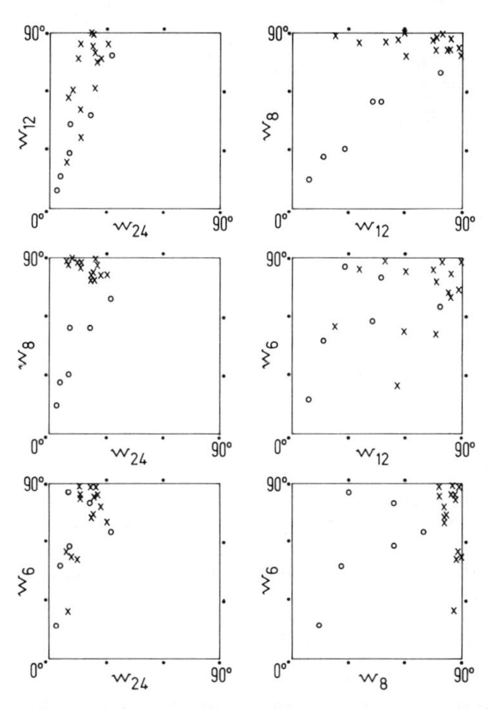

Figure 13

'Swings' plots (or (w_{T_1}, w_{T_2}) plots). Each plot is a regression of a w_{T_1} *vs* w_{T_2} (for EE stations alone). The 6 EE stations of Table 2 (i.e., having $w_8 < 70°$ in Figure 12) are marked by a small circle.

Similarly for the non-EE stations, Figure 14 shows even less correlations than for the EE stations in Figure 13. The conclusion is that, on the basis of the present data for 64 stations, no evidence can be provided on the possible role of local conductivity anomalies in the measured magnetic fields.

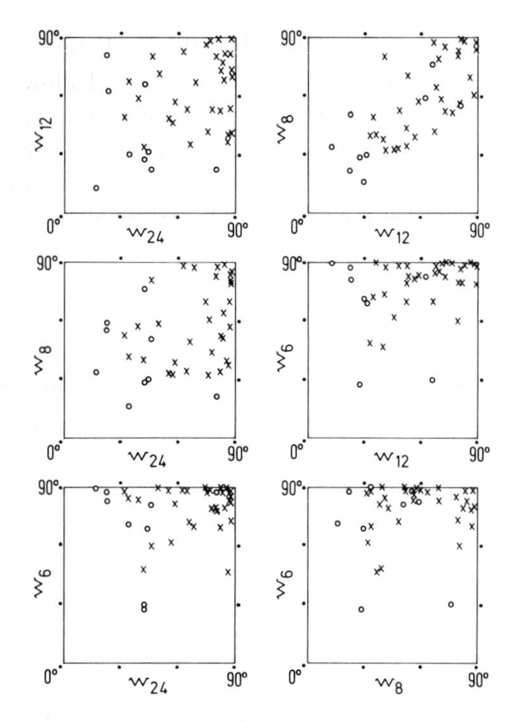

Figure 14
The same as Figure 13 but for the 42 non-EE stations. The 9 non-EE stations listed in Table 3 (and defined as having one of the $w_T < 30°$, $T = 24$, 12, 8 or 6 hours) are marked by a small circle.

9. *The Skin Depth of the Equatorial Electrojet-Associated Telluric Currents*

It appears that the equatorial electrojet has a strong influence on the polarization of the local **B**, and therefore on p_3. Since the skin depth for $T = 24$ hours is $S \sim 600$–800 km, it is reasonable to infer that the case of Figure 1c has to be considered.

If this is the case, the Parkinson plane at each EE station should be approximately tangent to a toroidal surface (a cylindrical surface in the case of a flat Earth) having its axis along the equatorial electrojet. It has not been possible for us, however, to use this information to estimate the height or intensity of the equatorial electrojet.

In fact, when the actual orientation of each Parkinson plane is considered in detail at each O_j for each T and for each month, it can be clearly recognized that the planes lie quite differently from the tangent plane to the ideal toroidal surface with an axis along the equatorial electrojet. The Parkinson plane appears to be approximately tangent to the Earth's surface. That is, it appears to be more sensitive to the shape of the Earth than to the geometry of the equatorial electrojet (the $\mathbf{B}_e$ field).

Therefore, it can be concluded that the case considered here is more similar to the situation outlined in Figure 1b rather than that in Figure 1c. This is consistent with the previous estimates (Section 1) on the role of shallow structures, like the oceans, on even the *Sq* field. That is, the telluric currents associated with the *Sq* field are flowing at a far shallower depth than the nominal skin depth $S \sim 600$–800 km. The N-S polarization of **B** for the EE stations is thus in fact a consequence of the strong $\mathbf{B}_e$ polarization and is not actually a skin-depth effect.

10. Conclusions

(i) The canonical GDS parameter p_3 is suitable for representing (locally) the distribution of the telluric currents on a planetary scale for each of the considered periods *T*. The parameter shows a good repetitiveness in time; therefore, it is not seriously affected by errors. However, 64 observatories do not appear sufficient for providing a reasonably complete picture of the global currents.

(ii) In general, the seasonal variations in the inducing external field $\mathbf{B}_e$ are far more relevant to producing the observed **B**'s than any eventual channelling within local conductivity anomalies for the stations studied.

(iii) The mid-oceanic ridges do not appear to affect the local **B**'s at any of the 64 observatories.

(iv) The equatorial electrojet strongly affects 21 out of the 64 stations considered here. The net result is a N-S polarization of the total observed field **B** at these stations associated with a similar polarization of the external field $\mathbf{B}_e$.

(v) The skin depth for the equatorial electrojet-induced telluric currents appears to be much shallower than the nominal value evaluated for the standard flat, homogeneous Earth model, consistent with the conclusions of several authors on the role of the oceans for *Sq* induction effects.

(vi) The Parkinson plane certainly reflects significantly more the geomorphological features of the conductivity structures underground than the $\mathbf{B}_e$ field polarization, even in a region which is strongly affected by the equatorial electrojet.

Acknowledgments

We thank Prof. W. D. Parkinson for providing his 1977 Report on *Sq* analysis of IGY magnetic observatory data.

References

Beamish, D., Hewson-Browne, R. C., Kendall, P. C., Malin, S. R. C. and Quinney, D. A. (1980a), *Induction in arbitrarily shaped oceans—IV. Sq for a simple case.* Geophys. J. R. Astr. Soc. *60*, 435–443.
Beamish, D., Hewson-Browne, R. C., Kendall, P. C., Malin, S. R. C. and Quinney, D. A. (1980b),

Induction in arbitrarily shaped oceans—V. The circulation of Sq-induced currents around land masses. Geophys. J. R. Astr. Soc. *61*, 479–488.

BEAMISH, D., HEWSON-BROWNE, R. C., KENDALL, P. C., MALIN, S. R. C. and QUINNEY, D. A. (1983), *Induction in arbitrarily shaped oceans—VI. Oceans of variable depth.* Geophys. J. R. Astr. Soc. *75*, 387–398.

BULLARD, E. C. (1970), *Geophysical consequences of induction anomalies* (Summary), J. Geomagn. Geoelect. *22*, 73.

FAINBERG, E. B., *Global geomagnetic sounding*, Preprint 50a (Izmiran, Moscow 1983) 66 pp.

FAINBERG, E. B. and ZINGER, B. Sh. (1981), *Electromagnetic induction in a nonuniform spherical model of the Earth.* Phys. Earth Planet. Interiors *25*, 52–56.

FAINBERG, E. B. and SANIN, S. I. (1981), *Effect of the world ocean on the results of global magnetovariational sounding.* Izv. Acad. Sci. USSR, Phys. Solid Earth *17*, 519–523.

FAMBITAKOYE, O. (1976), *Etude des effects magnétiques de l'electrojet équatorial.* ORSTOM, Paris, 202 pp.

FILLOUX, J. H. (1982), *Magnetotelluric experiment over the ROSE area.* J. Geophys. Res. *87*, 8364–8378.

FORBES, J. M. (1981), *The equatorial electrojet.* Rev. Geophys. Space Phys. *19*, 469–504.

GREGORI, G. P. and LANZEROTTI, L. J. (1980), *Geomagnetic depth sounding by induction arrow representation: A review.* Rev. Geophys. Space Phys. *18*, 203–209.

GREGORI, G. P. and LAZEROTTI, L. J. (1986), *Geomagnetic depth sounding, 1. The problem of separating external and internal origin geomagnetic fields* (in preparation).

GREGORI, G. P., LANZEROTTI, L. J., MELONI, A. and VALENTI, C. (1982), *Studies of the external origin component of Sq by 'canonical' GDS analysis.* Annales Geophys. *38*, 339–346.

HUTTON, V. R. S. (1976), *The electrical conductivity structure of the Earth and planets.* Rep. Prog. Phys. *39*, 487–572.

GUSTAFSSON, G. (1969), *A revised corrected geomagnetic coordinate system.* Ark. Geophys. *5*, 595–617 (see also Kiruna Geophysical Observatory Report no. 694, April 1969).

JONES, A. J. (1983), *The problem of current channelling: A critical review.* Geophys. Surveys *6*, 79–122.

LAW, L. K. (1983), *Marine electromagnetic research*, Geophys. Surveys *6*, 123–135.

LAW, L. K. and GREENHOUSE, J. P. (1981), *Geomagnetic variation sounding of the asthenosphere beneath the Juan de Fuca ridge.* J. Geophys. Res. *86*, 967–978.

MATSUSHITA, S. and CAMPBELL W. H. (1967), *Physics of geomagnetic phenomena*, 2 vol., 1398 pp.

PARKINSON, W. D. (1977), *An analysis of the geomagnetic diurnal variation during the International Geophysical Year*, Bulletin 173. Bureau of Mineral Resources, Geology and Geophysics, Canberra 1977) 196 pp.

ROBERTS, R. G. (1984), *The long-period electromagnetic response of the Earth.* Geophys. J. R. Astr. Soc. *78*, 547–572.

TAKEDA, M. (1985), *UT variation of internal Sq currents and the oceanic effect during 1980 March 1–18.* Geophys. J. R. Astr. Soc. *80*, 649–659.

TSYGANENKO, N. A. (1979), *Subroutines and tables for the geomagnetic field computations.* Materials for the WDC-B2, Moscow.

TSYGANENKO, N. A. (1981), *Numerical models of quiet and disturbed geomagnetic field in the cislunar part of the magnetosphere.* Annales Geophys. *37*, 381–394.

(Received 19th March, 1986, revised 15th July, 1986, accepted 16th July, 1986)

PAGEOPH, Vol. 125, Nos. 2/3 (1987) 0033-4553/87/030393-15$1.50+0.20/0

Conductivity Profiles from Global Data

B. A. Hobbs[1]

Abstract—Six magnetic storms investigated by Jady and Marshall (1984) are here analysed in the frequency domain to yield global response estimates in the range 0.09–0.75 cpd. These responses are presented as the potential ratio Q, depth of penetration c and apparent resistivity ρ_a. The data are compared to, and included with, other global response estimates and the combined data used to derive model earth conductivity profiles by linear, nonlinear and Monte-Carlo inversion techniques.

Key words: Electromagnetic induction, Earth conductivity, inversion.

1. Introduction

The most recently available data set appropriate to the Earth's global electromagnetic induction problem is that provided by Jady and Marshall (1984), more details of which are given by Marshall (1980). In both publications a representation of the Earth's electromagnetic response was determined from the data using time domain analyses. Paterson (1983) used the same data and extended the analysis into the frequency domain, employing an autocorrelation function approach. The result of that analysis was used by Jady *et al.* (1983) to infer earth conductivity. An alternative frequency domain approach is employed here which generates a response function incompatible with that of Paterson (1983) and it is shown why the latter requires correction.

The data of Jady and Marshall (1984) consist of hourly mean values from at least 58 magnetic observatories for six magnetic storms during the period June 1964 to August 1970 (Table 1). The number of hourly values describing each storm is variable and averages around 240 hours. For each hour Marshall (1980) made a careful time-domain analysis, that expanded the scalar magnetic potential function into spherical harmonics, and which separated these harmonics into parts of external and internal origin. The storm-time variations were found to consist predominantly of a P_1^0 harmonic and the convenient tables of external and internal P_1^0 coefficients listed in Marshall (1980) are the starting point for this present analysis.

---

[1] Department of Geophysics, University of Edinburgh, Mayfield Road, Edinburgh EH9 3JZ, U.K.

Table 1

Magnetic storms analysed by JADY *and* MARSHALL *(1984)*

Storm number	Commencement date
1	7 June 1964
2	17 April 1965
3	2 February 1969
4	10 February 1969
5	8 March 1970
6	16 August 1970

2. *Storm-Time Analysis in the Frequency Domain*

The response of the Earth to magnetic variations is usually expressed as a function of frequency, f, and this response may take many forms all of which are simply related. The storm-time data most immediately give rise to the potential ratio response measure

$$Q(f) = i(f)/e(f) \tag{1}$$

where $i(f)$ and $e(f)$ are Fourier transforms at frequency f of the time series of coefficients representing the internal and external potential functions respectively. For later comparison with the work of ROBERTS (1982), $Q(f)$ will be transformed to penetration depth $c(f)$ using

$$c(f) = a(1 - 2Q(f))/2(1 + Q(f)). \tag{2}$$

A further transformation to complex apparent resisitivity $\rho_a(f)$ will be made via

$$\rho_a(f) = \mu i \omega c^2(f). \tag{3}$$

A typical storm, that of 7 June 1964, is represented in Figure 1 by plotting the external P_1^0 coefficients as functions of time. It is clear from the beginning and end of the record that to avoid leakage a standard window should be applied. A useful window such as that of the minimum 3 point Blackman-Harris (HARRIS, 1978), if applied directly to the data of Figure 1 would remove much of the details of the storm commencement, since this is near the beginning of the record. In order to avoid such loss of information, the window needs to be centred near the storm commencement. In order to do this and to exploit the long length of available records (some 240 hours), data must be generated in advance of the start time. A convenient arrangement is to reflect the storm data about the beginning of the record and then apply the selected data window, centered around the storm commencement and of comparable length to the original record. Figure 1 shows the reflected data, a Blackman-Harris window spanning 256 hours and the windowed

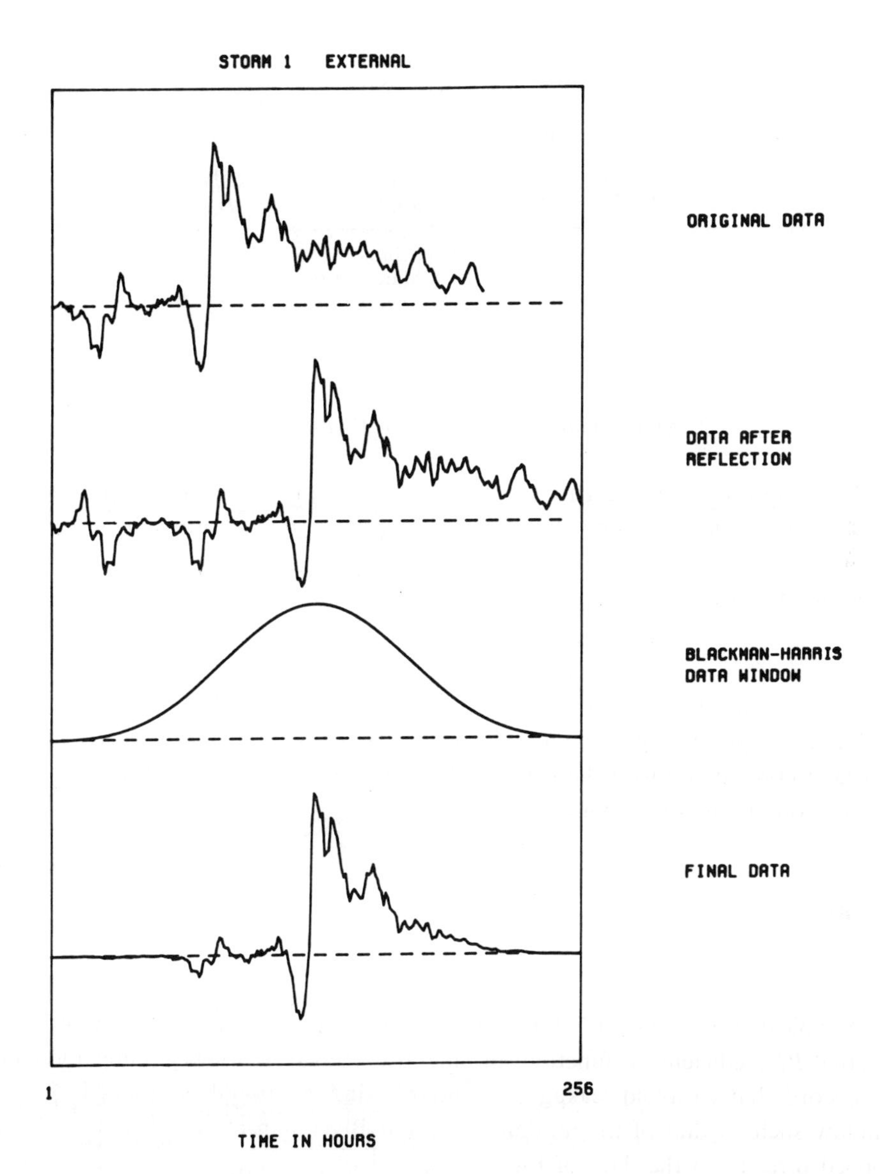

Figure 1
(a) External P_1^0 coefficients of storm 1 from MARSHALL (1980). (b) The same data after reflection about the beginning of the record and centering of the storm commencement. (c) A minimum 3-point Blackman-Harris data window spanning 256 hours. (d) The windowed data.

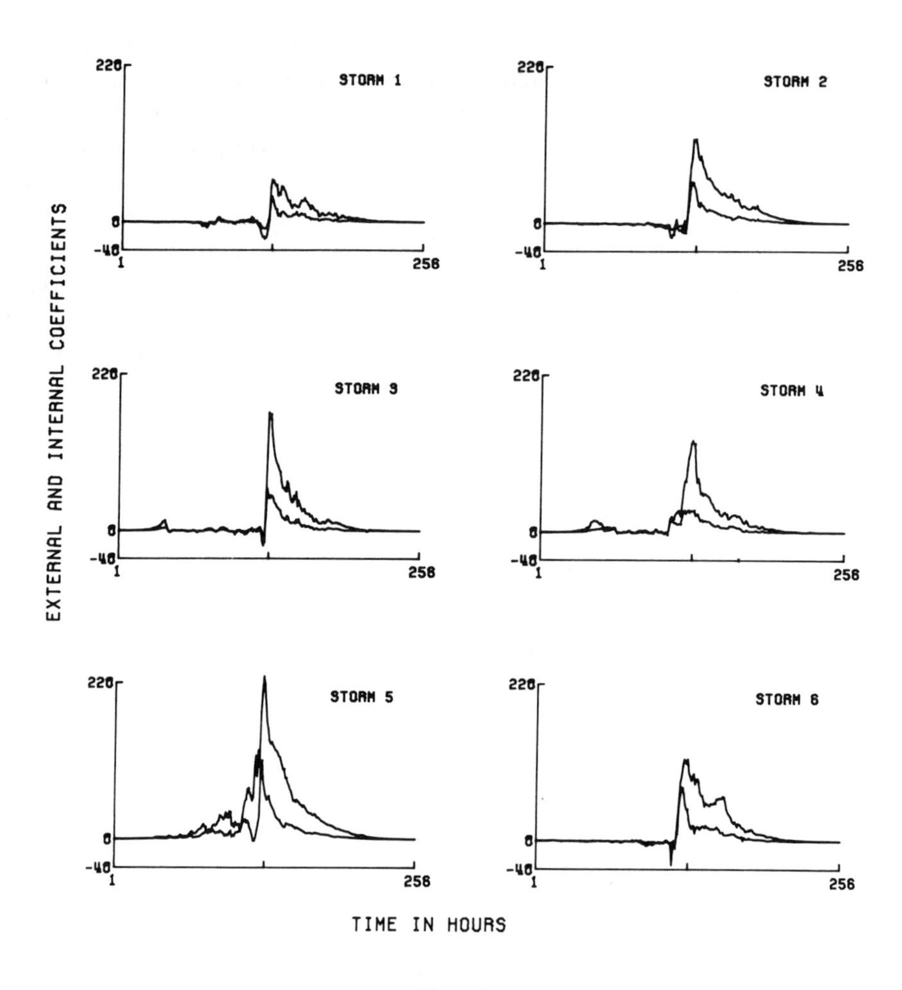

Figure 2
External and internal P_1^0 coefficients of all six storms after data reflection and windowing.

data for the external part of storm 1. Figure 2 shows the internal and external parts for all 6 storms after similar data reflection and Blackman-Harris windowing.

The storm-time data are now in a suitable form for the application of conventional Fast Fourier Transform techniques. Before such application the synthetic function

$$g(t) = a\,t\,e^{-bt} \tag{4}$$

was studied. With $a = 27$, $b = 0.1$ and t in hours this function resembles the major features of a storm such as storm 1. The Fourier transform of $g(t)$ is

$$G(f) = a/(b + 2\pi i f)^2. \tag{5}$$

From 256 hourly samples of the function $g(t)$, a standard FFT routine generates $G(f)$ accurately. To $g(t)$ was then added white noise with a variance representative of the six measured storms (MARSHALL, 1980). It was found that only the first 20 Fourier frequencies were accurately determined. These correspond to frequencies reaching 0.078 cph (the Nyquist frequency is 0.5 cph). It is therefore likely that meaningful data from the six storms will be restricted to at most this range.

To obtain the response estimate Q, the internal and external parts of each storm shown in Figure 2 were Fast Fourier Transformed and the ratio $i(f)/e(f)$ was formed at each Fourier frequency f. Figure 3 shows the amplitude and phase of this ratio for frequencies up to 0.1 cph (one-fifth of the Nyquist). The figure shows that no

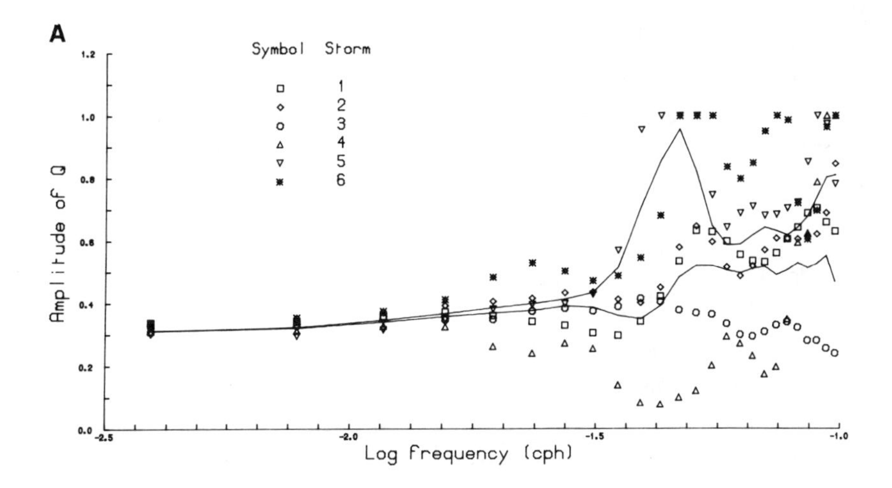

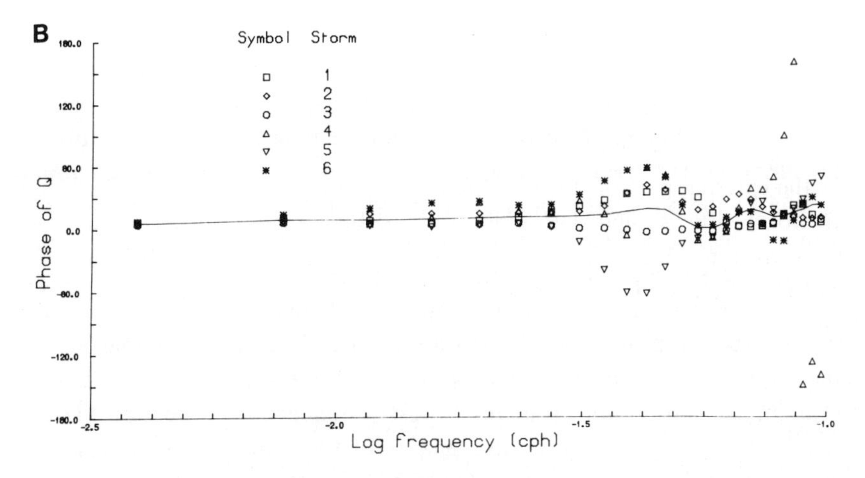

Figure 3
The potential ratio Q for each storm (A) amplitude (B) phase.

Table 2

Global response estimates

Frequency cpd	Q response Using all six storms				Q response Using storms 1,2,3,5			
	amp	δ(amp)	phase	δ(phase)	amp	δ(amp)	phase	δ(phase)
0.094	0.320	0.014	6.10	1.14	0.318	0.014	5.74	1.07
0.188	0.331	0.020	9.84	2.76	0.329	0.019	9.25	2.15
0.281	0.351	0.022	9.82	5.87	0.350	0.020	8.32	4.22
0.375	0.368	0.029	10.80	7.69	0.367	0.019	7.07	4.84
0.469	0.375	0.066	14.14	9.25	0.376	0.022	8.10	4.38
0.563	0.384	0.086	13.00	6.34	0.383	0.027	9.38	4.18
0.656	0.388	0.073	12.67	7.64	0.387	0.037	9.33	6.81
0.750	0.379	0.076	14.54	15.64	0.387	0.052	6.83	13.58

Frequency cpd	c response Using storms 1,2,3,5				Apparent resistivity Using storms 1,2,3,5			
	real	δ(real)	imag	δ(imag)	amp	δ(amp)	phase	δ(phase)
0.094	885.3	77.9	175.1	32.9	6.98	1.19	67.62	4.52
0.188	833.5	103.1	287.4	67.4	13.33	3.02	51.95	9.36
0.281	715.0	107.9	266.9	135.8	14.98	4.38	49.06	19.93
0.375	628.9	95.8	231.5	158.6	15.40	4.84	49.58	26.07
0.469	585.2	110.6	268.4	145.4	17.77	6.48	40.72	24.91
0.563	549.4	136.5	313.5	140.3	20.57	8.94	30.57	25.26
0.656	527.8	186.0	313.6	229.5	22.61	14.61	28.57	40.88
0.750	524.5	259.5	229.1	456.5	22.46	23.54	42.80	86.29

meaningful information is obtained beyond the first 8 discrete frequencies in accord with the above synthetic study. At the first 8 frequencies the phase seems well-determined but the amplitude of two storms, storms 4 and 6, depart substantially from the remaining four. An inspection of Figure 2 confirms this result. Near the storm commencement the internal part of storm 4 seems not to respond to the large external component and this is also apparent at a later time in storm 6. At the beginning of storm 6 the internal component is relatively higher than for any other storm.

MARSHALL (1980) used an alternative method to estimate Q and found for storms 3 and 5 the phase was unphysical. PATERSON (1983) concluded that storms 5 and 6 should be rejected (however it will be shown later that his analysis requires correction). According to the calculations presented here, storms 4 and 6 seem least representative. Further analyses in this paper will therefore concentrate on storms 1, 2, 3 and 5, although results for all 6 storms will be included for completeness.

The storm data are insufficient to allow band averaging and we are forced to extract as much information from the lowest 8 frequencies as possible. Treated as individual events the amplitudes and phases of the four (or six) storms may be averaged at each discrete frequency. Table 2 shows the means and standard deviations so determined. Using the four-storm averages together with Equations (2) and (3), means and standard deviations for the response measures c and ρ_a are also presented in Table 2.

3. *Comparison with Other Global Response Estimates*

A recent analysis of a global response function was made by ROBERTS (1982); who estimated $c(f)$ in the frequency range 0.014–0.333 cpd. His data set consisted of 14

A

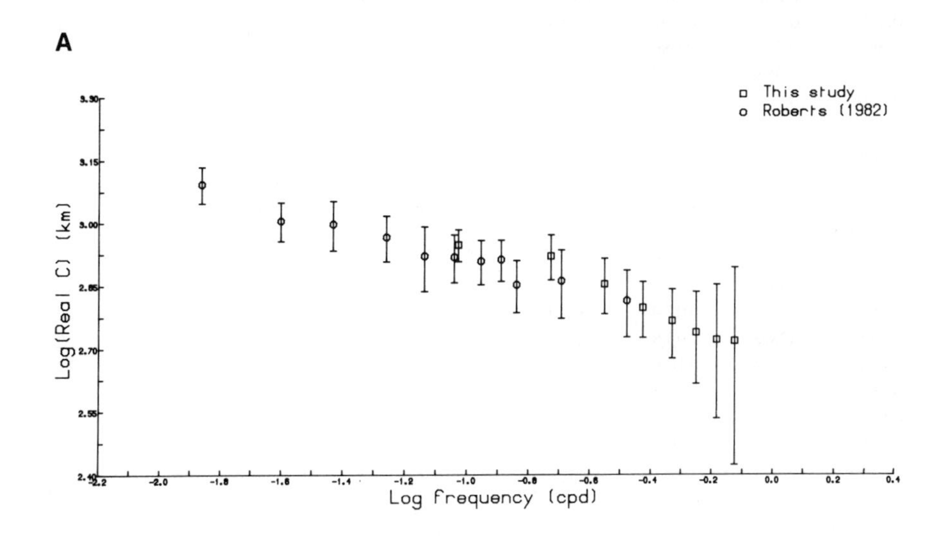

B

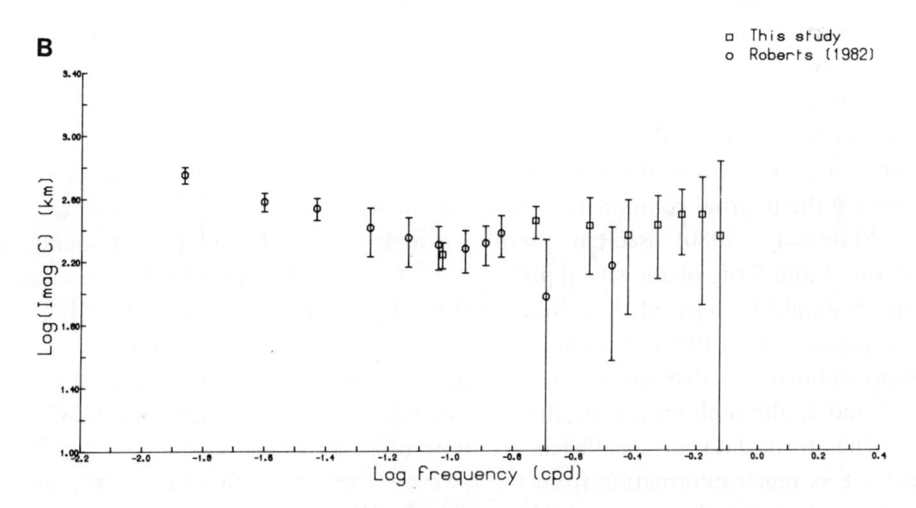

Figure 4

The penetration depth c from Table 2 combined with estimates from ROBERTS (1982). (A) Real part of c, (B) imaginary part of c.

years of hourly mean values from 12 observatories—a very different data set from that of MARSHALL (1980). Figure 4 shows the response estimate of ROBERTS (1982) and $c(f)$ from Table 2 above. There is striking agreement between these two differing data sets in both the real and imaginary parts of c, the imaginary part being, as is usually found, less well determined. (Note the different scales in Figure 4a and b).

BANKS (1969) estimated another form of the response measure which ROBERTS (1982) has converted to c. PATERSON (1983) also obtained c from the storm data.

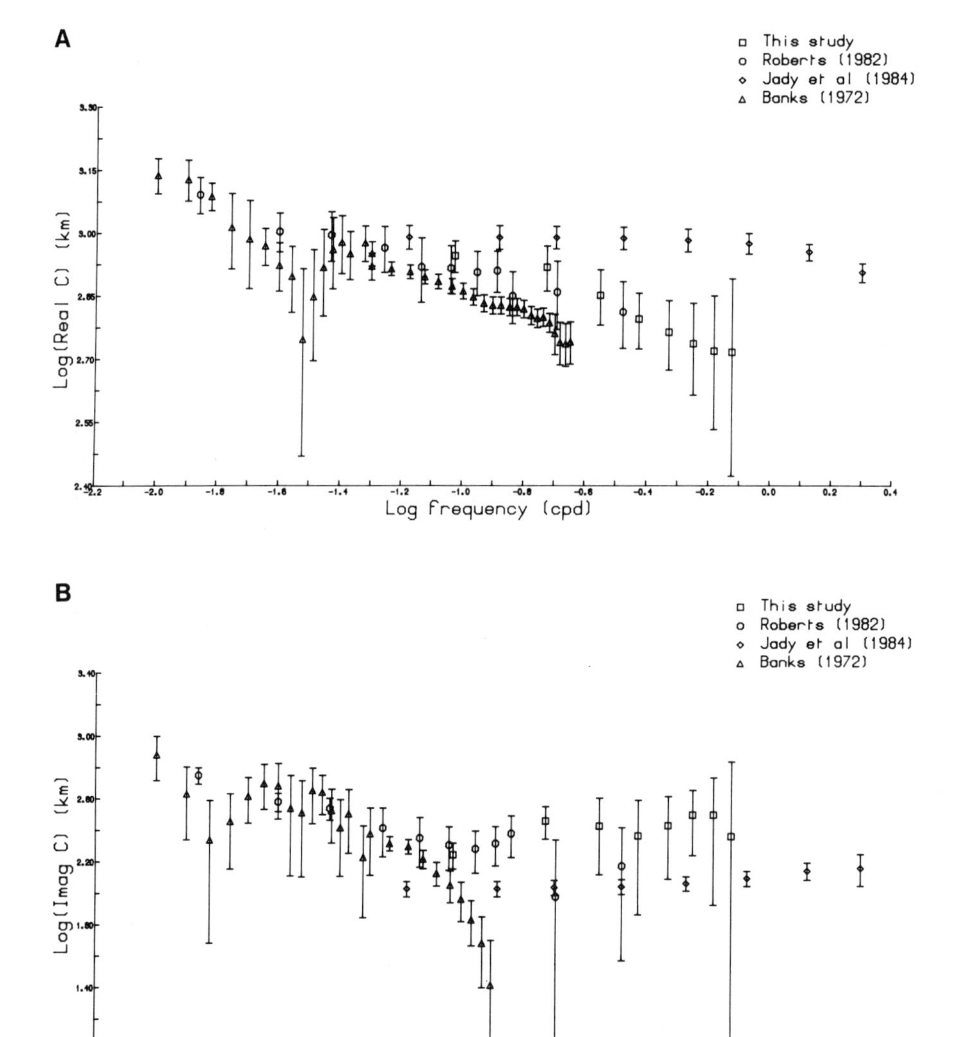

Figure 5
Comparison of the penetration depth c from four sources: (A) real part of c, (B) imaginary part of c.

The results of both analyses are added to Figure 4 to produce Figure 5. Here it can be seen that Banks' data is in reasonable agreement except for higher frequencies (where the imaginary part of c is negative and hence unphysical). However, the results of PATERSON (1983) are not in accord. A simple argument shows why this is so.

The response estimates published by JADY *et al.* (1983) are taken from PATERSON (1983), who analysed the decaying part of the six magnetic storms and applied a first order difference operator 'to detrend the data'. Unfortunately, a first order difference operator is a high pass filter, thus effectively filtering out the low frequency signals of interest. To this output was applied a 'smoothing by threes', which is a low pass filter. The combined effect is to replace the j-th term of a series $\{x_j\}$ by

$$\{x'_j\} = \{x_{j+1} - x_{j-2}\}/6. \tag{6}$$

The transfer function of this combined filter is

$$F(f) = (1/3)\sin(3\pi f/2f_c)\exp(i\pi f/2f_c) \quad 0 < f < f_c \tag{7}$$

where f is frequency and f_c is the Nyquist frequency. Thus the filter phase shifts the data and has a gain given by

$$|F(f)| = |(1/3)\sin(3\pi f/2f_c)|. \tag{8}$$

The gain is shown as the solid curve in Figure 6. Estimates of the global response function $c(f)$ were given by JADY *et al.* (1983) for f in the range $0 < f < 0.17f_c$ (a horizontal bar indicates these frequencies on Figure 6). Almost all the signal is filtered out in this range (88% rejection even in the best case $f = 0.17f_c$) and the phase is shifted by up to 10°. This has led to their incompatible result.

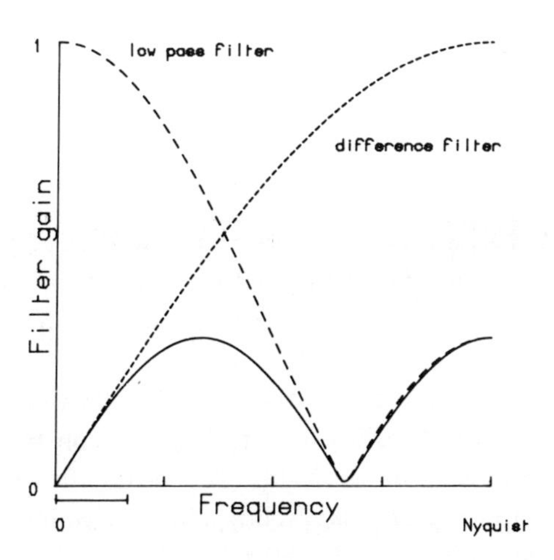

Figure 6
Gain of the filter employed by PATERSON (1983) (solid curve). Also shown are the gains for a first difference filter and for 'smoothing by threes'.

4. *Global Conductivity Estimates*

Global electromagnetic response data may be used to infer averaged values of conductivity as a function of depth by representing the Earth as spherically symmetric. It is important in this global problem, as it is in magnetotelluric and audiomagnetotelluric investigations, to use data spanning a large frequency range, several decades where possible. This present storm analysis yields estimates spanning less than one decade in frequency and has been generated for the purpose of enhancing existing response estimates rather than as an isolated set for inversion. When combined with the data of ROBERTS (1982), a data set with a reasonably useful frequency span is obtained. The inversion of the combined estimates (Figure 4) is briefly explored here by three different methods. In each method the model is a halfspace whose plane boundary represents the Earth's surface. Conductivity varies only with distance from this boundary. The plane geometry may be transformed to spherical geometry by employing the method of WEIDELT (1972). However, the conductivity solutions representing the Earth are almost identical in either plane or spherical geometry down to a depth of 1100 km, the penetration depth for this data. Thus the models given may be interpreted as those for a spherical earth.

4.1 *Nonlinear inversion*

PARKER (1980) shows that a model consisting of a finite number of δ-functions in conductivity embedded in an otherwise insulating halfspace forms the basis of a fundamental class which he denotes by D^+. The model whose responses fit a given set of data best will lie in D^+ and if this model is acceptable, measured by its χ^2 misfit, the data may be interpreted as one-dimensional. Acceptable models may be those whose value of χ^2 is less than the expected value plus say one standard deviation, that is

$$\chi^2 \leqslant N + \sqrt{2}N \tag{9}$$

for N independent data. For the data set of Figure 4, $N = 38$ thus models might be acceptable with $\chi^2 \leqslant 47$. The best fitting model in D^+ is found to have $\chi^2 = 7.49$, indicating clearly that it is possible to interpret the data with a one-dimensional conductivity distribution.

The class D^+ is of further use in estimating the effective depth of penetration below which the conductivity is arbitrary and hence unconstrained (PARKER, 1982). Figure 7 shows the relationship between this effective depth and the associated misfit χ^2. Accepting models with $\chi^2 \leqslant 47$ suggests the data determine conductivity down to a depth of approximately 1100 km.

A more physically satisfying class of models consists of horizontally layered structures in which the i-th layer has uniform conductivity σ_i and thickness h_i.

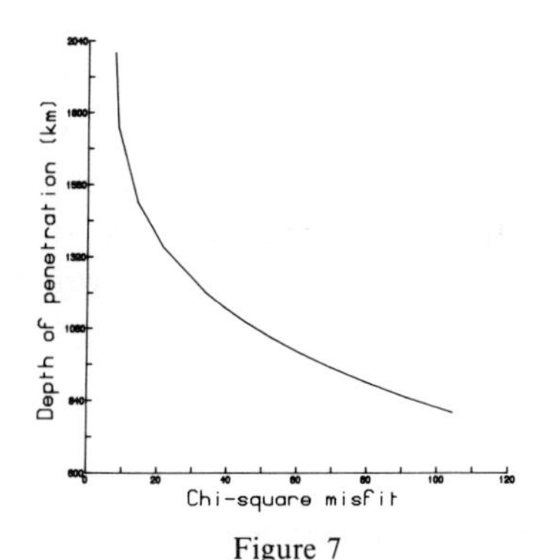

Figure 7
The relationship between misfit and penetration depth for models in D^+.

PARKER (1980) further restricts this class to that for which the product $\sigma_i h_i^2$ is the same for all layers and denotes this class H^+. Using the procedure of PARKER and WHALER (1981), a number of H^+ models were determined and two representative examples are shown in Figure 8. When the product $\sigma_i h_i^2$ is small the models resemble those in D^+ with highly conducting thin layers separated by thick regions of low

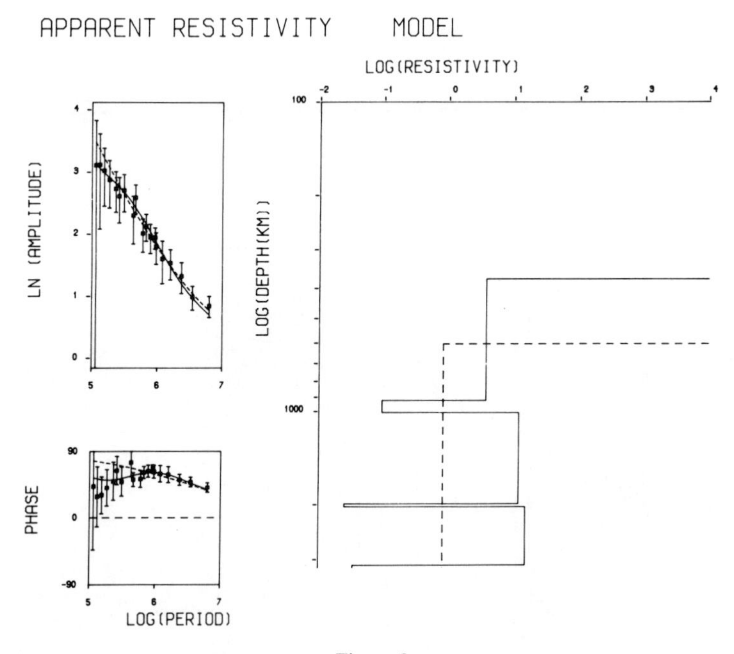

Figure 8
Two representative models (solid and dashed lines) from a range produced in H^+, together with their responses (solid and dashed curves) in terms of apparent resistivity and phase.

conductivity. Such models are associated with a small misfit and $\chi^2 = 8.90$ for that shown in Figure 8. Larger values of $\sigma_i h_i^2$ give rise to models with a few thick layers in which the conductivity variations are less extreme. A range of such models was explored and a two layer case is shown in Figure 8. Although $\chi^2 = 19.92$ for this model, a quite acceptable value, the fit does not seem somehow visually satisfying. It is easily possible to satisfy these data with two layer models that fit all the data points to within their standard errors, as will be shown later. Thus Parker's H^+ is perhaps too restricting.

4.2 *Linearised inversion with smooth conductivities*

An alternative to δ-functions and step functions in conductivity is to consider a smoothly varying profile with depth. One such model was determined by smoothing the point estimates obtained from a SCHMUCKER (1970) inversion. The Schmucker inversion provides profile information over the depth range 524–1240 km. Above and below this range uniform conductors are assumed and a simple search procedure generates the two conductivity values that yield the best fitting model. The model is shown in Figure 9 together with its response in terms of apparent resistivity and phase and the estimates of that response from Table 2. The model is an acceptable fit with

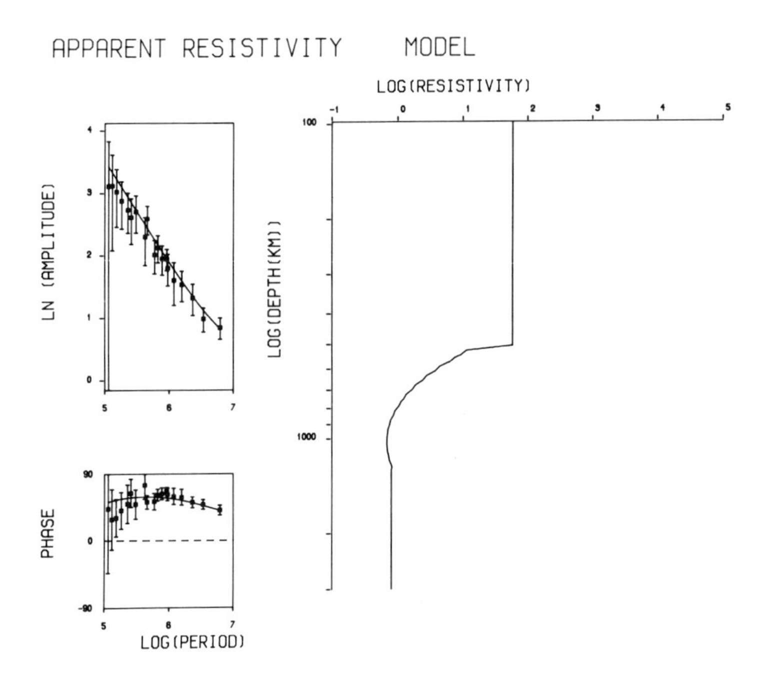

Figure 9
A smooth conductivity profile together with its response in terms of apparent resistivity and phase.

$\chi^2 = 14.70$. Had this not been the case, the model could have been improved using the method of HOBBS (1982).

Linearised inverse theory of BACKUS and GILBERT (1967), when applied to this smooth model, yields values of the averaged conductivity as a function of depth together with estimates of the resolution and errors of these averages. Above 600 km and below 1,000 km the solution is unacceptably poor. Between these depths the averaged conductivity is constrained quite well and nearly an order of magnitude increase with depth is indicated (Figure 10).

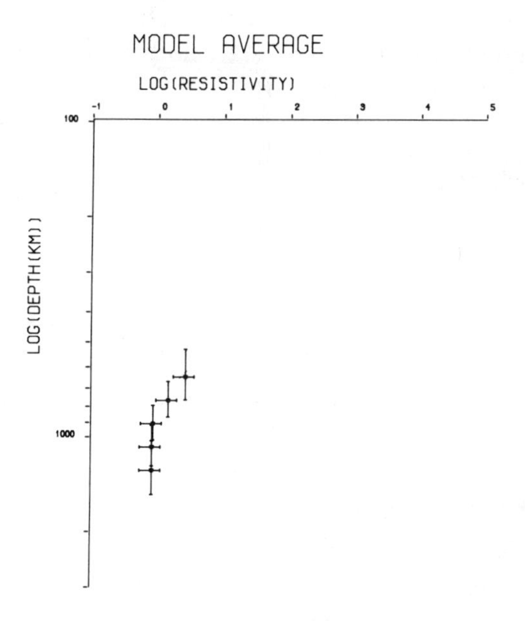

Figure 10
Model averages obtained from Backus-Gilbert inversion together with estimates of the depth resolution of these averages corresponding to a 35% error in their value.

4.3 *Monte-Carlo inversion*

For magnetotelluric and audiomagnetotelluric studies, Monte-Carlo methods may be used to explore acceptable model parameters. Using the apparent resisitivity data of Figure 9, DAWES (personal communication) calculated a variety of two-layer models yielding adequate data fits and these models are shown in Figure 11. The upper layer resistivity range is 25–45 Ωm and the underlying halfspace resistivity varies between 0.6 Ωm and 0.8 Ωm. The large increase in conductivity occurs within the depth range 637–722 km. Dawes also determined the best fitting two-layer model,

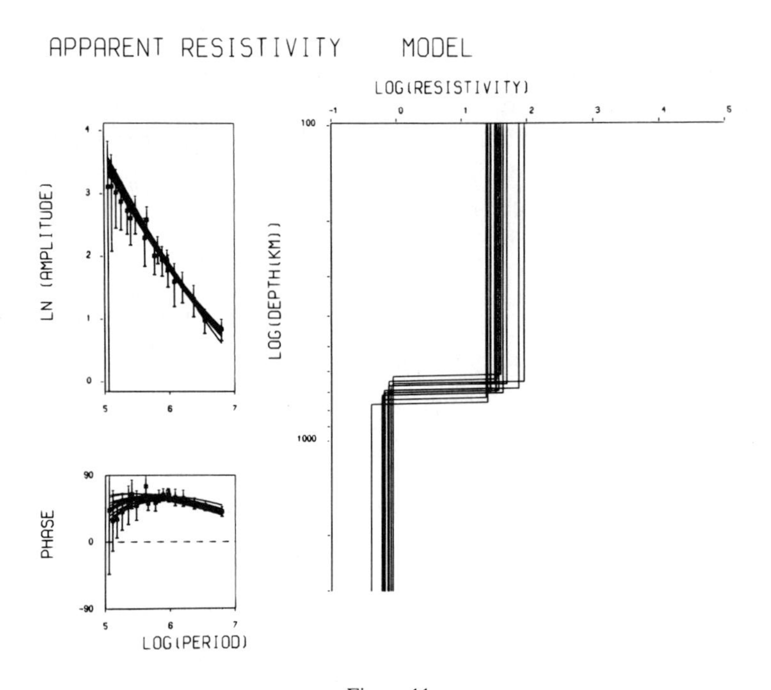

Figure 11
A range of two layer models determined in a Monte-Carlo inversion of apparent resistivity data together with their responses.

the resistivities being 25 Ωm and 0.6 Ωm with an upper layer thickness of 680 km.

The misfit for this model is $\chi^2 = 13.45$ and this type of model avoids the restrictions inherrent in H^+ solutions.

5. *Conclusions*

The analysis of storm-time data by conventional frequency domain methods yields transfer function estimates compatible with other recent results. Reliable estimates are only obtainable over a small frequency range, but these may be added to other global estimates for the purposes of inversion. Inversion of these data, combined with those of ROBERTS (1982), yielded layered, smooth and averaged models by a variety of techniques. All the models determined imply low conductivity in the upper few hundred kilometres with a rise in conductivity to approximately 1.25 Sm^{-1} at a depth of 600–700 km.

References

Backus, G. E. and Gilbert, F. (1967), *Numerical application of a formalism for geophysical inverse problems.* Geophys. J. R. Astr. Soc. *13*, 247–276.

Banks, R. J. (1969), *Geomagnetic variations and the electrical conductivity of the upper mantle.* Geophys. J. R. Astr. Soc. *17*, 457–487.

Harris, F. J. (1978), *On the use of windows for harmonic analysis with the discrete Fourier transform.* Proc. IEEE *66*, 51–83.

Hobbs, B. A. (1982), *Automatic model-finding for the one-dimensional magnetotelluric problem.* Geophys. J. R. Astr. Soc. *68*, 253–266.

Jady, R. J. and Marshall, R. T. (1984), *The analysis of geomagnetic storm-time variations.* Phil. Trans. R. Soc. Lond. A *310*, 365–406.

Jady, R. J., Paterson, G. A. and Whaler, K. A. (1983), *Inversion of the electromagnetic induction problem using Parker's algorithms with both precise and practical data.* Geophys. J. R. Astr. Soc. *75*, 125–142.

Marshall, R. T. (1980), *Geomagnetic storm-time variations and the determination of upper mantle conductivity.* Ph.D. thesis, University of Exeter.

Parker, R. L. (1980), *The inverse problem of eletromagnetic induction: existence and construction of solutions based upon incomplete data.* J. Geophys. Res. *85*, 4421–4428.

Parker, R. L. (1982), *The existence of a region inaccessible to magnetotelluric sounding.* Geophys. J. R. Astr. Soc. *68*, 165–170.

Parker, R. L. and Whaler, K. A. (1981), *Numerical methods for establishing solutions to the inverse problem of electromagnetic induction.* J. Geophys. Res. *86*, 9574–9584.

Paterson, G. A. (1983), *On the electrical conductivity of the Earth's upper mantle.* Ph.D. thesis, University of Exter.

Roberts, R. G. (1982), *The electrical response of the Earth and upper mantle conductivity.* PhD. thesis, University of Lancaster.

Schmucker, U. (1970), *Anomalies of geomagnetic variations in the south-western United States.* Bull. Scipps Inst. Oceangr. *13*.

Weidelt, P. (1972), *The inverse problem of geomagnetic induction.* J. Geophys. *38*, 257–289.

(Received 18th November, 1985, revised 4th June, 1986, accepted 10th June, 1986)

PAGEOPH, Vol. 125, Nos. 2/3 (1987) 0033-4553/87/030409-18$1.50+0.20/0

Models of Deep Electrical Conductivity Obtained From Data on Global Magnetic Variational Sounding

V. I. DMITRIEV,[1] N. M. ROTANOVA,[2] O. K. ZAKHAROVA[2] and M. V. FISKINA[2]

Abstract—An analysis of published papers containing magnetovariation sounding data (MVS) is carried out. A catalogue of the most trustworthy MVS parameters' values covering periods from 6 hours up to 11 years is compiled. With the help of averaged data of the magnetovariation sounding a planetary geoelectric profile is constructed meeting the requirements of up-to-date understanding of the physical processes in the Earth.

Key words: Global magnetic variational sounding, apparent resistance, inverse problem, planetary geoelectric profile.

1. *Introduction*

Global magnetic variational sounding, MVS, is an experimental determination of effective eletrodynamic Earth parameters such as the impedance, apparent specific resistance, etc., from the data on the natural geomagnetic field, Earth presumed spherically symmetrical. The data are obtained by frequency sounding; the variations in observations in the world network of magnetic observatories over a wide range of periods, from solar diurnal to eleven year, are processed by spherical harmonic analysis, SHA, (BERDICHEVSKY *et al.*, 1969). The MVS data are represented as global sounding curves,[3] GSC, which show the dependences of ρ_T and φ_T on the variation period T, where $\rho_T = |Z(\omega, n)|^2/\omega\mu_0$ and $\varphi_T = \text{Arg}\, Z(\omega, n)$ are used. Here ρ_T is the apparent resistivity, φ_T is the phase of spectral impedance $Z(\omega, n)$, and n is the sequential number of the spherical harmonic. The impedance is calculated either as the ratio of the amplitudes f_n^m and the difference of phases α_n^m of the internal and external parts of the scalar potential or as the ratio of the radical field component h_r to the tangential component h_θ.

Ever since the pioneering research of LAMB and SCHUSTER, new MVS data have been accumulated and processed, as in early papers by BERDICHEVSKY *et al.* (1970)

---

[1] MSU, Moscow, Lenin's Hills, 117234.

[2] IZMIRAN, Troitsk, Moscow District, 142092.

[3] These curves are in fact obtained from experimental points.

and ROKITYANSKY (1972). The most complete catalog of data available up to 1977 has been compiled in the research by FAINBERG *et al.* (1977) and later updated with SHA data (ROKITYANSKY, 1981; FAINBERG, 1983). This paper will analyze the MVS data available through 1984. The most credible parameters for periods spanning six hours to 11 years are cataloged in order to trace a new GSC where the scatter of experimental points is not very large. The effective parameters are estimated from mean values of the sample more accurately than in earlier research. This GSC is used to obtain a planetary geoelectric profile; the chief objective of deep geoelectricity studies.

2. *Global Sounding*

2.1. *Selection of most credible data*

Like the papers of BERDICHEVSKY *et al.* (1970), ROKITYANSKY (1972, 1981), FAINBERG *et al.* (1977), and FAINBERG (1983), this part of the paper will be an interpretation of the reported data subjected to spherical harmonic analysis according to the researcher's objectives and experimental potential. The significance of every group of MVS parameters depends on numerous factors which are hard to control, therefore the choice of data for the GSC is very important. We have analyzed about 60 articles published in 1919–1984 and a total of 298 values of MVS parameters. All of the parameters in each paper was analyzed so as to obtain most credible values. First, those MVS data were rejected which did not satisfy at least one of the inequalities $f_n^m > (n + 1)/n$, $|Z(\omega, n)| < \mu_0 \omega R/n + 1)$, or $|(h_Z/h_\theta) t_{\alpha n} \theta| > 1$, which follow from the theory of electromagnetic induction in a spherically symmetrical conducting ball. Then an adequate frame of reference for the SHA and harmonics, which would stably represent the spatial structure of the variation spectrum, was chosen for each case. The variability of MVS parameters, as a function of geomagnetic disturbance in the solar cycle, was recognized. The data of regional rather than global sounding were rejected as were the data of the so-called anomalous observation locations around which the field variations are strongly disturbed by local geoelectric inhomogeneities. Only MVS data on aperiodic variations obtained for the first zonal harmonic were used. Finally, we corrected obvious errors in the pioneering papers.

2.2. *A catalog of MVS parameters*

In the catalog composed of Tables 1–5 only the selected data are included, which account for almost half of the MVS parameters to be analyzed. In the figures + denotes the Schmucker continuum, ×—the Banks continuum, ·—the findings of spherical analyses, and ○—the mean values over the period.

Table 1

Final values of MVS parameters obtained from analysis of S_q-variations.

f_n^m	α_n^m	ρ_T	φ_T	Author(s)
		$\sqrt{T} = 147\ s^{1/2}, n = 5, m = 4$		
2.10	19.5	50.5	−61.0	BORISOVA *et al.* (1974)
2.29	22.0	50.5	−61.0	"
2.27	26.0	69.0	−58.0	"
2.15	30.2	55.2	−50.5	BERDICHEVSKY *et al.* (1970)
2.70	23.0	85.9	−67.5	CHAPMAN (1919)
2.20	27.0	66.5	−55.9	MALIN (1973)
2.80	15.0	80.6	−75.4	MALIN and GUPTA (1977)
2.10	27.0	61.2	−53.5	MALIN *et al.* (1975)
2.80	25.0	94.4	−67.0	"
3.05	9.2	88.8	−81.9	FAINBERG (1975)
1.76	29.0	48.4	−40.3	WINCH (1981)
2.00	25.0	52.1	−52.7	SCHMUCKER (1984)
		$\sqrt{T} = 170\ s^{1/2}, n = 4, m = 3$		
2.33	24.0	69.0	−60.0	BORISOVA *et al.* (1974)
2.19	26.0	66.5	−55.5	"
2.23	21.5	59.3	−60.0	"
2.22	12.7	53.2	−72.7	BERDICHEVSKY *et al.* (1970)
2.10	17.5	45.5	−61.9	PARKINSON (1974)
2.11	28.0	65.6	−50.8	AFRAIMOVICH *et al.* (1966)
2.50	21.0	73.6	−65.5	CHAPMAN (1919)
2.20	11.5	44.3	−72.1	MATSUSHITA and MAEDA (1965)
2.20	21.0	56.5	−60.0	MALIN (1973)
2.50	13.0	63.0	−73.9	MALIN and GUPTA (1977)
2.20	21.0	56.5	−60.0	MALIN *et al.* (1975)
2.50	19.0	70.5	−67.4	"
2.42	20.6	68.4	−64.6	FAINBERG (1975)
2.25	14.0	49.6	−69.5	HASEGAWA and OTA (1950)
2.10	2.7	62.8	−51.5	SCHMUCKER (1974)
2.31	15.3	54.6	−68.8	JADY (1974)
		$\sqrt{T} = 208\ s^{1/2}, n = 3, m = 2$		
2.43	21.5	67.5	−61.5	BORISOVA *et al.* (1974)
2.53	19.5	68.5	−65.5	"
2.32	16.0	52.2	−65.0	"
2.30	3.1	40.4	−84.5	BERDICHEVSKY *et al.* (1970)
2.16	18.0	45.7	−59.5	AFRAIMOVICH *et al.* (1966)
2.30	5.0	40.8	−81.7	BEN'KOVA (1941)
2.20	18.0	48.0	−60.6	CHAPMAN (1919)
2.30	14.0	48.1	−68.0	MATSUSHITA and MAEDA (1965)
2.40	16.0	56.4	−67.1	MALIN (1973)
2.40	13.0	52.7	−70.9	MALIN and GUPTA (1977)
2.40	16.0	56.4	−67.1	MALIN *et al.* (1975)
2.50	17.0	63.6	−67.5	"
2.24	19.0	51.8	−60.3	FAINBERG (1975)
2.20	23.0	57.0	−54.7	WINCH (1981)
2.43	10.0	51.5	−75.4	HASEGAWA, and OTA (1950)
2.20	16.0	45.0	−63.2	PARKINSON (1974)
2.25	15.1	46.6	−65.5	JADY (1974)
1.20	27.0	66.0	−50.8	SCHMUCKER (1974)

Table 1 *continued*

f_n^m	α_n^m	ρ_T	φ_T	Author(s)
		$\sqrt{T} = 294\ s^{1/2}, n = 2, m = 1$		
2.53	8.0	38.0	−76.5	Borisova *et al.* (1974)
2.53	10.5	39.5	−72.0	"
2.40	1.0	31.0	−76.0	"
2.68	5.4	44.7	−81.8	Berdichevsky *et al.* (1970)
2.54	18.0	48.8	−62.4	Afraimovich *et al.* (1966)
2.80	13.0	54.9	−72.4	Chapman (1919)
2.80	13.0	54.9	−72.4	Matsushita and Maeda (1965)
2.70	0.0	43.3	−90.4	Malin (1973)
3.00	6.0	59.3	−82.6	Malin and Gupta (1977)
2.70	13.0	50.1	−71.3	Malin *et al.* (1975)
2.53	7.1	37.1	−78.0	Fainberg (1975)
2.59	2.0	38.1	−86.7	Winch (1981)
2.34	5.0	27.1	−79.7	Ben'kova (1941)
2.40	7.2	31.0	−76.2	Parkinson (1974)
2.30	9.0	27.6	−71.33	Hasegawa and Ota (1950)
2.40	18.0	42.3	−59.3	Schmucker (1954)
2.70	13.4	50.5	−70.7	Jady (1974)

2.3. The global sounding curve

Figure 1 shows the GSC obtained from all the MVS data to be analyzed. The scatter of points is obvious. The GSC of Figure 2 was obtained from most credible MVS data.

The scatter in Figure 2 is visibly smaller but so is the sample of experimental ρ_T and φ_T. Recall that reduction of the scatter and the sample have opposing effects on estimating.

Our analysis of experimental data showed, however, that the estimation accuracy has increased. Thus the error in estimating the 'true' value on the sample boundaries has reduced by a factor of one and a half to two.

3. A Model of Deep Electrical Conductivity

Before proceeding to solve the inverse problem, let us first describe briefly the current geophysical picture of the mineralogical composition and physical interpretation of conductivity as a function of depth.

3.1. Choice of a model

Studies of rocks at high temperatures and pressures have revealed that to the depth of 400 km the mantle is pyrolytic, olivine being the chief component. As

Table 2

Final MVS parameter values obtained from analysis of D_{st} variations.

$\sqrt{T}$	f_n^m	α_n^{0m}	ρ_T	φ_T^0	Author(s)
339	2.86	11.6	44.0	−64.9	DEVANE (1977a)
480	2.90	10.3	22.9	−66.4	
679	3.03	7.6	12.9	−74.2	
960	3.23	5.5	7.7	−79.9	
1,358	3.28	2.7	3.9	−85.2	
1,920	3.29	1.3	2.0	−87.7	
480			20.0	−80.0	DEVANE (1977b)
480			30.0	−69.0	
680			11.0	−81.0	
680			14.0	−78.0	DEVANE, (1977b)
960			5.2	−85.0	
960			6.8	−83.0	
1,360			2.5	−87.0	
1,360			3.4	−85.0	
390			26.0	−66.0	DEVANE (1978)
440			22.0	−71.0	
510			21.0	−73.0	
630			16.0	−75.0	
890			9.2	−81.0	
930			7.6	−77.0	
1,040			7.0	−78.0	
1,200			5.5	−76.0	
1,470			4.2	−72.0	
416	2.41	40.7	25.0		LAGUTINSKAYA *et al.* (1975)
495	2.85	9.5	15.0	−66.9	
600			10.5		
624			17.0		
688			8.4		
455			28.0		FAINBERG *et al.* (1976)
515			18.3		
540			19.0		
600			16.4		
680			12.6		
760			11.5		
945			11.0		
1,380			6.5		
720			11.7		BERDICHEVSKY, *et al.* (1970)
882			9.7		
1,250			6.5		
1,765			3.3		
379			18.3		ROTANOVA, and CHIMIDDORZH (1977)

pressure increases, the olivine can take on a spinel-like structure, as shown initially by RINGWOOD and MAGOR (1970). This transformation is believed to occur at depths of 420 km and to be responsible for the anomalous increase in the seismic wave velocity and material density. A second phase transformation occurs at 670 km

Table 3

Final values of MVS parameters from analysis of 27 days variations.

$\sqrt{T} = 1{,}527\ s^{1/2}$		$\sqrt{T} = 1{,}082\ s^{1/2}$		$\sqrt{T} = 882\ s^{1/2}$		Author(s)
4.8		7.5		10.0		ECKHARDT *et al.* (1963)
4.0	−72.0	4.6	−66.0	6.4	−81.0	BANKS (1969)
5.34	−70.5					RIKITAKE (1951)
3.7	−71.0	4.6	−50.0	6.4		DZHODENCHUKOVA (1975)

where the velocity and density again increase, but this change is attributed to the increased silicone coordination number.

The behavior of electical conductivity σ during phase transitions is difficult to interpret because of the lack of reported experimental data. Still, the Ringwood

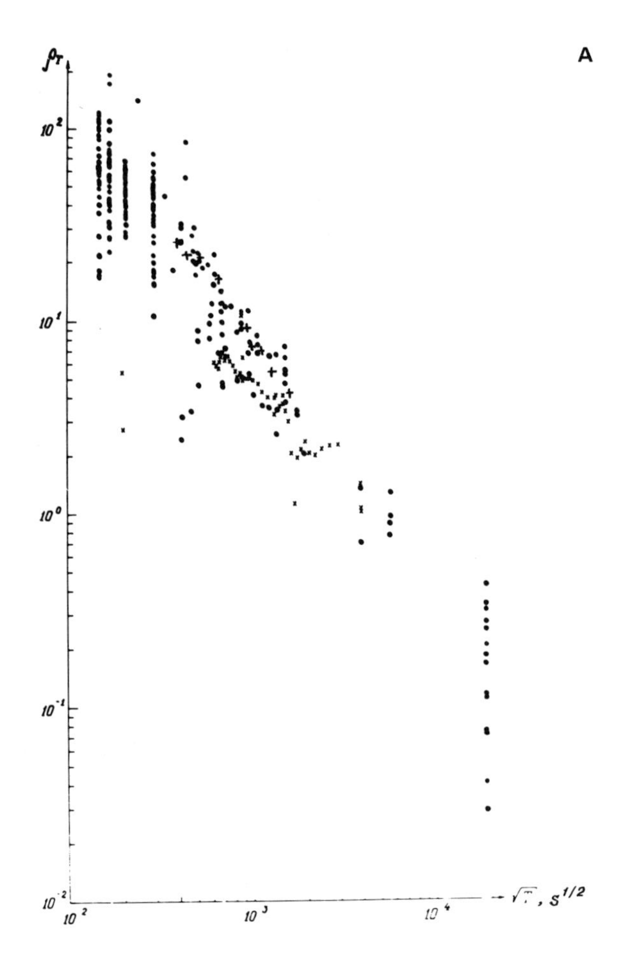

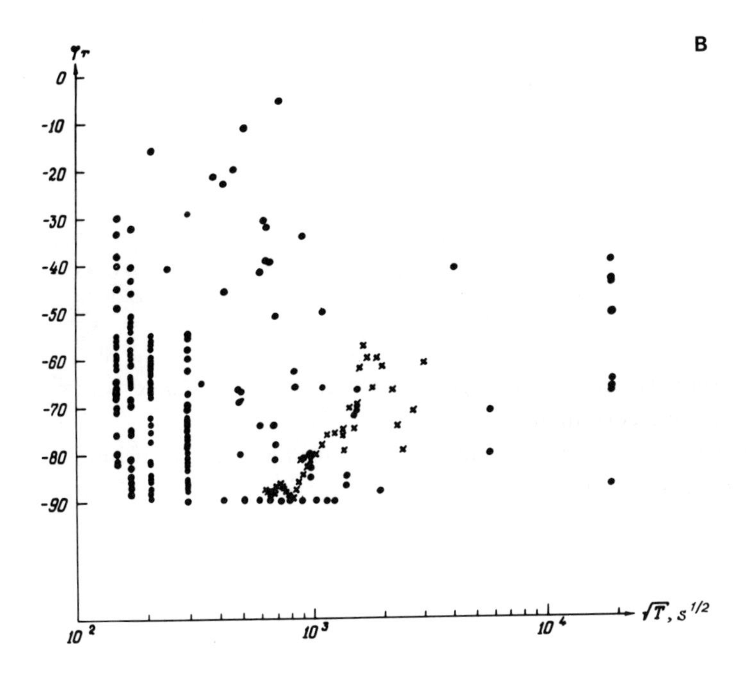

Figure 1
Global sounding data for all analyzed MVS results. ×—Banks continuum; ●—all other results.

Table 4

Final results of global sounding for half-year and one year periods.

Magnitude of h_r/h_θ	Argument of h_r/h_θ	ρ_T	φ_T	Author(s)
		$\sqrt{T} = 3{,}976\ s^{1/2}$, $n = 1$, $m = 0$		
0.45	49.0	1.03	−41.0	BANKS (1969)
0.52	49.0	1.37	−41.0	CURRIE (1966)
		0.80		ECKHARDT *et al.* (1963)
		$\sqrt{T} = 5{,}608\ s^{1/2}$, $n = 2$, $m = 0$		
1.78	170	0.89	−80	BANKS (1969)
212	10	1.27	−80	CURRIE (1966)
f_n^m	α_n^m			
7.29	73.3	0.95	−71	MALIN, and ISIKARA (1974)
9.52	41.0	0.75	−80	

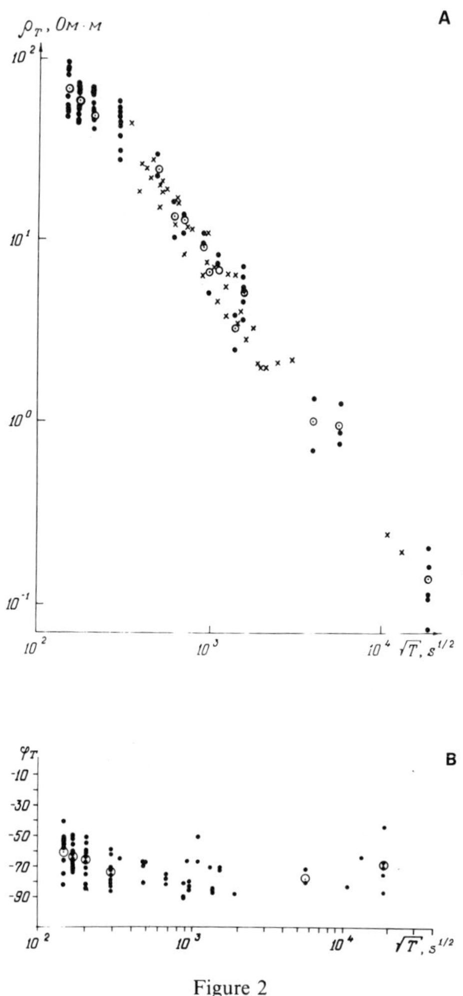

Figure 2
Global sounding data from most credible MVS parameters values. +—Schmuker continuum; ●—other results of spherical analyses; ⊙—data averaged over the period.

diagram (1977) and the experiment of AKIMOTO and FUJISAWA (1965), which demonstrated an increase of σ by two orders of magnitude in the case of a polymorphic phajalite—spinel transition, showed that electrical conductivity, as well as seismic propagation rate and density, can dramatically increase in the phase transition zone.

Another important factor in the choice of an electric conductivity model is physical interpretation in the light of studies at high temperatures and pressures. As temperature and pressure increase, olivines cease to be insulators and become semiconductors which can display impurity, ionic or electronic conductivity. The impurity type occurs at temperatures below 1,300 K and only in the first tens of kilometers, and so is not significant at mantle depths. The upper mantle shows the ionic type

Table 5

Final values of MVS parameters for an 11-year period.

f_n^m	α_n^{0m}	ρ_T	φ_T^0	Author(s)
		$\sqrt{T} = 18{,}625\ s^{1/2}$, $n = 1$, $m = 0$		
7.69	61	0.165	−69.3	YUKUTAKE (1965)
5.27	22	0.073	−74.2	HARWOOD, and MALIN (1977)
8.33	8	0.107	−76.6	
3.57	77	0.204	−43.6	ISIKARA (1977)
5.88	44	0.114	−66.6	COURTILLOT, and LE MOUEL (1976)
		$\sqrt{T} = 13{,}170\ s^{1/2}$, $n = 1$, $m = 0$		
5.47	38	0.195	−66	do
		$\sqrt{T} = 10{,}800\ s^{1/2}$, $n = 1$, $m = 0$		
6.31	12	0.247	−83.0	do

conductivity which plays a major role in the 'conducting layer' and its effect decreases against pressure at great depths (LYUBIMOVA, 1968). As a result, σ decreases towards the boundary with layer C but at about 400 km the conductivity increases again. This is attributed to an additional conductivity mechanism of olivine phase transitions, and ionic conductivity is replaced by electron. The lower mantle is an electronic conductor (ZHARKOV, 1983).

This reasoning confirms the adequacy of our conduction model (DMITRIEV *et al.*, 1977a). Therefore the new GSC will be interpreted by a parametric model (model 1) whose analytical expression is

$$\sigma(r) = \sigma_0 + \sigma_3 \exp\left[-\frac{E_1}{2kT(r)}\right]\cdot\frac{1}{2}\left(1 + th\frac{R - r - H_1}{Q_1}\right) + \sigma_4 \exp\left[-\frac{E_2}{2kT(r)}\right]\cdot\frac{1}{2}\left(1 + th\frac{R - r - H_2}{Q_2}\right) \quad (1)$$

where E is the activation energy and k is the Boltzmann constant,

$$\sigma_0 = \begin{cases} \sigma_1, 0 < R - r \leqslant 4 \text{ km} \\ \sigma_2, 4 < R - r < H_1 - Q_1 \end{cases}$$

H_1 and H_2 are middles of the first and second jumps
Q_1 and Q_2 are slopes of the jumps.

3.2. Solution of the inverse global sounding problem

The experimental data are a set of values of ρ_T which is represented as a discrete set $\{\rho_T^i(T_j)\}_{i=1}^l$, $j = \overline{1, L}$ where L is the number of periods and l is the number of values at every j-th period.

In plotting a $\sigma(r)$ curve which would best account for the data on ρ_T let us employ the procedure for solving the inverse problem (DMITRIEV *et al.*, 1977a) but with another functional $\varepsilon(\bar{P})$. The experimental values of ρ_T as a function of T change by four orders of magnitude, but we want to approximate them with the same relative accuracy. For this purpose DMITRIEV (1979) proposed using a transformant which reduces the order of ρ_T and transforms the functional of accumulated error to the form.

$$\varepsilon(\bar{P}) = \frac{1}{L}\sum_{j=1}^{L}\left\{\left[\ln \frac{\rho_T(\bar{P}, T_j)}{\rho^{\text{exper}}(T_j)}\right]^2 \cdot \ln \frac{T_{j+1}}{T_j}\right\}, \tag{2}$$

where $\bar{P}$ is the vector of unknown parameters in model (*I*) and $\ln (T_{j+l}/T_j)$ are in fact weighting coefficients.

New data on ρ_T and the procedure of solving the inverse problem lead to an optimal vector $\bar{P}$ for which $\varepsilon(\bar{P})$ adopted a minimal value. The geoelectrical profile represented by an optimal curve of $\rho_T^{\text{opt.}}$ has the form

$$\sigma(r) = \begin{cases} \sigma_1 = 1 \text{ Sm/m}, \quad 0 < R - r \leqslant 4 \text{ km}, \\ \sigma_2 = 10^{-3} \text{ Sm/m}, \quad 4 < R - r < H_1 - Q_1, \\ \sigma_3 \times \exp[-E_1/2kT]\frac{1}{2}\left(1 + th\dfrac{R - r - H_1}{Q_1}\right), \\ \qquad\qquad H_1 - Q_1 < R - r < H_1 + Q_1, \\ \sigma_3 = 0,7 \text{ Sm/m}, \quad H_1 = 660 \text{ km}, \quad Q_1 = 100 \text{ km} \\ \sigma_4 \times \exp[-E_2/2kT]\frac{1}{2}\left(1 + th\dfrac{R - r - H_2}{Q_2}\right), H_1 + Q_1 < R - r \\ \sigma_4 = 17,6 \text{ Sm/m}, \quad H_2 = 1{,}300 \text{ km}, \quad Q_2 = 260 \text{ km}. \end{cases} \tag{3}$$

DMITRIEV *et al.* (1977b) have shown that the parameters responsible for the second singularity in the curve or $\sigma(r)$ vary within a wide range. This suggests that this singularity is, rather than an abrupt change, a monotone increase of $\sigma(r)$ which stands for a wide area of adiabatic temperature distribution starting a depths of about 1000 km and ending at the second boundary layer in the vicinity of the core-mantle boundary. For this reason, the second singularity in model I is replaced by exponential increase. As a result we have model II,

$$\sigma(r) = \sigma_0 + \sigma_3 \times \exp[-E/2kT]\cdot\frac{1}{2}\left(1 + th\frac{R - r - H}{Q}\right) + \sigma_3 \times \exp\left(-l\frac{R - r}{R}\right), \tag{4}$$

where l = const. Following calculation, similar to the previous case, we have numerical values of the model parameters

$$\sigma(r) = \begin{cases} \sigma_1 = 1 \text{ Sm/m}, & 0 < R - r \leqslant 4 \text{ km} \\ \sigma_2 = 10^{-3} \text{ Sm/m}, & 4 < R - r \leqslant H - Q \\ \sigma_3 \times \exp\left[-E/2kT\right] \cdot \frac{1}{2}\left(1 + th\frac{R - r - H}{Q}\right), & \\ \qquad H - Q < R - r < H + Q, & \\ \sigma_3 = 0,9 \text{ Sm/m}, \quad 660 \text{ km}, & Q = 65 \text{ km}, \\ \sigma_3 \times \exp\left(-l\frac{R - r}{R}\right), & \text{H} + Q < R - r \\ l = 22 & \end{cases} \tag{5}$$

Let us consider two more models.

Model III, exponential increase of electrical conductivity with depth

$$\sigma(r) = \bar{\sigma}_0 + \sigma_1 \times \exp\left(-l\frac{R - r}{R}\right), \qquad \sigma_0 = \begin{cases} \sigma_{01}, 0 < R - r \leqslant 4 \text{ km} \\ \sigma_{02}, 4 < R - r \leqslant H \end{cases} . \tag{6}$$

The optimal parameters are $\sigma_{01} = 0.7$ Sm/m, $\sigma_{02} = 0.19 \times 10^{-2}$ Sm/m, $l = 55$, and $H = 140$ km.

Model IV is obtained from the simple dependence of electrical conductivity on temperature

$$\sigma(r) = \bar{\sigma}_0 + \sigma_1 \times \exp\left[-E/2k\left(T_0 + T_1 th\frac{R - r}{\alpha}\right)\right]. \tag{7}$$

In this model the unknowns are $\bar{P} = \bar{P}(\bar{\sigma}_0, \sigma_1, E, T_1, \alpha)$ which provide the functional minimum at $\sigma_{0_1} = 1$ Sm/m, $\sigma_{0_2} = 10^{-3}$ Sm/m, $\sigma_1 = 14.4$ Sm/m, $E = 2.1$ eV, $T_1 =$ 2,100 K, $\alpha = 730$, and $H = 430$ km.

An insight into the truth of the unique distribution of $\sigma(r)$, chosen by optimal models, is provided by the extent to which the theoretical ρ_T curve represents the experimental data; the model in which the functional $\varepsilon(\bar{P})$ takes on the lowest value is chosen. The numerical values of the functional $\varepsilon(\bar{P})$ for the four models are given in the first row of Table 6.

Model II thus provides the best representation of the experimental values of ρ_T and *a priori* data on the geoelectrical Earth profile. In it $\varepsilon(\bar{P})$ is the lowest.

In obtaining an optimal model of the electrical conductivity function, the experimental data were the mean values of ρ_T for each period. Using the mean values alone calls for confirmation that the optimal model is true, and above all, determining whether the optimal parameters vary widely for every model if the experimental data is a continuous variation of ρ_T.

To obtain a continuous analog, the initial values of ρ_T were approximated by splines and polynomial regression and the inverse problem was solved in the same way, by using parameteric models I–IV. Their parameters were found to change but slightly when digital values of ρ_T are replaced by their continuous analog.

The numerical values of the functional $\varepsilon(\bar{P})$ for the spline are given in the lower row of Table 6. The minimal value of the accumulated error is again provided by model II in which the electrical conductivity changes abruptly in the transition layer and monotonically increases in the lower mantle.

The distributions of $\sigma(r)$, as determined by models I–IV, are shown in Figure 3a as is the distribution of electrical conductivity obtained indirectly by ZHARKOV (1983). The curves ρ_T for the models are shown in Figure 3b.

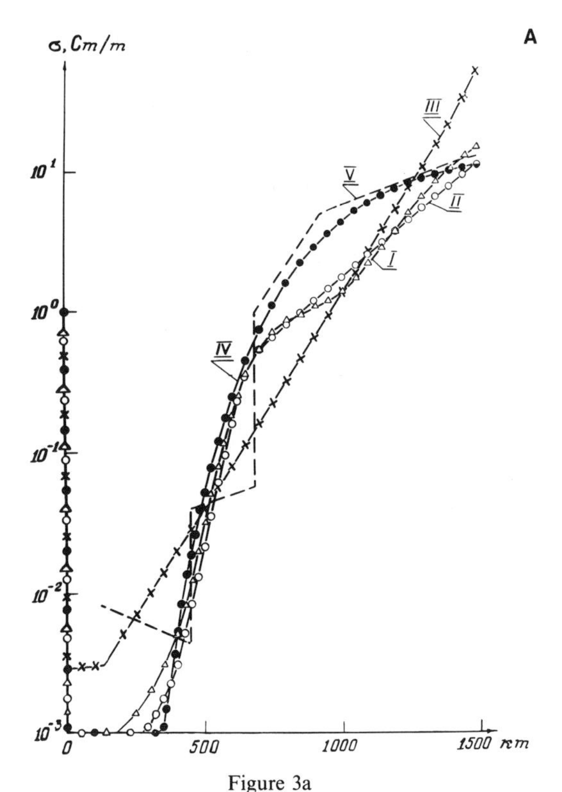

Figure 3a
Electrical conductivity distributions found by parametric models I–IV. V—distribution obtained indirectly by physical interpretation (ZHARKOV, 1983).

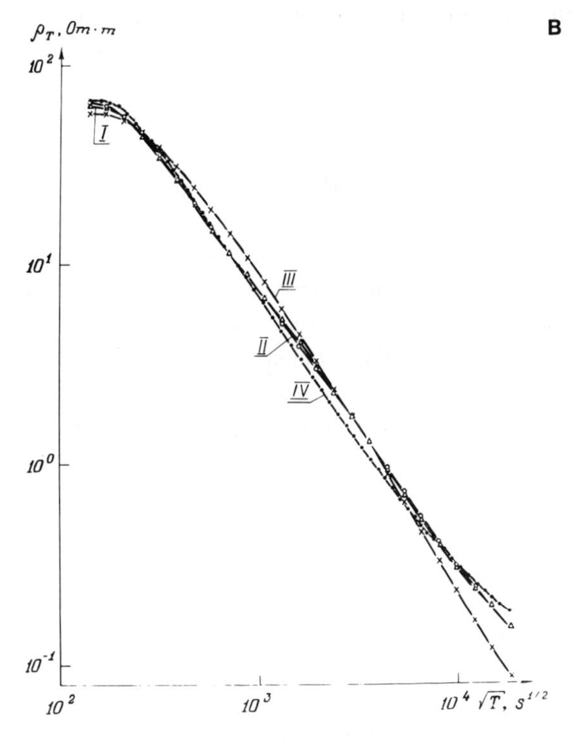

Figure 3b
Plots of apparent resistivity associated with the conductivity distributions of Figure 3a.

3.3. *On resolution of the parametric model*

Let us estimate resolution of the parameters introduced into formula (4). Transform the functional (2) in the following way:

$$\varepsilon(\bar{P}) = \sum_{j=1}^{L} \ln \frac{T_{j+1}}{T_j} (\ln \rho_T^{\text{exper}} - \ln \rho_T^{\text{theor}}(\bar{P}))^2 = \sum_{j=1}^{L} \frac{T_{j+1}}{T_j} (\ln \rho_T^{\text{exper}} - \ln \rho_T^{\text{opt}} + \ln \rho_T^{\text{opt}} - \ln \rho_T^{\text{theor}})^2 \leqslant \sum_{j=1}^{L} \ln \frac{T_{j+1}}{T_j} (\ln \rho_T^{\text{exper}} - \ln \rho_T^{\text{opt}})^2 + \sum_{j=1}^{L} \ln \frac{T_{j+1}}{T_j} (\ln \rho_T^{\text{opt}} - \ln \rho_T^{\text{theor}})^2. \tag{8}$$

In equation (8) the first sum is known in solving the inverse problem (Table 6); denoting the second sum as $\Delta(\bar{P})$. Let us see how the values of this functional vary with every parameter $\bar{P}$.

Table 6

Models	I	II	III	IV
$\varepsilon(\bar{P})$				
deviation from the mean	0.048	0.047	0.076	0.049
from the spline	0.003	0.002	0.044	0.015

To determine the range of each parameter, the numerical value of $\Delta(\bar{P})$ must be chosen. Let $\Delta(\bar{P}) = 0.002$ and assume that the geoelectrical profiles coincide with $\|\ln \rho_T^{\text{opt}} - \ln \rho_T^{\text{theor}}\| \leq 0.002$. The associated variation of all the parameters are shown in Table 7.

All optimal model parameters are determined somewhat stably even though the value of the functional, $\varepsilon(\bar{P}) = 0.004$, is twice the optimum. This confirms the truth of our model of electrical conductivity in which σ changes abruptly at depths of 400 to 600 km.

Table 7

Model parameters	σ	H	Q	l
Range	0.7 ± 0.17 Sm/m	660 ± 10 km	90 ± 30 km	24 ± 0.7 km

4. *Discussion*

The existence of a conductivity jump in transitional layer C has been ascertained in the classical Lamb and Lahiri-Price models. In these models conduction changed abruptly in transition from a conducting core to a nonconducting mantle and the data of solar-diurnal variations and geomagnetic storms placed this change in zone C.

In our research we have repeatedly noted the advantages of frequency sounding which make it possible to deal with conductivity models featuring a complex distribution law. Indeed, the numerical calculations of the functional of accumulated error have revealed that model II, where σ changes abruptly in layer C, provides a better description of experimental global sounding data than does simple model III (exponential change of σ with depth). This suggests that experimental change of electrical conductivity in layer C in contrast to the stepwise change in classical models.

The existence of an abrupt change is confirmed by the changed slope of the global ρ_T curve over periods around 10^3 $s^{-\frac{1}{2}}$, as is especially obvious in using data

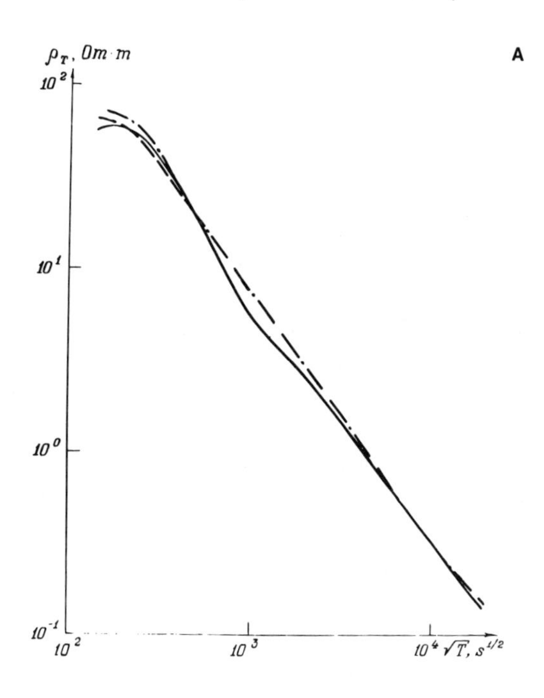

Figure 4a
Solid line—BANKS (1972) continuum spline; dash line—ρ_T curve for optimal model II; dash-and-dot line recognizes the effect of the World Ocean.

on ρ_T including the BANKS (1972) continuum. These data repeatedly were shown (FAINBERG, 1983) to be unduly low. For this reason the Banks data were not used in our plotting of the condition function. It would be natural to investigate how optimal model II would change if the Banks data are, after all, included in the initial data. The solid line in Figure 4a shows a spline obtained from data including the Banks continuum, the solid line in Figure 4b shows the associated distribution of $\sigma(r)$, and the dotted line was calculated in disregard of Banks data. The effect of the continuum results in a rise of the σ curve starting from depths of generally 550 km, but does not change the model to any significant extent.

In obtaining the function $\sigma(r)$ from parametric models the depths from four to about 500 km were assumed to feature a conductivity of 10^{-4} to 10^{-3} Sm/m. At depths of about 200 km an astenosphere where σ is rather high is assumed existent. To see whether the global sounding data can reveal the existence of the astenosphere, model II was modified by assuming the existence of a 'conducting layer'

$$\sigma(r) = 10^{-4} \exp [\gamma(R - r)]/[1 + \rho \times \exp \gamma(R - r) \qquad \text{for} \quad 4 < R - r < H_1 - Q_1,$$

where γ and β are constants; for this new model the primal problem was solved. Comparison of ρ_T values revealed that mean ρ_T values practically would not change

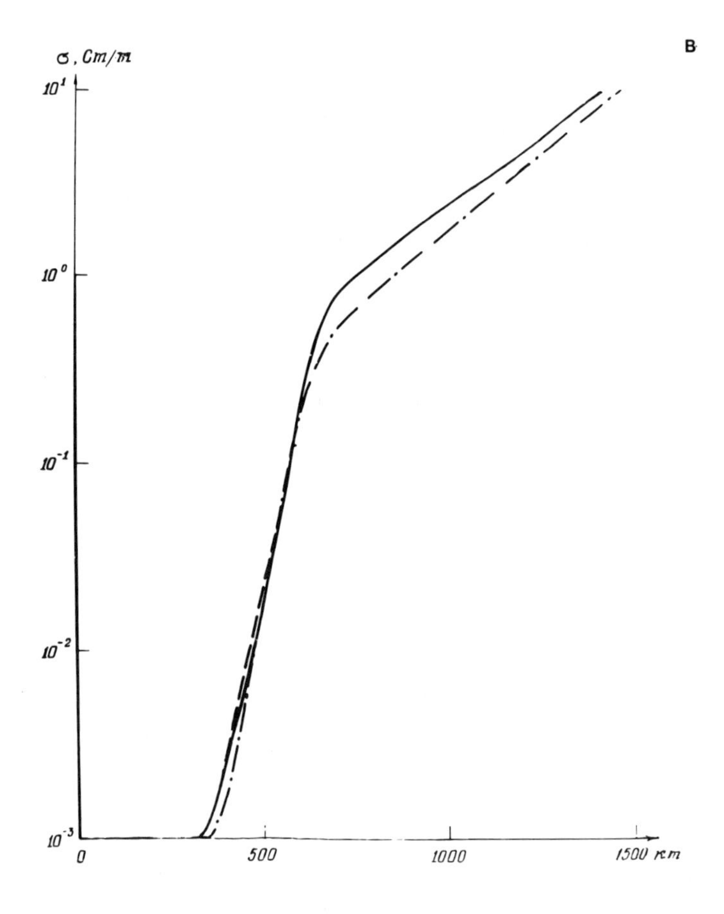

Figure 4b
Distributions of $\sigma(r)$ associated with the ρ_T curves of Figure 4a.

if such a layer exists; the spline values in the diurnal variation range changed by two to three percent. The global sounding curve obtained from diurnal variation periods thus does not show existence of an astenospheric layer.

Finally, let us see how the World Ocean would change the optimal parameters of the model. FAINBERG (1983) showed that this effect is seen as a 30 to 40 percent increase of ρ_T values in harmonics of S_q. We solved the primal and inverse MVS problems for model II by a 30 percent increase of the initial ρ_T values on the second to fourth harmonics. The curve ρ_T for this case is shown as the dash-and-dot line in Figure 4a and the associated distribution of $\sigma(r)$ as the dash-and-dot Figure 4b. The change of ρ_T results in a steeper jump ($Q = 60$ km) which is closer to the surface ($H = 600$ km).

REFERENCES

AFRAIMOVICH, S. R., BAZARZHAPOV, A. D., MISHIN, V. M., NEMTSOVA, E. I., OSIPOV, N. K., PLATONOV, M. L. and URBANOVICH, V. D. (1966), *Mean S_q fields inferred from the data from September 1958.* In *Geomagn. Issledovania.* Nauka (Moscow) *8*, 31–51.

AKIMOTO, S. and FUJISAWA, H. (1965), *Demonstration of electrical conductivity jump produced by the olivine-spinel transition.* J. Geophys. Res. *2*, 443–447.

BANKS, R. J. (1969), *Geomagnetic variations and the electrical conductivity of the upper mantle.* Geophys. J. R. Astr. Soc. *1–4*, 457–487.

BANKS, R. J. (1972), *The overall conductivity distribution of the Earth,* J. Geomag. Geoelectr. *3*, 337–351.

BEN'KOVA, N. P. (1941), *Quiet solar diural variations of terrestrial magnetism.* Proceedings of research bodies of the USSR Weather Service *1*, 75 pp.

BERDICHEVSKY, M. N., VANJAN, L. L. and FAINBERG, E. B. (1969), *Frequency sounding of the Earth by spherical analysis of electromagnetic variations.* Geomagnetism and Aeronamy *9*, 372–374.

BERDICHEVSKY, M. N., VANJAN, L. L., LAGUTINSKAJA, L. P., ROTANOVA, N. M. and FAINBERG, E. B. (1970), *Experience of frequency sounding of the Earth based on spherical analysis of geomagnetic field variations.* Geomagnetism and Aeronomy *10*, 374–377.

BORISOVA, V. P., BERDICHEVSKY, M. N., ROTANOVA, M. N. M., FAINBERG, E. B. and CHEREVKO, T. N. (1974), *New data on global magnetic variation sounding of the Earth.* In *Studies in Geomagnetism, Aeronomy and Physics of the Sun.* Nauka (Moscow) *30*, 194–205.

CHAPMAN, S. (1919), *The solar and lunar diurnal variation of the Earth's magnetism.* Phil. Trans. Roy. Soc. (London), Ser. A. *218*, 1–118.

COURTILLOT, V. and LE MOUEL, J. (1976), *On the long-period variations of the Earth's magnetic field from 2 months to 20 years.* J. Geophys. Res. *81*, 2941–2950.

CURRIE, R. G. (1966), *The geomagnetic spectrum—40 days to 5.5 years.* J. Geophys Res. *71*, 4579–4598.

DEVANE, J. F. (1977a), *Global electrical conductivity from a magnetic storm.* Acta Geod. Geophys. et Montanist. Acad. Sci. Hung. *12*, 369–375.

DEVANE, J. F. (1977b), *Electrical conductivity profiles and upper mantle structure.* In Rep. at Seattle IAGA Meeting, 12 pp.

DEVANE, J. F. (1978), *Monte-Carlo inversion of geomagnetic induction data.* In Rep. 4 Workshop on EMI, Murnau, 20 pp.

DMITRIEV, V. I., ROTANOVA, N. M., ZAKHAROVA, O. K. and BALYKINA O. N. (1977a), *Geoelectrical and geothermal interpretation of deep magnetovariational sounding results.* Geomagnetism and Aeronomy *2*, 314–321.

DMITRIEV, V. I., ROTANOVA, N. M., BALYKINA, O. N. and ZAKHAROVA, O. K. (1977b), *On the resolution of deep magnetovariational sounding curves.* Geomagnetism and Aeronomy *6*, 1092–1097.

DMITRIEV, V. I. and BALYKINA, O. N. (1977), *One application of spline-fluctuations to solution of the inverse magnetovariational sounding problem.* In *Numerical Methods in Geophysics.* MGU (Moscow), 105–121.

DZHODENCHUKOVA, A. K. (1975), *The results of global sounding of the Earth using the data on the spectrum of 27-day geomagnetic field variation.* Geomagnetism and Aeronomy *15*, 317–320.

ECKHARDT, D. H., LARNER, K. and MADDEN, T. (1963), *Long-period magnetic fluctuation and mantle electrical conductivity estimates.* J. Geophys. Res. *23*, 6279–6286.

FAINBERG, E. B. (1975), *Magnetovariation sounding of the Earth by a normal S_q-field.* Geomagn. Aeron. *1*, 179–181.

FAINBERG, E. B., FISKINA, M. V. and ROTANOVA, N. M. (1977), *Experimental data on the global electromagnetic sounding of the Earth.* In *Studies of the Space-Time Structure of the Earth's Magnetic Field.* Nauka (Moscow), 102–113.

FAINBERG, E. B., LAGUTINSKAYA, L. P. and SANIN, S. I. (1976), *Expansion of the field of individual D_{st}-variations into natural orthogonal components and study of deep Earth electrical conductivity.* In Fourth All-Union Workshop on Electromagnetic Soundings. MGU (Moscow), 98–101.

FAINBERG, E. B. (1983), *Global geomagnetic sounding.* Preprint, Moscow, *50a*, 66 pp.

HARWOOD, J. M. and MALIN, S. R. (1977), *Sunspot cycle influence on the geomagnetic field.* Geophys. J. Roy Astron. Soc., 605–619.

HASEGAWA, M. and OTA, M. (1950), *On the magnetic field of S_q in the middle and lower latitudes during the Second Polar Year.* Transactions of the Oslo Meeting, 1948, I.A.T.M.E. Bull. *13*, 426–430.

ISICARA, A. M. (1977), *Solar cycle controlled variation of the geomagnetic field.* Acta Geodaetica, Geophysica et Montanistica *1–3*, 397–405.

JADY, R. J. (1974), *The conductivity of spherically symmetric layered Earth models determined by S_q and longer period magnetic variations.* Geophys. J. Roy. Astr. Soc. 399–410.

LAGUTINSKAYA, L. P., ROTANOVA, N. M., FAINBERG, E. B. and DUBROVSKY, V. G. (1975), *Global magnetovariational sounding of the Earth from D_{st}—variation data.* Geomagnetism and Aeronomy. 5, 498–950.

LYUBIMOVA, E. A. (1968), *Earth and Moon thermics*, Nauka (Moscow), 279 pp.

MALIN, S. R. C. (1973), *Worldwide distribution of geomagnetic tides.* Phil. Trans. Roy. Soc. London, 551–594.

MALIN, S. R. C., CECERE, A. and PALUMBO, A. (1975), *The sunspot cycle influence on lunar and solar geomagnetic variations.* Geophys. J. Roy. Astron. Soc. 115–126.

MALIN, S. R. C. and GUPTA, J. C. (1977), *The S_q current system during the International Geophysical Year.* Geophys. J. Roy. Astron. Soc. 515–529.

MALIN, S.R.C. and ISIKARA, A. M. (1974), *Annual variation of the geomagnetic fields.* Geophys. J. Roy. Astron. Soc. 565–569.

MATSUSHITA, S. and MAEDA, H. (1965), *On the geomagnetic solar quiet daily variation field during the IGY.* J. Geophys. Res. *11*, 2435–2558.

PARKINSON, W. D. (1974), *The reliability of conductivity derived from diurnal variation.* J. Geomagn. and Geoelectr. 281–284.

RIKITAKE, T. (1951), *Electromagnetic induction within the Earth and its relation to the electrical state of the Earth's interior.* 3 Bull. Earthquake Res. Inst., Tokyo Univ. *29*, 61–69.

RINGWOOD, A. E. and MAGOR, A. (1970), *The system Mg_2SiO_4 ... Fe_2SiO_4 at high pressures and temperatures.* Phys. Earth Planet. Inter. 89–108.

ROKITYANSKI, I. I. (1972), *Geophysical methods of magnetovariation sounding and profiling.* Naukova Dumka, Kiev, 225 p.

ROKITYANSKI, I. I. (1981), *Inductional Earth sounding.* Naukova Dumka, Kiev, 296 pp.

ROTANOVA, N. M., CHIMIDDORZH, G. (1977), *New experimental determinations of apparent resistivity on the basis of D_{st}-variation data.* In *Studies in Time-Space Structure of the Geomagnetic Field.* Nauka, (Moscow), 159–166.

SCHMUCKER, U. (1974), *Erdmagnetische Tiefensondierung mit langperiodischen Variationen.* In Protokoll über das Kolloquium *Erdmagnetische Tiefensondierung* Grafrath/Bayern 11–13 März, Univ. München, 313–342.

WINCH, D. E. (1981), *Geomagnetic tides, 1964–1965.* Phil Trans. Roy. Soc. London 1473, 403 pp.

YUKUTAKE, T. (1965), *The solar cycle contribution to the secular change in the geomagnetic field.* J. Geomagn. and Geoelectr. *3–4*, 287–309.

ZHARKOV, V. N. (1983), *Internal structure of the Earth and plantes.* Nauka (Moscow), 415 pp.

(Received 22nd November, 1985, accepted 24th April, 1986)

PAGEOPH, Vol. 125, Nos. 2/3 (1987) 0033-4553/87/030427-31$1.50+0.20/0

The Upper Mantle Conductivity Analysis Method Using Observatory Records of the Geomagnetic Field

WALLACE H. CAMPBELL[1]

Abstract—The electrical conductivity of the Earth's upper mantle can be inferred from geomagnetic quiet-day, *Sq*, variations recorded at the world's observatories using the coefficients of a spherical harmonic analysis (*SHA*) that separate the external (source) and internal (induced) parts of the surface field. The conductivity profile determined from such an analysis can be sensitive to special characteristics of the quiet field itself as well as the separation techniques employed. This review of the Sq-analysis features critical to a conductivity derivation is pictorially presented along with the equations for application of the SCHMUCKER (1970) technique to the *SHA* coefficients for a conductivity determination. Three examples illustrate the use of these equations with different *Sq* models.

Key words: Electric conductivity, upper mantle, Sq, geomagnetic daily variations

Introduction

The purpose of this paper is to review a method that can be used to estimate the electrical conductivity structure of the Earth's upper mantle. This method requires a source of natural field variations with wavelengths of sufficient size to penetrate about 100 to 600 km into the conducting Earth. The method also requires the surface measurements of the field to be separated into the external and internal component parts for each individual wavelength obtained from a constituent frequency decomposition of the observed field variation. In the first part of the paper, I will discuss the observed field and its separation analysis. In the last part of the paper, I will describe the conductivity determination.

The quiet daily field variations, called *Sq* for 'solar quiet-time', provide a natural signal source with frequencies appropriate to upper mantle conductivity studies. The mathematical technique for an external/internal field separation of the *Sq* field has been adequately described in the literature (cf. CHAPMAN and BARTELS, 1940; MATSUSHITA, 1967; PARKINSON, 1983; and CAMPBELL and SCHIFFMACHER, 1985). Rather than repeat here the mathematics of this separation, I have instead chosen to

[1] U.S. Geological Survey, Mail Stop 964, Denver Federal Center, Box 25046, Denver, Colorado 80225.

describe the process with detailed illustrations and to indicate those characteristics of particular importance to Earth-conductivity studies.

The conductivity determination follows the procedure originally presented by SCHMUCKER (1970, 1979), which has been shown by WEIDELT *et al.* (1980) and JONES (1983) to be equivalent to the 'Niblett' and 'Bostick' depth-conductivity profile transformations. I will provide the mathematical details and indicate the limitations of the 'transfer function' that translates the separated external and internal field coefficients into values of conductivity and depth. Several sample computations will be given to illustrate the method.

Quiet Geomagnetic Field Change Source

The focus of this section is on concepts and illustrations rather than upon mathematical formulas. The purpose is to develop a level of understanding that will help the reader avoid the type of equation enchantment that may obscure an understanding of the important physical processes and occasionally lead to grave analysis errors.

Figure 1 is an interesting record of the variation in the H (magnetic northward) component of field for two consecutive days near the March equinox. Local Time (LT) is about five hours earlier than the Universal Time (UT) indicated. March 23 is obviously active; March 24 is clearly quiet; there is apparently a linear recovery of the active period field level through the quiet day. We could use almost any measure of the range of variations during short increments of the day to select the 'quiet' conditions. The quasilogarithmic K_p is a convenient three-hour index of this type; its linear counterpart is the A_p index (MAYAUD, 1980). On the 23rd, the lowest K_p value was 3−, the highest was 7*o*, and the day's A_p value was 67. On the 24th, the lowest K_p value was 0*o*, the highest was 2+, and the day's A_p value was 2. Some

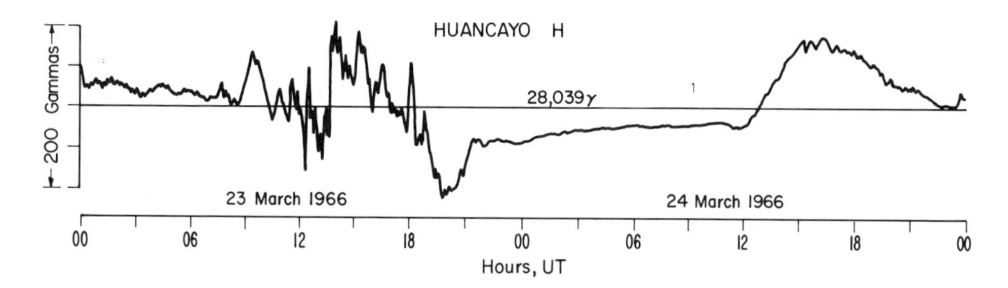

Figure 1
Example of active (23 March) and quiet (24 March) variations of the magnetic northward, H, component of field at a standard observatory (Huancayo, Peru). The average field level (baseline) for the period and the variation scale, both in gammas (nanoteslas), are indicated.

Table 1

*Quiet year activity levels**

Quiet year	Percentage of $K_p < 3$	Percentage of $K_p < 0$	$\bar{A}_p$ (Annual av.)
1934	85.7	39.7	7
1945	77.5	26.3	10
1955	74.9	17.9	11
1965	84.9	32.8	8
1976	70.5	14.2	13

* Values from Meyers and Allen (1977).

authors prefer to select quiet days by a limiting value of the day's A_p (e.g., $A_p = 10$). Others take a fixed number (e.g., five) of the quietest days (judged by the day's A_p) for a given month, whatever the values may be. I have preferred to select those days for which no K_p index exceeded a limiting level (e.g., 2+) because active records can cause unrealistic conductivity computations. The methods of selecting quiet days are often responsible for the variations in results between authors. Despite some difficulties in assuming that '*p*' means a true 'planetary' characteristic of K_p, the index has always proved to be a useful selection tool for determining geomagnetically quiet days.

Field-variation measurements at the world observatories are sensitive indicators of a number of physical changes that transpire between the Sun and the Earth's surface. The large amount of literature on this subject (cf. CAROVILLANO and FORBES, 1983) need not be summarized here. It is sufficient to note that the Sun has an approximate 11-year activity cycle that has been documented in detail by sunspot numbers (*SSN*) back to 1700 (Figure 2). In the mid-1800's, shortly after the application of photographic recording techniques to magnetograms, the solar activity cycle was discovered in geomagnetic registrations. Maxima and minima in field variations were found to occur on or about a year after the corresponding extreme in *SSN*. For example, Table 1, showing the percentages of low K_p values during the recent geomagnetically quiet years, can be compared to 56.2 percent and 47.6 percent of $K_p < 3$ in active years 1958 and 1974, respectively. Geomagnetic data taken on years near the *SSN* minima are typically best for representing *Sq* in its purest form.

The principal cause of the quiet day geomagnetic field variations is the ionospheric dynamo current created when there is a force on the ionized region of the atmosphere in the presence of the Earth's main field. Selective characteristics of the collision frequencies of the ionized atmospheric particles make the *E*-region electrons near 100 km the most suitable current carriers. A force of these electrons is created by the day-to-night thermotidal changes in the atmosphere as the Sun rises and falls daily through the year and by some upper atmospheric winds of global scale. Figure 3 exhibits a typical average March *D* component daily *Sq* variation at midlatitudes

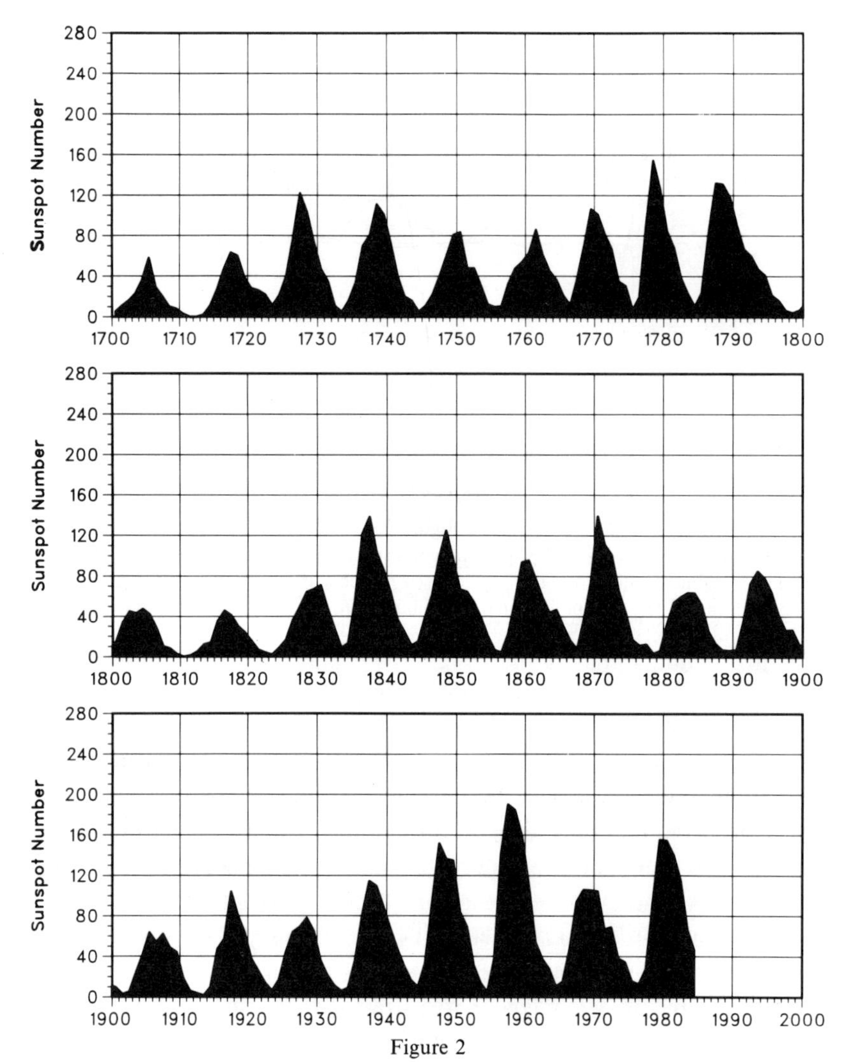

Figure 2

Yearly mean sunspot numbers for the period 1700 to 1984 (from WILKINSON, 1986).

(noon at Dallas is near 18 hr *UT*). The angle χ between the Sun and the zenith measured from the observatory determines the Chapman factor $(\cos\,\chi)^{0.5}$ that is proportional, in a first approximation, to the *E*-region ionization. Note the corresponding onset and subsistence of *Sq*(*D*) and the Chapman factor. This ionization control of the *Sq*, therefore, depends upon local time, geographic latitude, and the serial day-number of the year. However, even small amounts of solar acctivity can change the ion density at *E*-region levels, therefore the selection criteria for quiet days is very important.

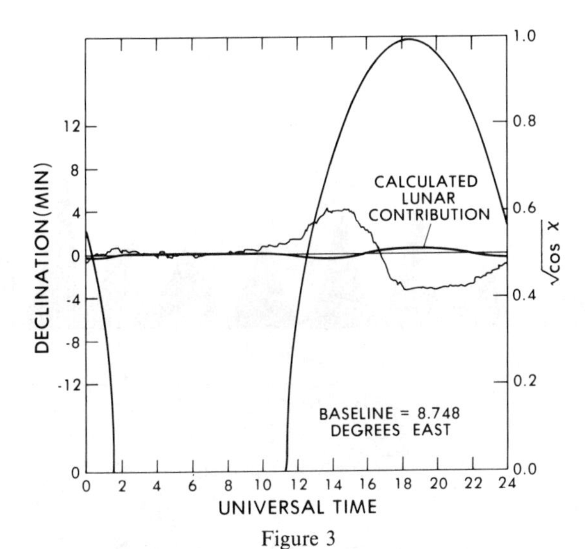

Figure 3

Quiet daily variation (irregular trace), *Sq*, of the magnetic declination, *D*, component at Dallas, USA, for 21 June 1986. The Chapman ionization factor, $\sqrt{\cos \chi}$ (in which χ is the solar zenith angle is shown by the parabolic-shaped curve. The calculated lunar contribution to the daily field variation is also indicated (from CAMPBELL and SCHIFFMACHER, 1985).

The two major forces for the dynamo currents are also arranged best by geographic latitude and season. The thermotidal motion of the atmosphere that drives most of the observed *Sq* depends upon the sunlight. Figure 4 illustrates the seasonal and geographic latitude variation of the daily solar radiation reaching the Earth's atmosphere. Note the North–South asymmetry because the Earth's orbital position is closer to the Sun during the Southern Hemisphere summer. The summer-to-winter difference in dynamo currents should be greater in the Southern Hemisphere than in the Northern Hemisphere because of this asymmetry. The solar radiation at polar latitudes has a low angle and high refraction that decreases its heating effect there. Also, the observed lag of surface temperatures (to about a month after the summer solstice, because of the surface heat accumulation) is not shown in this type of figure. The second driving force is the horizontal wind near 100 km, of the type illustrated in Figure 5 (FESEN *et al.*, 1986). The wind patterns are dependent upon a great number of factors such as the solar radiation, the global pressure and temperature distribution, the Earth's spin, and the solar-terrestrial activity level. Noon is at 0° longitude in the figure. Note the westward component of wind near the equator and the eastward component at other latitudes. The oppositely directed Earth's field of the two hemispheres causes this wind direction bias to have an oppositely directed effect upon the dynamo current in the two hemispheres.

From the above considerations, the source currents would seem to be organized by season and geographic latitude. However, these ionospheric dynamo currents

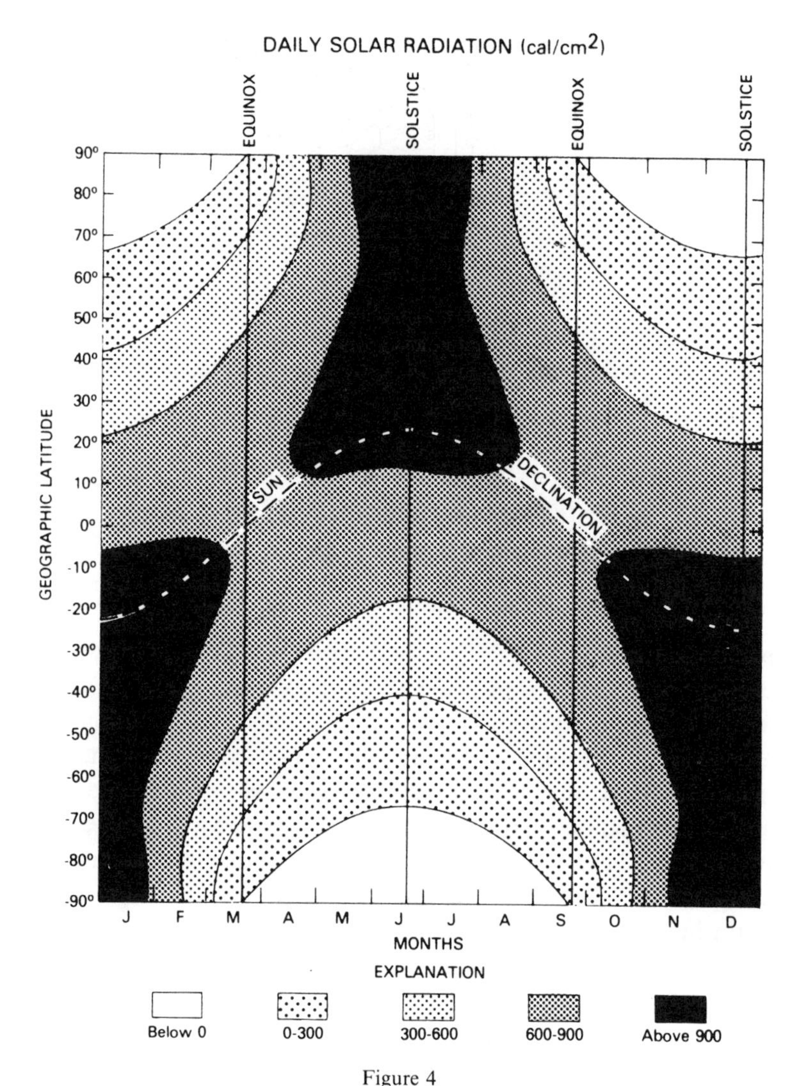

Figure 4
The solar radiation at the top of the atmosphere as given in the Smithsonian Meteorological Tables, 1958. The solar declination and seasonal demarcations are also shown. Darker shading indicates more radiation. Values are in calories per square centimeter

also depend upon the *E*-region conductivity. This property is not a simple one for an ionized gas in a magnetic field. The conductivity depends upon the direction of the driving force with respect to the Earth's magnetic field direction and magnitudes of all three: field, force, and electron density. To a good approximation, the field can be represented by its dipole term. Unfortunately, however, the dipole is tilted about 11° from the geographic axis (Figure 6) giving the dynamo currents some geo-

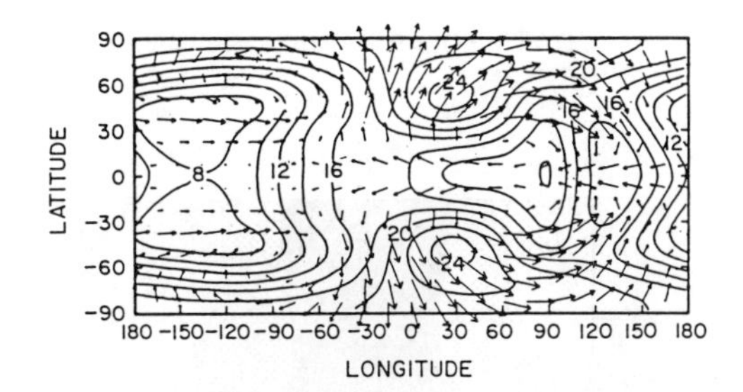

Figure 5
The 1200 *UT* temperatures (solid curves) and winds (arrows) at about 135 km altitude for equinox solar cycle minimum conditions. The maximum wind arrows are 24 m/s (from FESEN *et al.*, 1986).

magnetic arrangement bias. This feature of the *Sq* source causes a problem in the subsequent analysis. A truncated series of terms will be needed to represent the currents. Less error is encountered in a limited-term representation when the analysis is organized in a coordinate system that is natural to the source phenomenon.

Both geographic and geomagnetic arrangements have been tried. In addition, a third, dip coordinate system is used by some researchers; these dip latitudes depend upon the observed field direction local to the measuring site (whereas the dipole system depends upon an analysis in which the full global distribution of sites have contributed). The dip coordinate system produces an uneven spacing of latitudes over the Earth's surface; this feature could introduce some ambiguity in the depth-conductivity determinations that assume a normal spherical coordinate system. Most researchers find the *Sq* better organized with respect to a geomagnetic coordinate system (e.g., Figure 6) because the field alignment is particularly important at the dip equator, where the horizontal field encourages a strong electrojet current, and at the auroral zone latitudes, where the ionization is affected by the magnetospheric (geomagnetic) field structure. Although the dipole orientation provides some symmetry with respect to the field direction, the main field total strength shows an asymmetry, with low values southeast of Brazil and high values near northern Canada, northern Siberia and south of Australia (Figure 7). Both the field strength and its direction with respect to the E-region forces determine the tensor conductivity and resulting *Sq* currents.

Three other special effects contribute changes to the quiet-field measurements. In order of relative amplitude, these are the lunar tides, the magnetospheric distortion, and the *IMF* sector reaction. The semidiurnal lunar tides have a period of about 13 hr 25.2 min. The dynamo currents driven by these tides exhibit field levels, during the daytime hours, usually less than 10 percent of the amplitudes of the *Sq* daily variation at the same site (MATSUSHITA and CAMPBELL, 1972). When it is possible to determine the lunar contributions (Figure 3), they should be removed from the analysis. The 50.4 min daily lag of the lunar contribution, behind the solar daily

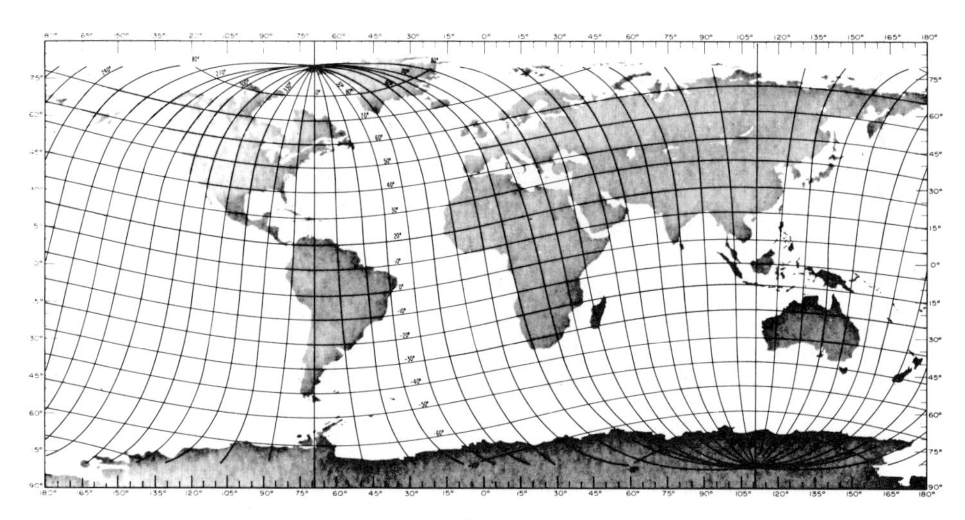

Figure 6
Global map of geomagnetic coordinate system (curved lines), for the quiet year 1965.

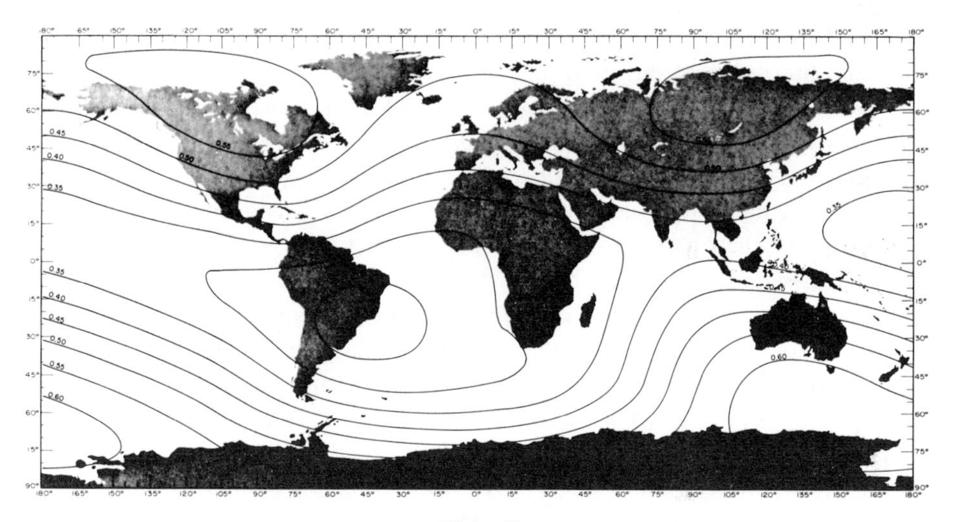

Figure 7
Global map of total geomagnetic field intensity, B, contours in 10^5 gamma units, for the quiet year 1965.

variations, makes a different contribution to the total quiet-field change each day; thus, if a number of days are averaged to form the typical Sq for a given month, the lunar contribution, changing daily, represents a noise in Sq that will add error to a conductivity determination if it is not removed.

The tear-drop shape of the magnetosphere is largely determined by the Earth's main field and its encounter with the solar wind of fields and particles. The distortion of the Earth's external field has been represented by the field of a magnetopause current system that varies with season and solar activity. As the Earth spins, in quiet solar-terrestrial conditions, a station at the Earth's surface should experience a diurnal change in the field from these magnetopause currents. OLSEN (1970) has modeled the amplitude of these diurnal fields to be about 10 gammas at the Earth's surface. There is considerable doubt that the effect is that large in quiet periods because the sharp cutoff of *Sq* at ionospheric sunset and sunrise (Chapman factor = 0 in Figure 3) is not a magnetospheric feature. The relative spectral composition of the quiet day variations for the ionospheric dynamo and the magnetospheric distortion models should be slightly different.

The quiet daily field at polar latitudes reacts to the interplanetary magnetic field (*IMF*) from the Sun (CAMPBELL and MATSUSHITA, 1973; CAMPBELL, 1976) as a result of interactions at the Earth's magnetotail. The effect is represented by a diurnal variation that shifts phase by about 90° between toward and away conditions which occur typically every week or two. The *IMF*-field response can be considered a perturbation of the magnetopause current model described above. MATSUSHITA *et al.* (1973) found traces of this effect in the quiet-day field changes at low latitudes. As we shall see below, the spherical harmonic analysis coefficients are relatively insensitive to variation near the poles. The *IMF* contributions to *Sq* are quite small outside the polar cap but, if not removed, would represent a small (several percent) noise in the *Sq* determination.

Auroral region geomagnetic activity related to solar disturbances is associated with Birkland (magnetospheric field aligned) and ionospheric currents. When strong, the fields from the Birkland currents may be detectable at mid-latitude locations. Ionospheric electrojet currents driven by the Birkland system can flow to lower latitudes in the conducting *E*-region. Even on the quietest of days, some small westward auroral electrojet current can be observed at high latitudes. For *Sq* modeling, an attempt is usually made to remove small disturbance effects from the original field measurements in *SSN* minimum years by data smoothing techniques. For active-year analysis, high-latitude currents have been eliminated from the data set by special extrapolation of 60°-field values to the pole (MATSUSHITA and MAEDA, 1965).

Figure 8 illustrates some of the complex array of phenomena that can contribute to the establishment of the ionospheric tensor conductivity. The factors contributing to the ionospheric source current, external to measurements made anywhere on the Earth's surface, are illustrated in Figure 9. When such a source current changes in time, it induces a flow, in the electrically conducting Earth, a secondary current at a depth, direction, and strength determined by the conducting structure of the subsurface region. The surface observations of the magnetic field variation are, therefore, a mixture of external (source) and internal (induced) parts whose separation and relationship will allow the researcher to infer the Earth's conductivity behavior. The information gained from this electrical property, complementary to that obtained

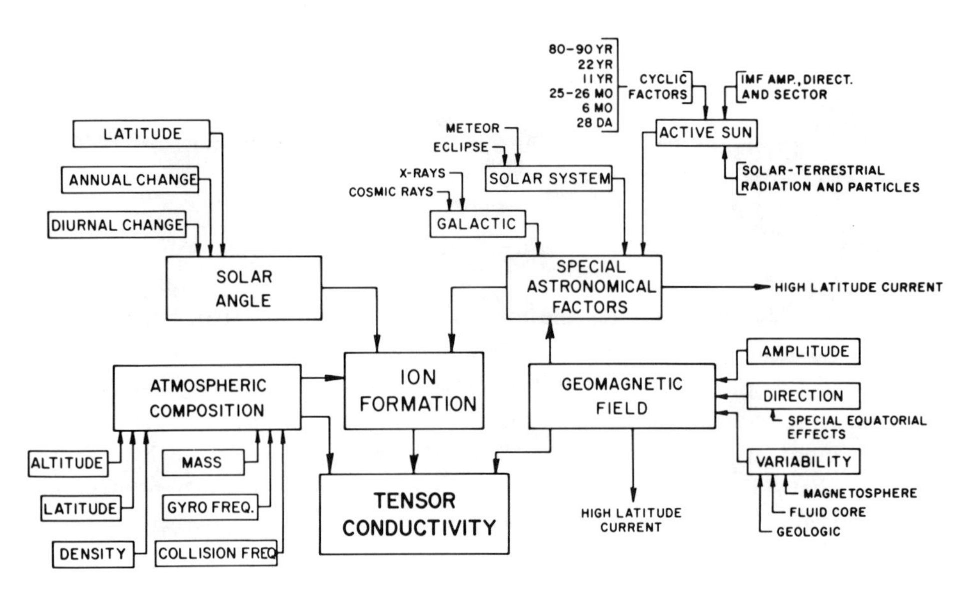

Figure 8
Outline of contributions to tensor conductivity determinations.

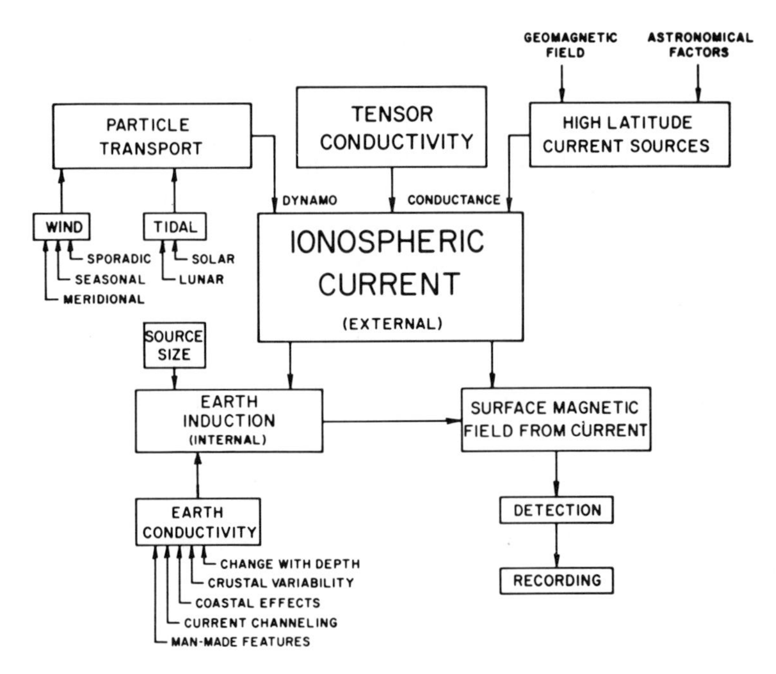

Figure 9
Outline of contributions to ionospheric current and surface magnetic field recordings.

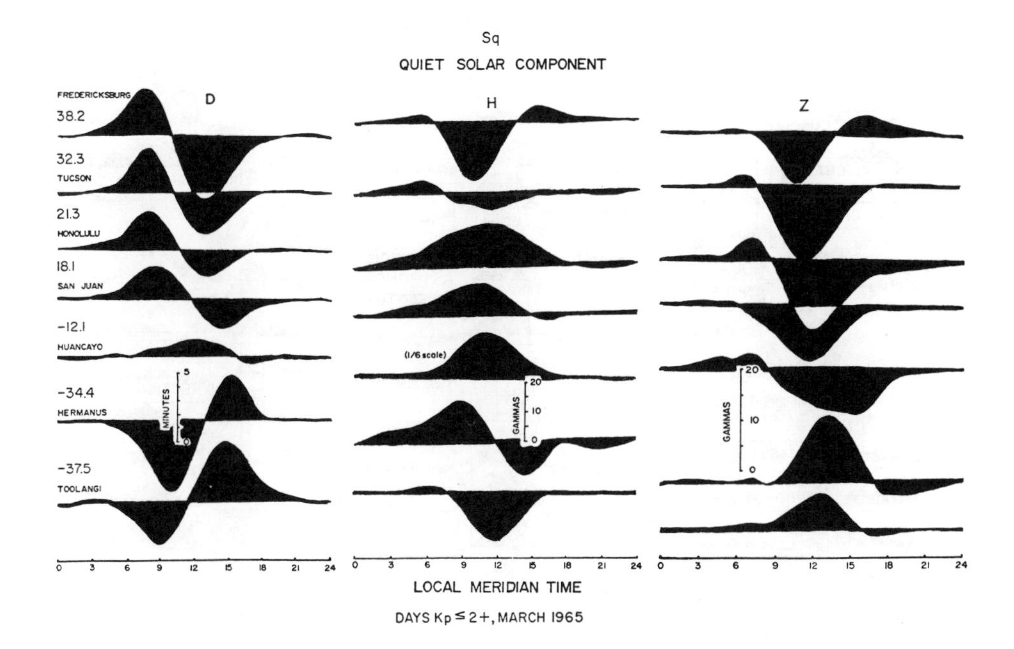

Figure 10
Field variations, *Sq*, of the magnetic northward, *H*, angle eastward, *D*, and downward, *Z*, components for seven observatories (indicated at left with geographic latitude) for average of quiet days in March 1965, when all K_p activity indices were less than or equal to 2+. Note that the large *H* variations at Huancayo are shown as 1/6th actual size.

from the seismic and gravity observations, is important for accurately describing the composition of the inaccessible Earth.

Sq *Characteristics of Importance to Conductivity*

Some special global features of the quiet-day-field variations have been recognized since the first observatories were established in the Southern Hemisphere over a hundred years ago (cf. Hobarton magnetograms in WALKER, 1866). In opposite hemispheres, the *H* components show the same direction of variation, whereas the *D* and *Z* components exhibit opposite variations (Figure 10). This feature can be expected to result from the dynamo current effect of the Earth's main field opposite direction in the two regions. The Southern Hemisphere field maxima and minima occur about an hour or two later than the corresponding values of the Northern Hemisphere observatories (Figure 10). It is believed that this time lag is largely a reaction of the eastward ionospheric wind components in the oppositely directed main field (Figure 5).

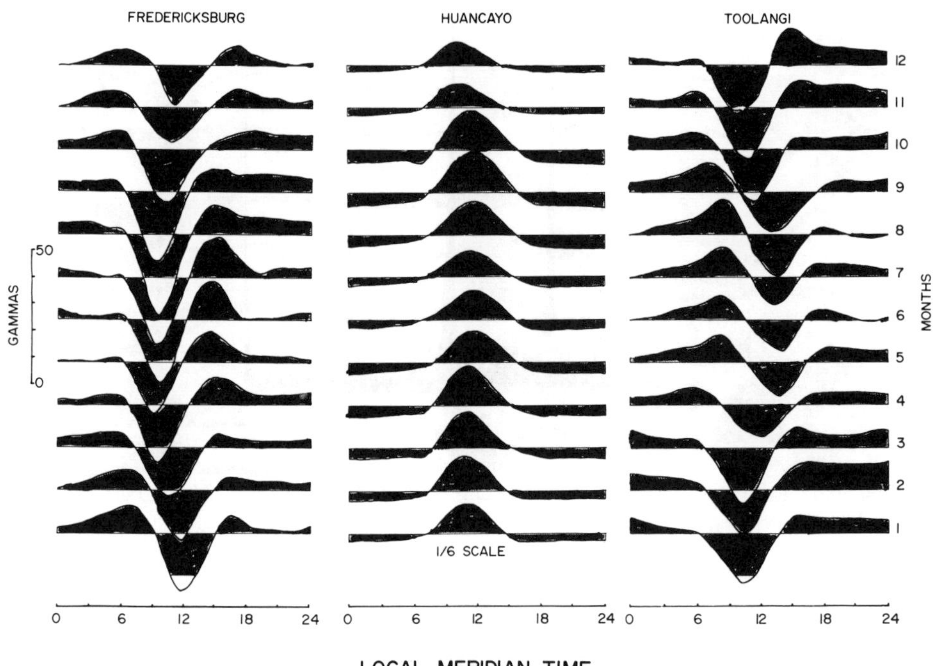

Figure 11

Month-by-month variations of $Sq(h)$ at Fredericksburg (USA), Huancayo (Peru), and Toolangi (Australia) observatories. The large Huancayo values near the geomagnetic equator, are shown as 1/6th actual size. Fredericksburg and Toolangi are at similar geomagnetic north and south mid latitudes, respectively.

Both the amplitude and the phase of the quiet daily variations at a given station change through the year (Figure 11). The amplitudes show marked seasonal changes, largely a result of more intense Sun: semiannual (equinoctial) maxima near the equator; annual (summer solsticial) maxima elsewhere. The annual phase change through the year shifts the extrema from later hours in winter to earlier hours in summer. I assume that this effect is more related to the seasonal changes of global winds than to the thermal tides.

A spectral analysis of the quiet-day-field variations always shows strong peaks at periods of 24, 12, 8, and 6 hours (Figures 12 and 13). The relative amplitudes of the peaks may vary through the year, for different components, or for different locations. On occasion, shorter period components are distinguishable above the noise; a high frequency part of the spectrum is necessary to represent the sharp cutoff of Sq at sunrise and sunset.

Figure 14 illustrates the decomposition of a Sq variation into the four significant

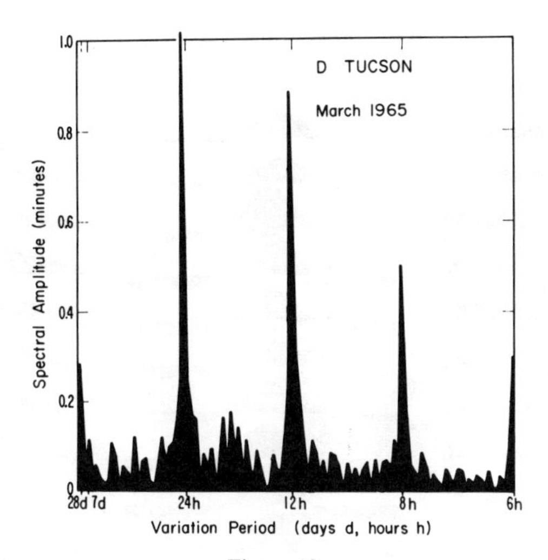

Figure 12
Example spectral amplitude composition of the D field-component variation for March 1965, at Tucson (USA). Values are indicated for periods from 28 days to 6 hours. To convert to power spectral density (gamma^2-sec), multiply the square of the illustrated amplitude by 2.45×10^6.

Fourier components as the value of m changes from 1 to 4 for the 24-, 12-, 8-, and 6-hour field-variation periods. Amplitudes, c, and phases, α are given for a 360° cycle of Φ occurring each 24-hour day. The A and B amplitudes of the equivalent cosine and sine coefficients are also shown. The important aspect of this Fourier representation is that the complex waveform of Sq can be replaced by a superposition of a series of sinusoidal oscillations of systematically decreasing wave lengths; each of these oscillations can be treated as a separate physical entity in the induction process.

Any linear trend in the data (cf. March 24 of Figure 1) needs to be removed from the recorded Sq variation before the Fourier analysis is performed or an error in the sinusoidal amplitude coefficients may be introduced. If the sample number is large within the period of the shortest Fourier component (e.g., a 2.5 min data rate that produces 144 samples for the 6-hr Fourier component), it is usually sufficient to determine the variation end values from the nearby data points and tilt the entire record appropriately. However, when the sample number is small with respect to the shortest analysis period (e.g., one-month sampling for the semiannual change), special linear trend removal techniques need be employed to insure that the full sinusoidal component can be recovered.

There are two methods for obtaining a global pattern of the Sq changes. One is to select a particular UT time, e.g., 1200 UT (noon at 0° longitude), and then to determine the best values of the Sq that fit all global observatories at that moment. In this method, the surface effect of the quiet ionospheric current system is sampled about the Earth at a particular instant of time. Extrapolations between observatory

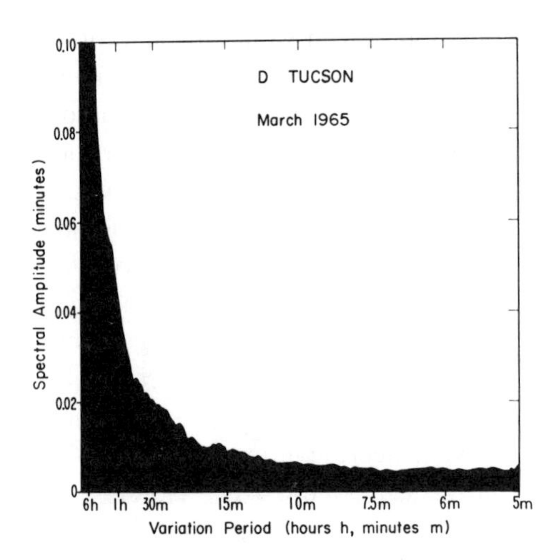

Figure 13
Example of spectral amplitude composition of the D field-component variations for March 1965, at Tucson (USA). This is the same data sample as Figure 12, only values for the shorter spectral periods from 6 hours to 5 minutes are displayed here.

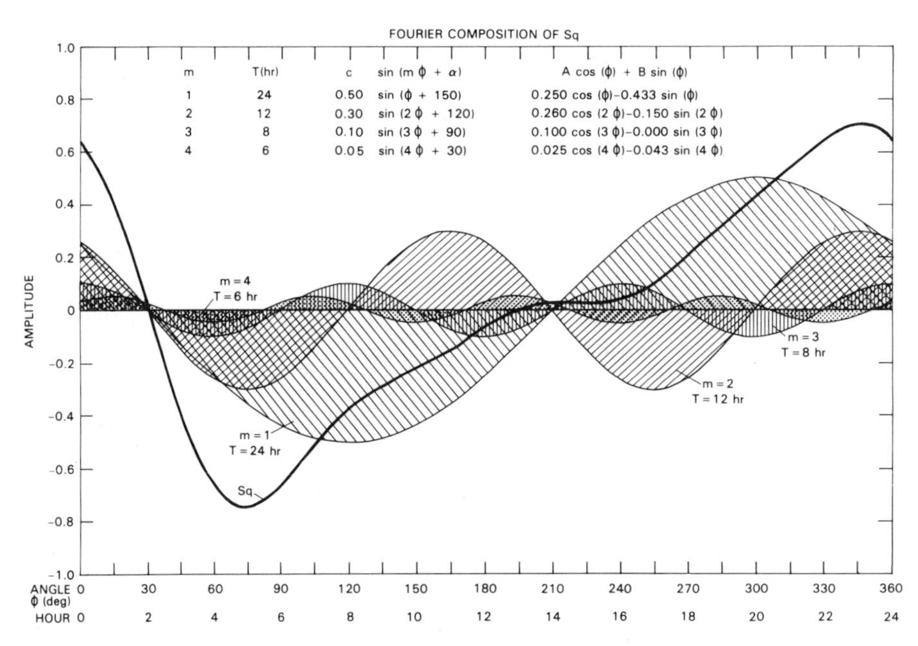

Figure 14
Example of Fourier analysis of Sq field variation (heavy solid curve for a sample day 0 to 24 hrs) indicated as angle Φ° and with amplitudes given on a scale of 0 to 1. Fourier components with $m = 1$ to 4 (for 24, 12, 8, 6, hr periods, T) are indicated by the amplitude, C, and phase angle, α, representation as well as by the amplitudes (A and B) of the cosine and sine representation.

locations provide an estimation of field values for all latitudes and longitudes. Subsequent analysis, to separate the external (source) and internal (induced) contributions for conductivity determinations, would provide a conductivity profile of the 'average' Earth weighted by the upper mantle characteristics beneath the particular observatories that contribute most to the fitting functions.

The second method assumes that during the very quiet solar-terrestrial conditions the stable ionospheric current system is fixed in position with respect to the Sun. As the Earth spins 360° under this fixed system, an observatory senses the 24-hour change of *Sq*. Then, data from a group of observatories distributed in latitude over a narrow sector of the Earth may be used for a spherical analysis (cf. MATSUSHITA and MAEDA, 1965). Such procedure restricts the analysis region to a matched pair of Northern and Southern Hemisphere continents. CAMPBELL and ANDERSSEN (1983) introduced a further refinement in conductivity resolution by using observatories in only a half sector and then modeling the opposite half (i.e., providing a boundary condition of the opposite hemisphere) from the expected field reversals and seasonal differences (Figures 10 and 11) that would match an oppositely directed *Sq* current vortex. Their technique was roughly equivalent to the creation of a sphere surrounded by a unique-source current pattern and having an internal, spherically symmetrical, conductivity depth profile that varied in latitude in a manner that represented the effect of the grouped stations of one continental study region.

Smoothed latitudinal distributions of the Fourier coefficients, representing an *Sq* obtained by a carefully estimated trace of the field changes versus latitude or by some more reproducible automated technique, seem to produce similar external/-internal field separations. MATSUSHITA and MAEDA (1965) used the former method for each of the three sectors (zones) into which they divided the Earth's observatories. Those authors smoothed separately the amplitude and phase values. CAMPBELL and SCHIFFMACHER (1986), working with the cosine and sine coefficient representations, obtained 2.5°-latitude amplitudes by fitting a parabola to the equatorial station coefficient values, fitting line segments between other values, mirroring values about the end regions, and performing a five-point smoothing over all fitted values.

Sq analyses have been carried out as an annual average, broken into three seasons (centered on the equinoxes, June solstice, and December solstice), taken as the monthly average, and measured on a selected day. CAMPBELL (1982) introduced the method of modeling any day of the year by determining the annual and semiannual Fourier components of the best monthly *Sq* coefficients. Figure 15 illustrates the typical fitting procedure. Seasonal changes in the *Sq* source current pattern can affect the selection of fitting polynomials that, in turn, can respond to latitudinal differences in a conductivity profile. This problem is to be discussed below in more detail.

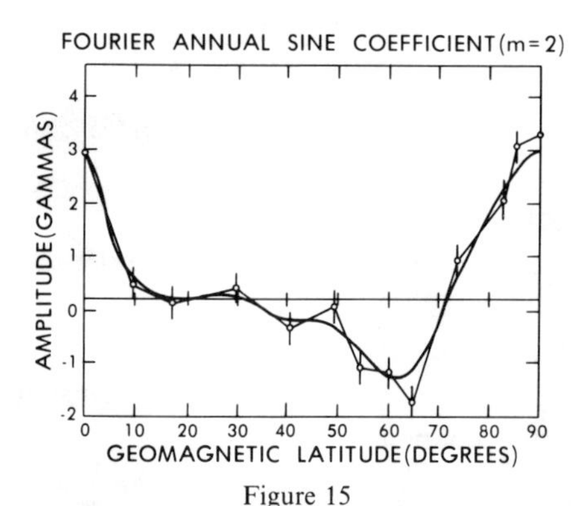

Figure 15
Example of smoothing over latitude for the Fourier sine coefficients of the 12-hour ($m = 2$) annual field variations of H at North American stations (indicated by circles).

External and Internal Separation of the Sq *Fields*

The three directional field components of *Sq* represented over the surface of the spherical Earth are obtainable from a scalar magnetic potential function, V. The spherical harmonic analysis fits this function by sinusoidal oscillations along circles of latitude and by Legendre-type polynomial oscillations along circles of longitude. The waves must close on themselves in a differentiable fashion. In a spherical coordinate system, the northward component of field is obtained from a derivative of V with respect to the colatitude angle, θ; the eastward component, from a derivative with respect to the longitude angle, Φ; and the vertical component from the derivative with respect to the radial direction. When the potential function is separated for external and internal sources, the corresponding external and internal parts of the surface-field measurements are obtainable. Just as the *Sq* fields change with time-of-day, location, and day-of-the-year (Figures 10 and 11), so the potential function undergoes corresponding changes.

Figure 16 is an example of a potential function, V, that was determined at various latitude and time increments in an analysis modeled from an American Hemisphere study for the 21st of August in a quiet year. The displayed amplitudes have been divided by the Earth's radius (R) to give a scale in gammas (nanoteslas). Longitude has been converted to local time. On the left is shown the daily change at a given latitude that appears to vary much like that of a *Sq* field. On the right is the latitude change of V at specified hours; note the nodes at the poles and low variation amplitudes near the equatorial latitudes. The daily variations of V show a phase change in the opposite hemispheres that are necessary for the oppositely directed ionospheric current vortices in the north and south. Maximum amplitudes of V occur near the mid latitudes (these would be the locations of the external current foci). The potential functions are clearly larger in the summer (Northern) Hemisphere mid

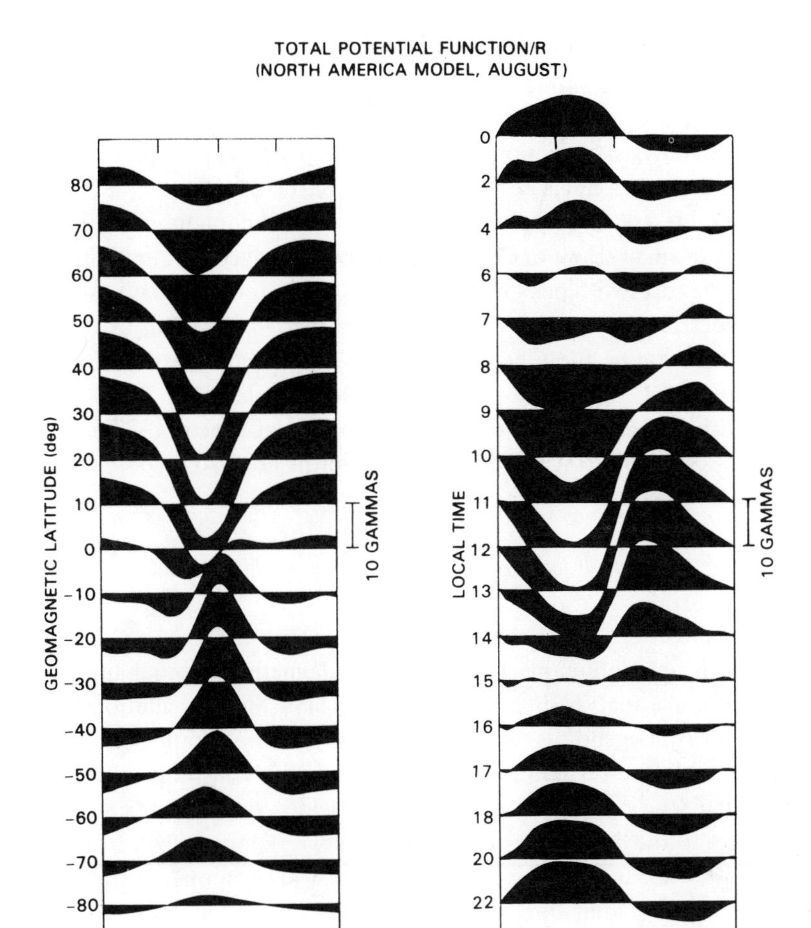

Figure 16
Geomagnetic latitude and longitude (in local time) variation of the scalar potential function, V (divided by the Earth radius R), for selected latitudes (left set) or local times (right set) for a field model of North America in August during a quiet year.

latitudes. All these characteristics are important for an understanding of the selection of Legendre polynomials that appear in the conductivity determination.

A series of polynomials is used to fit the scalar potential function over the spherical Earth's surface. These are essentially a Fourier series of m harmonics along circles of fixed latitudes (illustrated by the variations at left of Figure 16) and Schmidt normalized associated Legendre polynomials, $P_n^m(\theta)$, along the great circles of longitude (i.e., fixed time, illustrated by the variations at right of Figure 16). The indices n and m are called the degree and order, respectively. Fitting of the Legendre functions involves a superposition of a number of terms whose amplitude coefficients need be determined in a fashion similar to the Fourier series fitting with $m = 1$ to 4 (cf. Figure 14). There are $(n - m + 1)$ wave oscillations around a great circle of longitude. We have analysis restrictions, such as n must be greater than or equal to m, and the Legendre series must be truncated at a particular value of n (e.g., $n = 8$ or 12). Figure 17 illustrates some of the unique $P_n^m(\theta)$ characteristics that we must consider when fitting V if we are to understand the derived conductivities. If $n - m$ is even, as in the top two blocks of this figure, the functions can be best fit to values that are symmetric with respect to the equator. If $n - m$ is odd, as in the bottom two blocks of the figure, the functions can best fit to values that are antisymmetric about the equator. It is this $(n - m)_{\text{odd}}$ case that would show the opposite-hemisphere phase shift characteristic of the scalar potential function (Figure 16). Low-equatorial values of V during the equinoctial months are also favored by these $(n -$

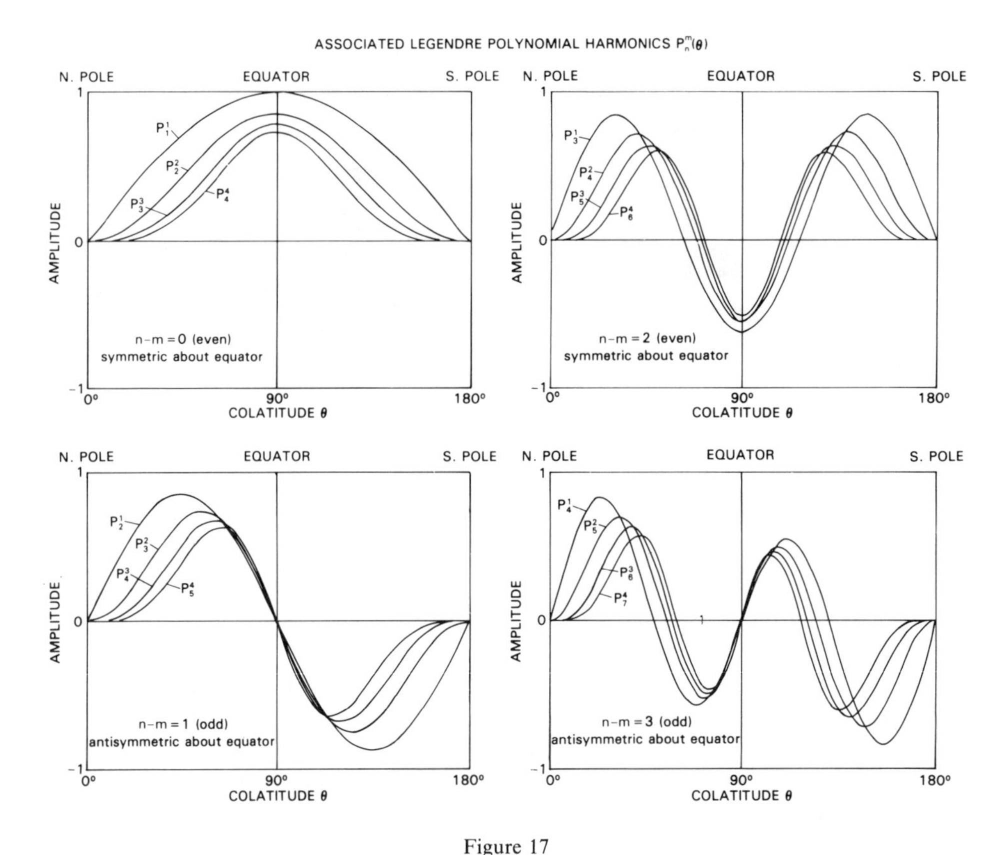

Figure 17

Examples of the associated Legendre polonomial harmonics, P_n^m, variation with latitude for selected values of n and m. The four sets are separated for similar values of $(n - m)$. There are $(n - m + 1)$ waves in 360°.

$m)_{odd}$; on other months, some $(n - m)_{even}$ terms need be included to build values of V at the equator. Because the seasonal changes in ionospheric current focus have their counterpart changes in maxima of the potential function (Figure 16), the relative amplitudes of the P_n^m with $(n - m)_{odd}$ must have seasonal changes as principal representors of V.

If we simply indicate the associated Legendre polynomial values of the last figure and the Fourier coefficients as + when they are positive and as − when they are negative, a simplified global representation of spherical harmonics can be drawn as in Figure 18. Note that there will be m waves (Fourier type) around a latitude circle and $n - m + 1$ waves (Legendre type) around a great circle of longitude. Checkerboard-like patterns are called 'tesseral' harmonics; pattern, like sections of oranges are 'sectoral' harmonics; and stationary patterns around circles of latitude are 'zonal' harmonics. Note that for Sq variations, there will be no values of $m = 0$ (an ampli-

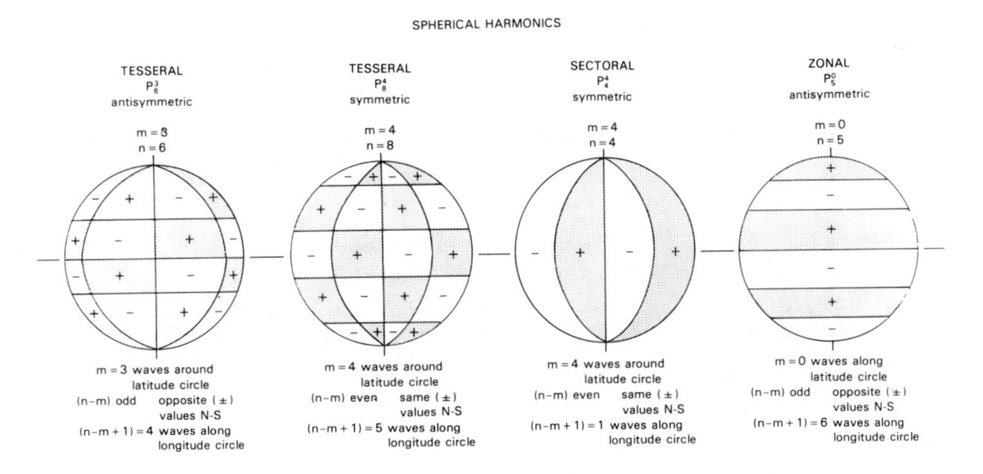

Figure 18

Illustrative map of positive and negative regions (cf. Figure 17) for spherical harmonic representations of $P_n^m \cos(m\Phi)$.

tude with no daily change). The opposite direction of the main field in the Northern and Southern Hemispheres for the similar thermotidal motions of the ionosphere drive oppositely directed current vortices in the two hemispheres whose potential functions are, as a result, largely antisymmetric with respect to the equator and are, therefore, best fitted by the $(n - m)_{odd}$ tesseral harmonics (although other harmonics will be needed to provide the full representative fitting).

The spherical harmonic analysis technique involves the representation of the potential function V over the surface of the Earth by two rapidly converging series. One series has terms with powers of radial distance (to the surface of analysis) that increase in single steps with the term number. The other series has terms with powers of reciprocal radial distance that also increase in single steps with term number. With increasing radial distance for a given set of *SHA* coefficients, the first series would provide larger and larger contributions to the potential (and the field) determination; whereas, the second series would provide decreasing contributions. The first series, therefore, represents sources external to the analysis surface (approaching a source increases the field), and the second series represents sources internal to the analysis surface (receding from a source decreases the field). When the best fit of the potential function about the Earth's surface is obtained, pairs of *SHA* terms with similar n (degree) and m (order) indices represent the external and internal parts of individual harmonic waves that can be separately analyzed for their physical properties in a way comparable to the separate handling of the Fourier components in a plane-wave analysis.

With the determination of the coefficients of the two harmonic series of terms that are truncated at low amplitudes, the separated H, D, and Z components of the external (source) and internal (induced) parts of the surface *Sq* field are resolved. Figure 19 illustrates the appearance of the separated quiet-field variations for a midlatitude site in an equinoctial period. From the separated fields, an equivalent external current system at 100 km can be determined; Figure 20 illustrates this *Sq* source current behavior through the months of a quiet year. There are 10^4 Amperes between contours and a midnight zero level was assumed for the display. A generally similar vortex occurs in the Southern Hemisphere but with the current direction reversed (because of the main field direction) and an annual change shifted by 6 months (for seasonal differences). Note the shift in latitude of the current vortex through the year. There is a corresponding shift in the latitude location of the potential function maximum (Figure 16) representing each month's fields. The 12 sets of *SHA* coefficients required to illustrate this month-by-month change would show the annual and semiannual characteristics of the originally observed field. Conductivities determined from these coefficients must reflect such month-by-month change if there are latitude-dependent variations in the depth profiles of the conductivity between locations of the dominant amplitudes of the fitted potential functions.

Figure 21 is a diagram of the analysis procedure for determining the conductivity from quiet daily variations of field. In fitting the station observations of field changes to the external and internal spherical harmonic series, two major smoothing processes occur. The first involves station-by-station characterization of the *Sq* daily change plus the annual and semiannual changes by Fourier harmonic terms; the appropriate coefficients are smoothed and extrapolated to all global locations. The second involves the representation of this global pattern by the spherical harmonic terms. To evaluate the variance in the two processes, hourly values of field can be determined at the three steps of representation (observation, Fourier, and *SHA*), and the 'error' between these values are found. Such computation is only a comparison of the three representations of the quiet field, not necessarily an error in the general

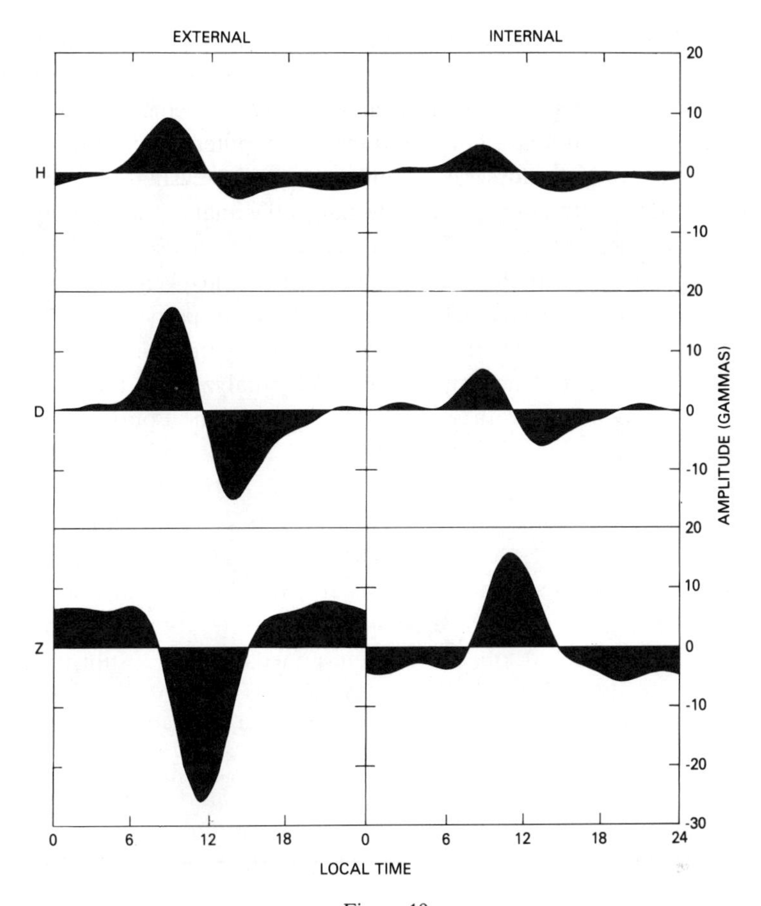

Figure 19
Example of separated external and internal *Sq* field components *H*, *D*, and *Z* (representing the magnetic northward, eastward, and downward directions in gammas) for North America at 30° latitude on 21 March of a quiet year.

sense; for example, a smoothing may actually remove some of an unwanted irregular local anomaly effect at a station and improve the conductivity depth profile determination.

Once the external-internal separation has been obtained, it is relatively straightforward to apply the appropriate transfer functions and obtain paired values of conductivity and depth. The more difficult problem is to determine what region of the Earth is best represented by the resulting profiles. Specifically, it is necessary to determine just what region of the Earth dominates the characteristics of the potential functions that are to be fitted by the spherical harmonic procedure; the actual conductivity of the Earth's upper mantle is not spherically symmetric. Analysis cases for

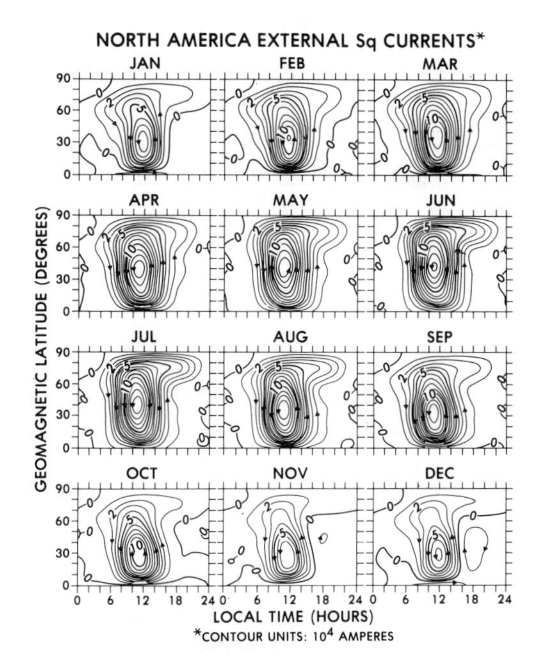

Figure 20
Equivalent ionospheric source current for *Sq* daily variations of field in the continental region of North America for separate months of 1965. Each pattern, in local time versus geomagnetic latitude coordinates, shows the equivalent current contours with 10^4 ampere steps. A midnight zero current level was assumed.

three observatory distributions will illustrate the problem: global, sectoral, and half sectoral.

For a global distribution of observations, the typical study involves a determination of the *Sq* current system at a particular Universal Time. Near the equinoctial periods of the year, the potential functions, V, will be mostly antisymmetric about the equator so the $(n - m)_{\text{odd}}$ Legendre polynomial harmonics will prevail and the mid-latitude regions near the current foci in the two hemispheres will most influence the selection of the constituent harmonic polynomials. The conductivity-depth determination is blind to which hemisphere contains the region of interest. The actual conductivity profiles of the north and south hemisphere mid latitudes may be quite different near the midday longitudes for the *UT* data sample; the profile obtained from the harmonic coefficients would be some intermediate value as the result of an intermediate polynomial fitting to the potential function. For the dominant $(n - m) = 1$ terms, if current systems with the foci closer to the equator in location require higher values of m to fit V (Figure 17), then the $n = 2$, $m = 1$ terms may be of insufficient size to be included and the $n = 3$, $m = 2$ terms may be more

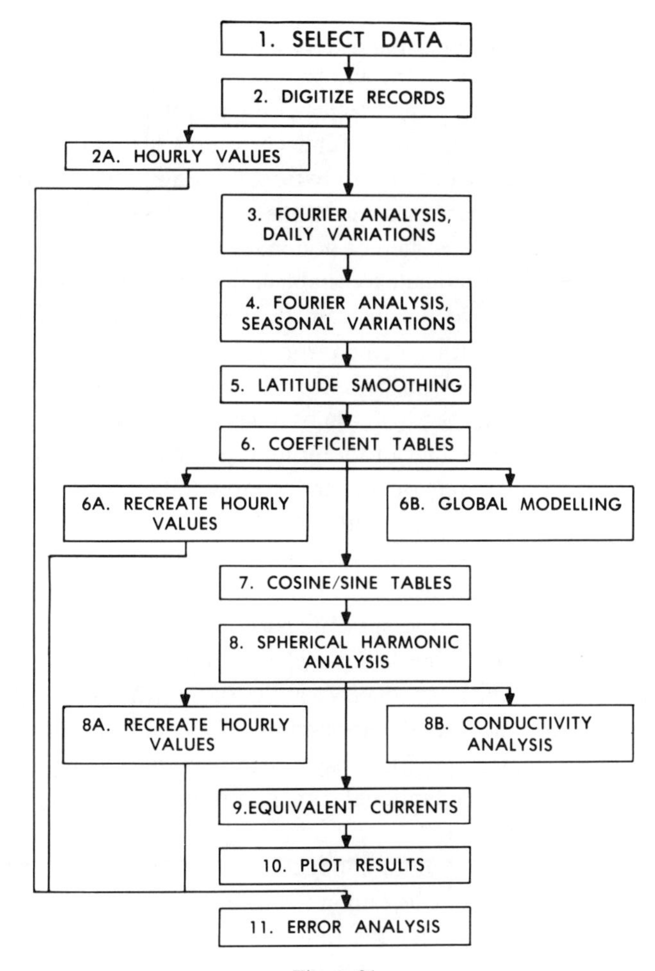

Figure 21
Flow diagram for spherical harmonic analysis of *Sq* daily variations in the geomagnetic field to derive an upper mantle conductivity (from CAMPBELL and SCHIFFMACHER, 1985).

important. Near the solsticial periods of the year, more of the $(n - m)_{\text{even}}$ Legendre polynomials (in addition to the $(n - m)_{\text{odd}}$ that are necessary to establish the oppositely directed *Sq* current vortices in the two hemispheres) must be included for the series representations of the summer-winter hemispherical differences in *Sq*. A change in the selection of the hour of *UT* analysis would bring about a different selection of the fitting polynomials and different conductivity results.

For a sectoral arrangement of observatories, the typical assumption is made that the 24-hour *Sq* variation in local time represents 360° longitude of sampling as the Earth spins under a current system fixed with respect to the Sun. The spherical harmonic functions necessary for the field distribution will have Legendre polynomials that try to match the potential functions representative of both the Northern and Southern Hemisphere parts of the analysis sector. Although the researcher hopes that the average conductivity characteristics of the sector will dominate the picture, the usual irregular station distribution (and corresponding differences in the true conductivity profiles) can only provide a mixture of fitting polynomials that describe neither hemisphere. As the seasons change, one might expect changes in the fitting polynomials that emphasize the different dominant regions contributing to the conductivity determination. It is difficult to establish the latitude most representative of a particular external-internal coefficient pair. If analyses are carried out each month of the year, the 12 conductivities obtained from a given n, m set will vary in a cyclic fashion following the year's cycle in behavior of the potential function.

In the half-sector-model method, an opposite hemisphere half-sector Sq is created from appropriately modified values of the primary region (daily variation of D and Z are reversed in amplitude direction and the months shifted by one-half year). Such modeling restricts the location determination to one hemisphere but introduces another difficulty. The opposite hemisphere duplication of the primary region each six months causes a six-month cycle in the conductivity values to occur with the representation of similar station regions. The contribution of semiannual Sq characteristics modifies this six-month cycle somewhat. The $(n - m)_{\text{odd}}$ coefficients represent Sq current features whose symmetry causes the north and south hemisphere areas dominating the conductivity determination to be equidistant from the equator. The $(n - m)_{\text{even}}$ coefficients represent unsymmetrically located regions in the two hemispheres dominating the polynomials; an interpretation of conductivities from such coefficients is more difficult.

Even if the Earth's upper mantle conductivity were spherically symmetric, the source field location has another effect upon the computed conductivity depth profile. To appreciate this feature, consider the $(n - m) = 3$, P_7^4 polynomial in Figure 17 and assume two current system cases: one has a potential function requiring a large positive value near 45°, the other has a potential function requiring a large negative value near 12° (assume both are appropriately antisymmetric with respect to the equator). The relationship of the external to internal fields can be different for the two cases because the effective wavelength for the induction process is different near 45° from that near 12° for P_7^4. That means, as the source current changes location, the conductivity-depth values determined for a particular n, m index can change. In effect, the location of the source current system can modify the sampling depth in the conductivity profile determination. Therefore, even for laterally homogeneous profiles, when fields are analyzed throughout the course of a year, the changing Sq patterns may provide some variation to the computed sampling depth for a given n, m.

Conductivity Determination

In the method outlined by SCHMUCKER (1970) for profiling the Earth's substructure, formulas were developed that provide the depth (d) and conductivity (σ) of apparent layers (substitute conductors) that would produce surface-field relationships similar to the observed components. These profile values,

$$d_{n,m} = z - p \qquad \text{(km)} \tag{1}$$

and

$$\sigma_{n,m} = 5.4 \times 10^4 (\pi p)^2 \qquad \text{(siemens/meter)} \tag{2}$$

need to be determined for each n,m set of SHA coefficients using the real (z) and imaginary (p) parts of a complex induction transfer function, C_n^m, given as

$$C_n^m = z - ip \qquad \text{(km)} \tag{3}$$

SCHMUCKER (1970) showed that the transfer function is obtained from the ratio of the field components, for a given n, m, as

$$C_n^m = \left[\frac{R(dP_n^m/d\theta)}{n(n+1)P_n^m(\theta)}\right]\left(\frac{Z}{X}\right)_{n,m} \qquad \text{(km)} \tag{4}$$

or

$$C_n^m = i\left[\frac{-mR}{n(n+1)\sin\theta}\right]\left(\frac{Z}{X}\right)_{n,m} \qquad \text{(km)} \tag{5}$$

where X, Y, and Z are the northward, eastward, and into the Earth field components in gammas, R is the Earth's radius in kilometers, θ is the colatitude at the field measurement location, and P_n^m is the Schmidt normalized associated Legendre polynomial. To use the above conductivity-depth formulation, it is now necessary to find the individual n, m SHA terms of the three field components at the field-measurement location. With the fitted potential function (V) given as

$$V = R \sum_{m=1}^{M} \sum_{n=m}^{N} [A_n^m \cos(m\Phi) + B_n^m \sin(m\Phi)]P_n^m = V(\theta, \Phi)_{\text{ex}} + V(\theta, \Phi)_{\text{in}} \tag{6}$$

in which the cosine (A) and sine (B) coefficients of the expansion are taken to be

$$A_n^m = [(a\text{ex})_n^m + (a\text{in})_n^m] \qquad \text{and} \qquad B_n^m = [(b\text{ex})_n^m + (b\text{in})_n^m] \tag{7}$$

for the external (ex) and internal (in) parts given by

$$V(\theta, \Phi)_{\text{ex}} = R \sum_{m=1}^{M} \sum_{n=m}^{N} [(a\text{ex})_n^m \cos(m\Phi) + (b\text{ex})_n^m \sin(m\Phi)]P_n^m, \tag{8}$$

and

$$V(\theta, \Phi)_{\text{in}} = R \sum_{m=1}^{M} \sum_{n=m}^{N} [(a\text{in})_n^m \cos(m\Phi) + (b\text{in})_n^m \sin(m\Phi)]P_n^m. \tag{9}$$

Then the $(a\text{ex})_n^m$, $(b\text{ex})_n^m$, $(a\text{in})_n^m$, and $(b\text{in})_n^m$ represent the cosine (a) and sine (b) external and internal *SHA* coefficients determined in the field separation (cf. MATSUSHITA, 1967; CAMPBELL and SCHIFFMACHER, 1985). Typically the maximum order, M, is taken as 4 and the maximum degree, N, is a value between 8 and 12.

CAMPBELL and ANDERSSEN (1983) showed that Equations (1) and (2) may be determined directly from the *SHA* coefficients with the expressions

$$z = \frac{R}{n(n+1)} \left\{ \frac{A_n^m[n(a\text{ex})_n^m - (n+1)(a\text{in})_n^m + B_n^m[n(b\text{ex})_n^m - (n+1)(b\text{in})_n^m]}{(A_n^m)^2 + (B_n^m)^2} \right\} \tag{10}$$

and

$$p = \frac{R}{n(n+1)} \left\{ \frac{A_n^m[n(b\text{ex})_n^m - (n+1)(b\text{in})_n^m] - B_n^m[n(a\text{ex})_n^m - (n+1)(a\text{in})_n^m]}{(A_n^m)^2 + (B_n^m)^2} \right. \tag{11}$$

The ratio of the internal to external parts of the magnetic surface field, S_n^m, was given by SCHMUCKER (1970) as

$$S_n^m = \frac{R/(n+1) - C_n^m}{R/n + C_n^m} \tag{12}$$

a complex number of the form

$$S_n^m = u + iv = \frac{r_{\text{in}} e^{-i\alpha_{\text{in}}}}{r_{\text{ex}} e^{-i\alpha_{\text{ex}}}} \tag{13}$$

in which r is the amplitude of the wave and α is the associated phase angle. CAMPBELL and ANDERSSEN (1983) found that this function may be simply written in terms of the external and internal *SHA* coefficients as

$$u = \frac{(a\text{ex})_n^m (a\text{in})_n^m + (b\text{ex})_n^m (b\text{in})_n^m}{[(a\text{ex})_n^m]^2 + [(b\text{ex})_n^m]^2} \tag{14}$$

and

$$v = \frac{(b\text{ex})_n^m (a\text{in})_n^m - (a\text{ex})_n^m (b\text{in})_n^m}{[(a\text{ex})_n^m]^2 + [(b\text{ex})_n^m)^2} \tag{15}$$

For a given n, m the argument of S becomes

$$\arg(S) = (\alpha_{\text{ex}} - \alpha_{\text{in}}) = \tan^{-1}\left(\frac{v}{u}\right). \tag{16}$$

S is typically the ratio of small numbers whose uncertainties can lead to computational difficulties when the arg (S) is small, negative, or close to 90°. I find it advisable to restrict the conductivity computations to the values

$$L < \arg(S) < H, \tag{17}$$

with values of $L = 9°$ to $11°$ and $H = 80°$ to $85°$.

There is also a restriction on the transfer function (Equation (3)) argument for a given n, m:

$$0° \geqq \arg(C) = \tan^{-1}(-p/z) \geqq -45°, \tag{18}$$

described by SCHMUCKER (1970), but note a typographical error in the identification of this value following Equation (24) of his paper.

Low amplitude *SHA* coefficients are eliminated from the analysis in various ways. Some authors simply do not introduce *SHA* coefficients unless one of the external-internal pair exceed an estimated error level. I use

$$[(A_n^m) + (B_n^m)^2]^{0.5} \geqq G_m, \tag{19}$$

where G_m is the cutoff amplitude, in gammas, that is different for each m because the Fourier components of the *Sq* potential function usually have lower amplitudes at higher m. For the quiet-year data, I have used values in the ranges 1.5 to 1.2 for G_1, 1.0 to 0.6 for G_2, 0.5 to 0.4 G_3, and 0.4 to 0.2 for G_4 without significant differences in the conductivity results within those ranges. I believe that the preferred ratio of amplitudes is between 1:1/2:1/3:1/4 and 1:2/3:1/3:1/6 for quiet-year *Sq*.

Let us now look at some conductivity determination made from *SHA* coefficients. One of the early analyses was that reported in Chapman and Bartels (1940, page 690) for data of the 1902 quiet year. First note that we must convert from the present notation to their e and i values using

$$\begin{aligned} (a\text{ex})_n^m &= e_{na}^m; \qquad (b\text{ex})_n^m = e_{nb}^m \\ (a\text{in})_n^m &= i_{na}^m; \qquad (b\text{in})_n^m = i_{nb}^m \end{aligned} \tag{20}$$

which, in turn can be computed for their Table 2, page 690, values using their Equations (18), page 691. Only the $(n - m) = 1$ values are given to $m = 4$ for the mean equinox period. Figure 22 shows the conductivity determination using the Schmucker technique with their *SHA* coefficients of *Sq*. The value near 475 km seems unreasonably low. The analysis is rather remarkable considering the fact that the global *SHA* coefficients were based on data from only five stations.

The conductivity values that can be computed from the *SHA* coefficients of MATSUSHITA and MAEDA's (1965) analysis of 18 N–S American sector stations in the three, four-month seasons (Equinox, June Solstice, and December Solstice) of 1958, a year of extreme solar terrestrial activity (cf. Figure 2), is illustrated in Figure 23. Values of G_1 to G_3 were taken as 1.5, 0.75, 0.5, and 0.375 and computation carried out for $(n - m)_{\text{odd}}$. The values do not form a coherent grouping for a number of probable reasons: because of activity, field values above 60° were taken as an extrapolation between the 60° value and a zero value at the pole; the 15 selected days of each season group may have contained nondynamo current systems; a dip coordinate system was employed for the study; and both north and south hemisphere data were included in the anlaysis.

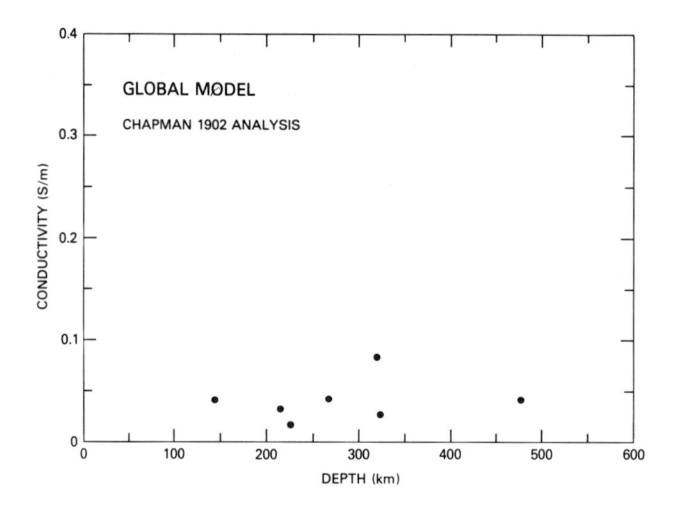

Figure 22
Conductivity values computed with formulas listed herein from Chapman's 1902 separation (CHAPMAN and BARTELS, 1940) of external and internal *SHA* coefficients for a quiet year.

The conductivity values shown in Figure 24 are from a study of *Sq* records in the 1965 quiet year. Data from 13 North American region observatories were analyzed month-by-month on days when all K_p intervals were less than 2+. Lunar variations were removed. Latitude smoothing was accomplished. Values for G_1 to G_4 were taken as 1.2, 0.6, 0.4, and 0.3 gammas. Only conductivity values for $(n - m)_{\text{odd}}$ were

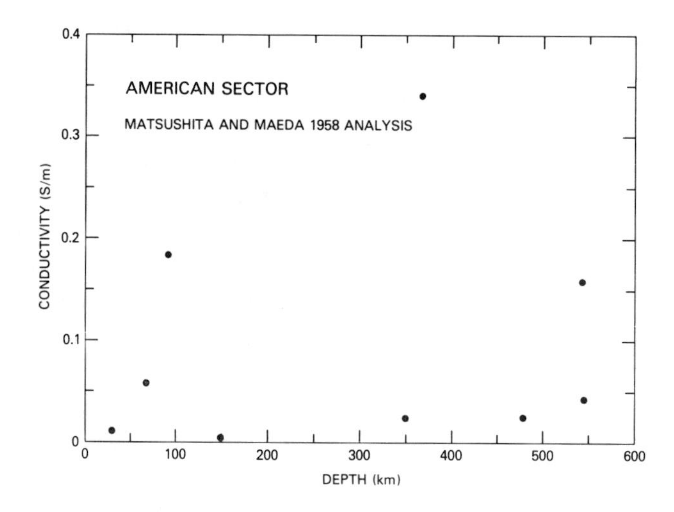

Figure 23
Conductivity values computed with formulas listed herein from MATSUSHITA and MAEDA's (1965) separation of external and internal *SHA* coefficients for the active year, 1958.

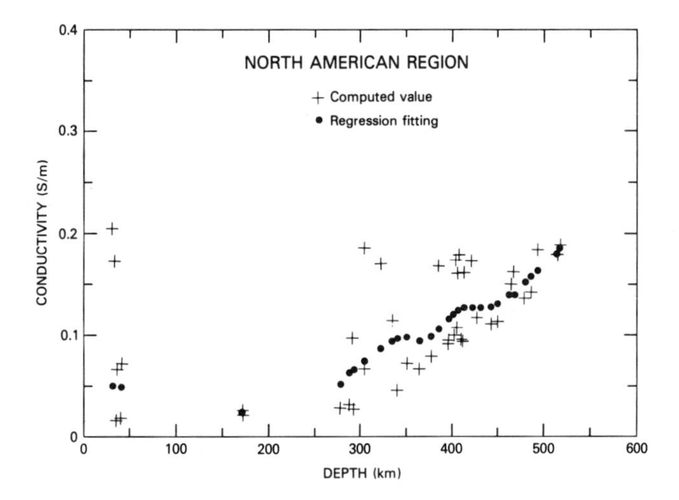

Figure 24
Conductivity values computed with formulas listed herein, following the method outlined in Figure 21, for a North American region with quiet data of 1965. Points indicated by '+' are the direct computations. Points indicated by '*o*' are the values from a locally weighted regression fitting of the direct computations.

computed. The direct computations provide an indication of a highly conducting layer near 30 to 40 km. There is a general rise in conductivity with depth for the values above 170 km.

CAMPBELL and ANDERSSEN (1983) obtained a 'best profile' from a scatter of computed values by fitting an exponential curve to the distribution, excluding points outside a 88-percent-confidence interval about the exponential, and then redetermining a best exponential fitting. They selected an exponential form because an average composition should show a conductivity that increases with depth in that fashion as a result of the known temperature dependence (cf. SHANKLAND and WAFF (1977)). CAMPBELL and SCHIFFMACHER (1986) employed a locally weighted regression method of CLEVELAND (1979) which makes no functional shape assumptions but does require a reasonably large number of values to be fitted. An illustration of this regression fitting is also shown in Figure 24. A heavy weighting coefficient of 0.5 was applied to separate regression fittings with each group of $(n - m)_{odd}$ having different m. Such weighting favors similar wavelength components. Next, a second regression fitting of the above fitted values was accomplished with a light weighting coefficient of 0.35. This weighting provides some smoothing to the profile features distinguished by the various wavelength components. A distinctive profile (Figure 24) results; however, both the original station distribution and the unique feature of the Legendre polynomial fitting of the potential function determine what part of the North American region is represented. I believe that the conductivities of Figure 24 are most influenced by the 25° to 45° geomagnetic latitude, focus region for the observatories of that data set.

If the noise level is small and if the crustal anomalies can be overlooked at an observatory site, it should be possible to extract the regionally (*SHA*) determined external source part of the field and identify the residual as the internal-induced contribution. Then for separated ($n - m$) parts, a conductivity profile unique to the observatory location may be found. A procedure of this type may further decrease some of the location ambiguity remaining in the conductivity analysis procedure described above.

A measure of the conductivity profile error in the locally weighted regression model can be obtained from the distribution of computed values about the fitting. For the Figure 24 example, the absolute deviation in conductivity has a mean value of 0.029 S/m with 95% confidence limits of 0.015 and 0.044 S/m. Depth and deviation were not correlated.

Acknowledgments

I wish to thank Edward R. Schiffmacher for help in preparing some of the figures. The magnetograms used in the North American analysis of Figure 24 were supplied by World Data Center A for the Solar Terrestrial Physics. The *SHA* coefficients of the MATSUSHITA and MAEDA (1965) analysis were provided by S. Matsushita.

REFERENCES

CAMPBELL, W. H. (1976), *Polar cap geomagnetic field responses to solar sector change.* J. Geophys. Res. *81*, 4731–4743.

CAMPBELL, W. H. (1982), *Annual and semiannual changes of the daily quiet variations (Sq) in the geomagnetic field at North American locations.* J. Geophys. Res. 87, 785–796.

CAMPBELL, W. H. and ANDERSSEN, R. S. (1983), *Conductivity of the sub-continental upper mantle: An analysis using quiet-day geomagnetic records of North America.* J. Geomag. Geoelectr. *35*, 367–382.

CAMPBELL, W. H. and MATSUSHITA, S. (1973), *Correspondence of solar field sector direction and polar cap geomagnetic field changes for 1965.* J. Geophys. Res. *78*, 2079–2087.

CAMPBELL, W. H. and SCHIFFMACHER, E. R. (1985), *Quiet ionospheric currents of the Northern Hemisphere derived from geomagnetic field records.* J. Geophys. Res. *90*, 6475–6486; see also corrections in J. Geophys. Res. *91*, 9023–9024, 1986.

CAMPBELL, W. H. and SCHIFFMACHER, E. R. (1986), *Quiet ionospheric currents and Earth conductivity profile computed from quiet time geomagnetic field changes in the region of Australia.* Aus. J. Phys., 40, (in press).

CAROVILLANO, R. L. and FORBES, J. M. (1983), *Solar-terrestrial physics; principles and theoretical foundations.* D. Reidel Pub. Co., 859 p.

CHAPMAN, S. and BARTELS, J. (1940), *Geomagnetism.* University Press, Oxford, 1049 p.

CLEVELAND, W. S. (1979), *Robust locally weighted regression and smoothing scatter plots.* J. Am. Statistical Assn. *74*, 829–833.

FESEN, C. G., DICKINSON, R. E. and ROBLE, R. G. (1986), *Simulation of the thermospheric tides at equinox with the National Center for Atmospheric Research thermospheric general circulation model.* J. Geophys. Res. *91*, 4471–4489.

JONES, A. G. (1983), *On the equivalence of the 'Niblett' and 'Bostick' transformations in the magnetotelluric method.* J. Geophys. Res. *53*, 72–73.

MATSUSHITA, S. (1967), *Solar quiet and lunar daily variation fields*, In Physics of Geomagnetic Phenomena (eds S. Matsushita and W. H. Campbell). Academic Press, New York, 302–424.

MATSUSHITA, S. and CAMPBELL, W. H. (1972), *Lunar semidiurnal variations of the geomagnetic field determined from the 2.5-min data scalings.* J. Atmos. Terr. Phys. *34*, 1187–1200.

MATSUSHITA, S. and MAEDA, H. (1965), *On the geomagnetic solar quiet daily variation field during the IGY.* J. Geophys. Res. *70*, 2535–2558.

MATSUSHITA, S., TARPLEY, J. D. and CAMPBELL, W. H. (1973), *IMF sector structure effects on quiet magnetic field.* Radio Sci. *8*, 963–972.

MAYAUD, P. N. (1980), *Derivation, meaning, and use of geomagnetic indices.* Amer. Geophys. Union, Washington, D.C., 154 p.

MEYERS, H. and ALLEN, J. H. (1977). *Some summary geomagnetic activity data, 1932–1976, NGSDC Data Fact Sheet No. 1.* World Data Center A. NOAA, Boulder, Colorado, USA, 12 p.

OLSON, W. P. (1970), *Variations in the Earth's surface magnetic field from the magnetopause current system.* Planet. Space Sci. *18*, 1471–1484.

PARKINSON, W. D. (1983), *Introduction to Geomagnetism.* Scottish Academic Press, Edinburgh, 433 p.

SCHMUCKER, U. (1970), *An introduction to induction anomalies.* J. Geomag. Geoelectr. *22*, 9–33.

SCHMUCKER, U. (1979), *Erdmagnetische Variationen und die electrische Leitfähigkeit in tieferen Schichten der Erde.* Sitzungsberichte u. Mitteilungen der Braunschweigischen Wissenschaftlichen Gesellschaft *4*, 45–102.

SHANKLAND, T. J. and WAFF, H. S. (1977), *Partial melting and electrical conductivity anomalies in the upper mantle.* J. Geophys. Res. *82*, 5409–5417.

WALKER, E. (1866), *Terrestrial and Cosmical Magnetism.* Deighton, Bell, and Co., Cambridge.

WEIDELT, P., MÜLLER, W., LOSECKE, W. and KNÖDEL, K. (1980), *Die Bostick Transformation*, In Protokoll über das Kolloquium Elektromagnetische Tiefenforschun (eds. V. Haak and J. Homilius). Berlin-Hannover, pp. 227–230.

WILKINSON, D. C. (1986), Personal communication. National Geophysical Data Center, World Data Center A, NOAA, Boulder, Colorado, USA.

(Received 25th July, 1986, revised 12th September, 1986, accepted 15th September, 1986)

PAGEOPH, Vol. 125, Nos. 2/3 (1987) 0033-4553/87/030459-05$1.50+0.20/0

Limitations in the Use of Spherical Harmonic Methods for Deep Conductivity Determinations

W. D. PARKINSON [1]

Abstract—Diurnal variations supply only four frequencies with sufficient power to be useful, giving an overall frequency ratio of only 4. Also only permanent observatories supply sufficient accurate data, and their distribution on the Earth is very irregular. More importantly, the assumption implicit in deep conductivity determinations, that the conductivity is a function of depth only, breaks down because of near surface conductivity anomalies.

Key words: Earth's interior, Earth's conductivity, diurnal variation, GDS.

The first quantitative determinations of the conductivity of the Earth's interior came as a by-product of the work of Schuster. He showed that the sense of the vertical component, in relation to that of the horizontal components, required an external origin for the diurnal variation field. However a knowledge of the horizontal components, required an external origin for the diurnal variation field. However a knowledge of the horizontal components is sufficient to specify the vertical component exactly if the origin is entirely external. When SCHUSTER (1889) tried this, he found that the vertical component was too small. There must also be a part of the field of internal origin whose vertical component is opposite in direction to that of the external field. He deduced that this internal component was due to eddy currents induced by the external field. Schuster remarked 'This we might have expected'. What was probably not expected was the magnitude of the internal part, for an Earth composed entirely of the kind of rocks that are common at the surface would not give rise to such a large internal field. There is evidently an increase in conductivity with depth. It is interesting to note that Schuster considered the increase in conductivity with temperature in semiconductors to be the likely cause of this conductivity increase. Thus the possibility of determining the conductivity of the interior of the Earth came about.

The technique used by Schuster to determine the expected magnitude of the vertical component was that introduced half a century earlier by Gauss. Each

[1] Geology Department, University of Tasmania, Box 252-C, Hobart, Australia 7001.

sinusoidal wave of the diurnal variation, as derived by a Fourier analysis, is analysed into a series of surface spherical harmonics. It is found that the analysis of the wave of period (1 day)/k can be moderately well represented by the spherical harmonic of the form

$$P^k_{k+1}(\cos\theta)\cdot\cos(k\varphi + \psi) \quad (1)$$

where P is the associated Legendre function, θ the colatitude, φ the longitude and ψ a phase angle.

Schuster was able to use only four observatories, and of these he relied mainly on Lisbon, Greenwich and Bombay. The same technique was used by CHAPMAN (1919) who made use of 21 observatories. He confirmed Schuster's conclusion that in order to account for the phase and magnitude of the vertical component it was necessary to assume that the conductivity of the Earth is not uniform but increases with depth. This presents the challenge of determining the distribution of conductivity as a function of depth within the Earth. With some notable exceptions this task has been tackled mainly by refinements of the spherical harmonic method used by Schuster. A glance at plots of this distribution according to various authorities (e.g. GARLAND, 1979) indicates that only slight advance on the work of Schuster and Chapman has been made. It is interesting to inquire why this is so.

Even a casual inspection of magnetograms from various observatories reveals that the diurnal variations are strongly dependent on latitude but that the longitude dependence is mainly one of dependence on local time. That is to say, to a first approximation, the diurnal variation field depends on latitude and local time. But which latitude; geographic, geomagnetic or magnetic? Unfortunately, none of these. If we omit consideration of sites near the magnetic equator, MAEDA (1953) has shown that we can define a pole somewhere between the geographic and geomagnetic poles such that, using this as a reference for latitude, the above approximation surpasses that using any other reference pole. But when we come to lower latitudes, within 5 degrees of the magnetic equator, a new phenomenon appears, namely the equatorial electrojet. The treatment of this has always been a problem. To represent it precisely requires a long series of spherical harmonics. We could simply ignore it and confine our analysis to nonequatorial sites. We shall then derive a smoothed field that omits the electrojet. But there must be return currents at some latitude. According to SUZUKI (1973) they flow at low temperate latitudes and so must have some effect on the *Sq* pattern. Actually if the electrojet is omitted and the return current included in the analysis, the resulting field is non potential. The parameter that controls electromagnetic induction for an oscillation of period T in a body of conductivity σ and permeability μ is the skin depth $(T/\pi\mu\sigma)^{\frac{1}{2}}$ which is proportional to the square root of the period. Now the longest period we can derive from *Sq* is 24 hours; the shortest is 6 hours and the amount of power in the 6 hour wave is very small. Thus we have a range of periods extending over a factor of 4. Translated into skin depth this amounts to a factor of 2. The objective is to measure the conductivity throughout the mantle. Therefore the limitation of period range is a very serious one. It is significant that the most useful determinations of the conductivity to great depth have been made by abandoning the diurnal variations as a source of data and analysing the much smaller continuous spectrum derived from erratically occurring storms and substorms.

Another, and probably more important limitation in the conventional manner of determining conductivity distributions from *Sq*, is the network of observatories from which the data come. Schuster used 4 sites, Chapman 21 and MATSUSHITA and MAEDA (1965) used 69. This sounds like a remarkable improvement. However, the important matter is the size of the largest area devoid of observatories rather than the number of observatories. Chapman's analysis left a gap of 3.2 steradians or 25% of the Earth's surface and the largest gap in the distribution of observatories used by MATSUSHITA and MAEDA (1965) is 1.3 steradians or 10% of the Earth's surface. The situation is less grave than this indicates because the dependence on longitude is much weaker than on latitude. If we consider the distribution of observatories purely in latitude then Chapman's leaves a gap of about 20 degrees. One method of handling the weak dependence on longitude is to divide the Earth into longitude zones and consider the field as a function of latitude within each zone. MATSUSHITA and MAEDA (1965) used three zones. If we examine each of their zones, again we find gaps of about 20 degrees.

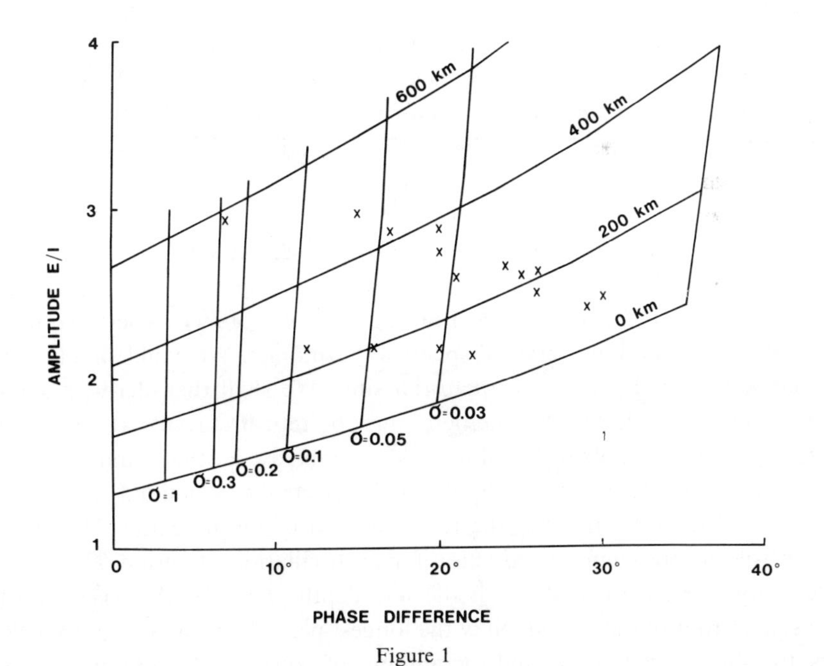

Figure 1
Amplitude and phase of the ratio of external to internal potentials of *Sq* at sites of the north-west India survey. The superimposed grid indicates the corresponding conductivity (in S m^{-1}) and depth of the uniform core model.

Finally, and perhaps most importantly, there is the assumption that the conductivity is a function of depth only. There are many causes of conductivity inhomogeneities in the crust and upper mantle; the oceans being the most widespread. It seems inevitable that any global conductivity distribution we derive in the near future from diurnal variations will be biased in favour of the continents and against the oceans. Many observatories are near coastlines and the ratio of vertical to horizontal components is influenced by the presence of the nearby ocean. This has been demonstrated for the east and west coasts of both Canada and Australia (LILLEY, 1979; LILLEY and PARKER, 1976).

Quite apart from oceans there are many conductivity anomalies that can influence the field as recorded at a magnetic observatory. The survey made in north-

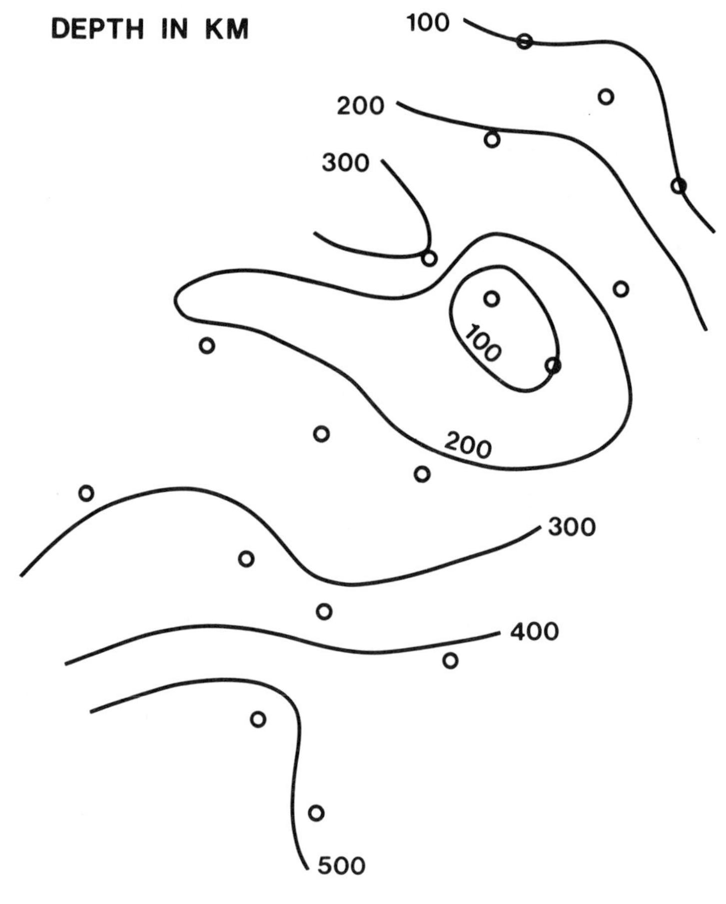

Figure 2
Contours of depths derived from Figure 1.

west India by ARORA *et al.* (1982) illustrates this point. Following the method developed by SCHMUCKER (1970) we can use the amplitudes of the north and east components of the 12-hour wave to determine an effective latitude. Assuming that the 12-hour wave is sufficiently well expressed by equation (1), we can determine the ratio of external to internal fields at each site. These are complex numbers which allow us to act upon their amplitudes and phases. To acquire some insight into the distribution of conductivity with depth, we can use the uniform core model (see e.g., RIKITAKE, 1966). The complex ratio of external to internal fields allows us to determine the depth and conductivity of the equivalent conductor in the uniform core model. This is illustrated in Figure 1 in which the amplitude and phase differences are ordinate and abscissa. A grid of conductivity and depth allows us to determine immediately the corresponding uniform core model. The scatter of points shows the varying models required for the sites of this array. Figure 2 is a contour map of the depths derived from Figure 1. The influence of the conductivity anomaly near the north of the array can be clearly seen.

To summarise, there are three limitations on the use of spherical harmonic analysis of diurnal variations for the determination of the global conductivity. They are (i) limited period range, (ii) irregular and biased observatory distribution, and (iii) the presence of surface and near surface conductivity anomalies.

REFERENCES

ARORA, B. R., LILLEY, F. E. M., SLOANE, M. N., SINGH, B. P., SRIVASTAVA, B. J. and PRASAD, S. N., *Geomagnetic induction and conductive structures in north-west India.* Geophys. J. R. Astr. Soc. *69*, 459–475.

CHAPMAN, S. (1919), *Solar and lunar diurnal variations of terrestrial magnetism.* Phil. Trans. Roy. Soc. Lond. *A218*, 1–118.

GARLAND, G. D. (1979), *Introduction to Geophysics.* W. B. Sauners & Co., Philadelphia, p. 261.

LILLEY, F. E. M., (1979), *Magnetic daily variations compared between the east and west coasts of Canada,* Canad. J. Earth Sci. *16*, 585–592.

LILLEY, F. E. M. and PARKER R. L. (1976), *Magnetic daily variations compared between the east and west coasts of Australia.* Geophys. J. Roy. Astr. Soc. *44*, 719–744.

MAEDA, H. (1953), *On the residual part of the Sq field in the middle and lower latitudes during the IPY.* J. Geomag. Geoelect. *5*, 39–51.

MATSUSHITA, S. and MAEDA, H. (1965), *On the geomagnetic solar quiet day variation field during the IGY.* J. Geophys. Res. *70*, 2535–2558.

RIKITAKE, T. (1966), *Electromagnetism and the Earth's Interior.* Elsevier, Amsterdam, p. 222.

SCHMUCKER, U. (1970), *An introduction to induction anomalies.* J. Geomag. Geoelect. *22*, 9–33.

SCHUSTER, A. (1889), *The diurnal variation of terrestrial magnetism.* Phil. Trans. Roy. Soc. Lond. *A180*, 467–518.

SUZUKI, A. (1973), *Return flow of the equatorial electrojet.* J. Geomag. Geoelect. *25*, 249–258.

(Received 22nd October, 1985, revised/accepted June, 1986)

PAGEOPH, Vol. 125, Nos. 2/3 (1987) 0033-4553/87/030465-18$1.50+0.20/0

MTS Studies on the Upper Mantle Conductivity in China

LIU GUODONG [1]

Abstract—Based on the data from more than 200 MTS sites distributed within different areas of the Chinese continent, general characteristics of upper mantle conductivity have been described. At least two conductive layers have been found in the upper mantle of some areas. The first is thin with a resistivity of a modicum to few tens Ωm; the second one is thicker with a resistivity of one to few Ωm. Nearly 300 heat-flow values indicate that there exists an exponential correspondent relationship between a depth of the upper mantle conductive layer with a thickness and an average value of heat flow. Based on the above results, the top depth map of this upper mantle conductive layer has been outlined for parts of the Chinese continent. This conductive layer is basically consistent with the low velocity zone in the upper mantle, and Cenozoic tectonism and current seismicity are significantly related to the variation of depth of the conductive layer in the upper mantle. The possible origins of the conductive layers in the upper mantle have been discussed here.

Key words: Magnetotelluric, heat-flow, upper mantle conductivity, Chinese continent.

1. General Neotectonics of the Chinese Continent

The Chinese continent located between the subduction zone of the west Pacific plate and the collision zone of the Himalayas can be divided into a western part and an eastern part in terms of geological and geophysical data. The boundary of the two parts, north-south geotectonic zone, adjoins an eastern longitude of 102° (South) to 107° (North).

In the western part there are the Tibetan Plateau; the Cenozoic basins of Talimu, Chaidamu, and Zhungeer; the ranges of Kunlun-Aerjin-Qilian; and Tianshan. The thickness of Cenozoic deposits in these basins generally reaches 6,000–7,000 m. The geological trend is mainly NWW direction and neotectonic movements are very strong. According to the relevelling data, the Tibetan plateau had a violent uplift with average rate of 2–10 mm per year from north to south, and the Talimu, Chaidamu, and Zhungeer basins sank with average rates of 1–3 mm, 2–4 mm, and 5–10 mm per year, respectively. Strong earthquakes have frequently taken place beneath the Tibetan plateau and the margins of the Cenozoic basins (Figure 1).

[1] Institute of Geology, State Seismological Bureau, Beijing, China.

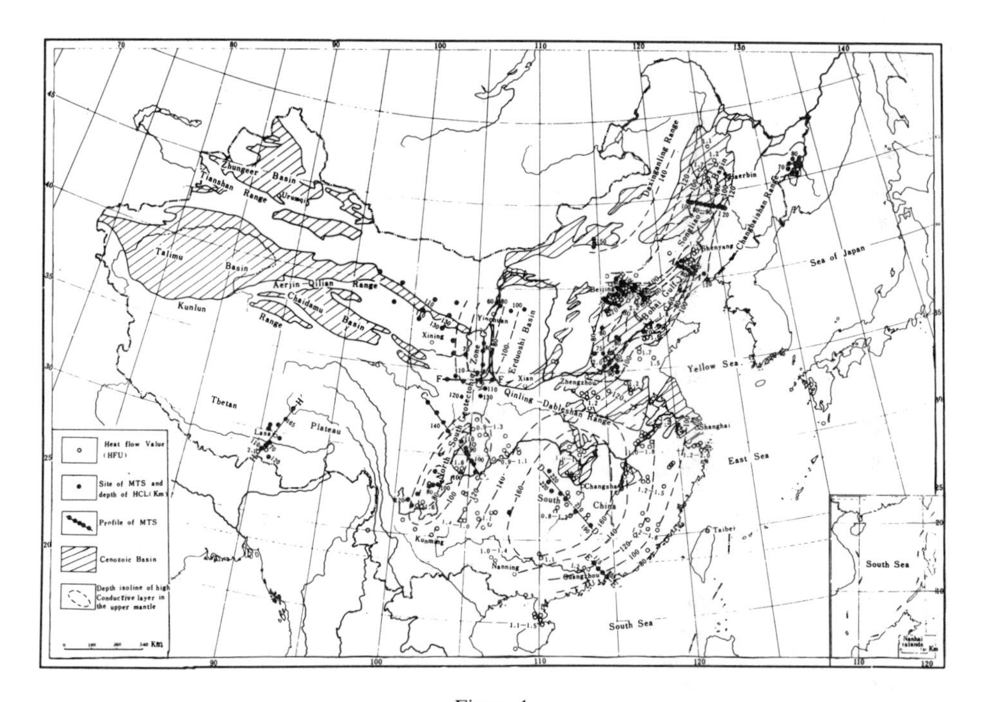

Figure 1
Distribution of Cenozoic basin, MTS sites with the depth of upper mantle conductive layer in km heat flow sites with values in HFU. (In areas where the sites are too dense only representative sites are shown and depth isoline of upper mantle conductive layer.)

In the eastern part of the Chinese continent, even though the geological trends are mainly NNE, the features of neotectonics are very different in diverse regions, for example:

(1) South and southwest China have been slowly uplifting since the later Mesozoic period and the seismicity has been very weak, with the exception of the southeastern coast where the neotectonism and seismicity are strong.

(2) The north China area called North China Platform has two different geologic regions. The Erduoshi basin in the western region wholly subsided during the Mesozoic period and has wholly uplifted since the Cenozoic period. Currently it has a slow uplift rate of 1–2 mm per year with no seismicity. However, the grabens were generated around the basin during the Cenozoic period and the thickness of Cenozoic deposits ranges from 2,000 to 8,000 m in the grabens, the seismicity is very strong and sinking activity is outstanding along the surrounding grabens. The North China Plain and the surrounding mountains in the eastern region belong to the Cenozoic rift system (Liu Guodong, *et al.*, 1982). Strong earthquakes have frequently taken place around the area of the rift system and beneath the Bohai Gulf.

(3) The northeast China area includes the Songliao basin in the middle part, Daxinganling range in the western part, and Changbaishan range in the eastern part. The Songliao basin was a rift during the Mesozoic period and has Mesozoic sediments 5,000–6,000 m thick while the Cenozoic sediments are only 200–400 m thick. The Daxinganling and Changbaishan ranges were uplifting during the Cenozoic period.

The north-south geotectonic zone, the boundary between western and eastern parts of the Chinese continent, is the strongest seismic zone, in which a series of rift valleys grew at different times during the Cenozoic period.

2. *MT Sounding in China*

An analogue and digital MT system was developed by the Institute of Geology in the 1970's, the state seismological bureau cooperating with the Geophysical Prospecting Bureau of Petroleum Ministry. The magnetic sensors are inductive coils with a core of mu-metal. The lower frequency limit of this MT system is 0.00025 Hz. China also has bought some MT equipment from other countries. Measurements have been made for more than 500 MT sites on the Chinese continent, 200 MT sites are for the study on the conductivity of crust and upper mantle, and more than 300 MT sites are mainly for the study of the sedimentary basins. The majority of MT sites for deep research are distributed in North China and the northern segment of north-south geotectonic zone and their surroundings, while only a few were placed in South China, Northeast China, and Tibet areas (Figure 1).

The available period of MT signal spectra for deep sounding is usually less than 3,000 seconds, therefore we only have upper mantle conductivity information from them. The recording duration for one MT site is usually 3–5 days in the field. The high quality MT signal segments were selected for data processing. The MT data, at first, were processed with a high-pass digital filter and convolved with a Hanning window to reduce truncation errors, and then were transformed into a spectrum by Fourier transformation. Thereafter, the frequency responses of the instrument were corrected, the average power spectra in the given frequency bands were calculated; the tensor impedance elements, the direction of the principal axes, the 'skewness' rations, the E-predictions, and phasor coherences were estimated. Finally, the data satisfying the acceptance criteria were chosen to calculate the apparent resistivity curves in E-polarization ($\rho_{\parallel}$) and H-polarization ($\rho_{\perp}$) (SIMS *et al.*, 1971). Modified analysis of band limited MT signal pairs and the cross-power method were also used for estimating the tensor impedance elements in some cases (JIN GUANGWEN *et al.*, 1982; GOUBAU *et al.*, 1978).

There are differences between an apparent resistivity curve in E-polarization and an apparent resistivity in H-polarization in most of the MT sites. These differences are mainly caused by an inhomogeneous conductivity in the shallow layers. In this case the curves in E-polarization were customarily used to study the conductivity of upper mantle, because they have a minimum distortion on the long periods due to inhomogeneity of the shallow layers. Therefore, the curves in E-polarization were generally fitted to one-dimensional models, while two-dimensional inversions for a few MT profiles have been done for the purpose of checking the validity of the one-dimensional inversion.

Nearly 300 heat flow values in the Chinese continent have been published to date (CHEN MOXIANG *et al.*, 1982; ZHANG RUHAI *et al.*, 1982; WANG TIYANG *et al.*, 1985; WU QIANFAN *et al.*, 1985; WANG JUN and LIU GUODONG, 1985), among them 120 measurements were distributed within the North China Plain and the surrounding mountings, 160 measurements within South and Southwest China, and some 20 measurements within other regions in China. The accuracy of most of the heat flow values in South and Southwest China is usually low, because the thermal conductivities of the rocks were presumed from rock cores recorded from the one bore hole in which the geothermal gradient had been measured (WANG JUN and LIU GOUDONG, 1985).

Since there exists a close relationship between the upper mantle conductive layer depth and the heat flow, the heat flow values will be equated and discussed in this work. Nearly 200 MT sites, with a depth for the upper mantle conductive layer and 300 heat flow values, are shown in Figure 1.

3. The Upper Mantle Conductivities of Some Areas in China

1. The North China Plain and surrounding mountains

This area is situated in the eastern part of the North China platform. The North China Plain is a Cenozoic rift system which starts at Shenyang in the north and terminates in the south around Zhengzhou, with a total length of about 1,000 km. The maximum width of the system is 400 km at the middle, with widths of 50–100 km at the south and north, and a trend of NNE. The terrain inside the rift system is low lying, covered by Bohai Gulf in its furthest east and northeast parts. This rift system consists of a series of depressions and uplifts. The thickness of Cenozoic deposits in the depressions may reach more than 6,000–7,000 m, the maximum thickness is 12,000 m in the Bohai Gulf depression, and the accumulated thickness of Cenozoic basalts is about 1,000–2,000 m in the depressions. Relevelling data shows the North China Plain rift system still has strong differential movement. In the Bohai Gulf area, probably with Bohai Gulf as its center, there is a violent sinking zone; over the past 14 years, the average relative drop in the Bohai Gulf coast has reached 50–150 mm. The surrounding mountains of the North China Plain rift system have maximum elevations of 2,000–3,000 m and have been uplifting since the early Cenozoic, experiencing a violent uplift over the past 14 years with an average relative uplift of 20–60 mm. The North China Plain rift system is a region of strong seismic activity in the eastern part of the Chinese continent. In the years since historical records have been kept, 51 earthquakes of magnitude more than 6 have occurred. In the ten years between the Xingtai earthquakes of 1966 and the Tangshan earthquakes of 1976, 15 earthquakes of magnitude greater than 6 have occurred. More of the strong earthquakes occur along the border of the rift system. Particularly, there is a strong seismic zone with direction NWW called the Zhangjiakou-Bohai seismic zone, extending from the north margin of the rift system to the center of the Bohai Gulf (LIU GUODONG, 1985).

Eighty MT sites for deep research have been measured in the North China Plain rift system and surrounding areas, with more than 30 of the sites distributed within the Lower Liaohe depression, which is in the northern branch of the rift system (Figure 1). The typical apparent resistivity curves are shown on Figures 2a and 2b. The results of the one-dimensional inversion for E-polarization curves indicate that there exists two conductive layers in the upper mantle. The first conductive layer has a depth of 35–40 km, a thickness of about 4–5 km, and resistivity of about 3–4 Ωm. The depth of the Mohe in the Lower Liaohe depression is 30–32 km, found by deep seismic sounding (LU ZAOXUN, 1985). Therefore, the first conductive layer of the upper mantle is a thin layer and occurs just under the Mohe discontinuity. The second conductive layer has a depth of 80–90 km and a resistivity of 1–2 Ωm, its

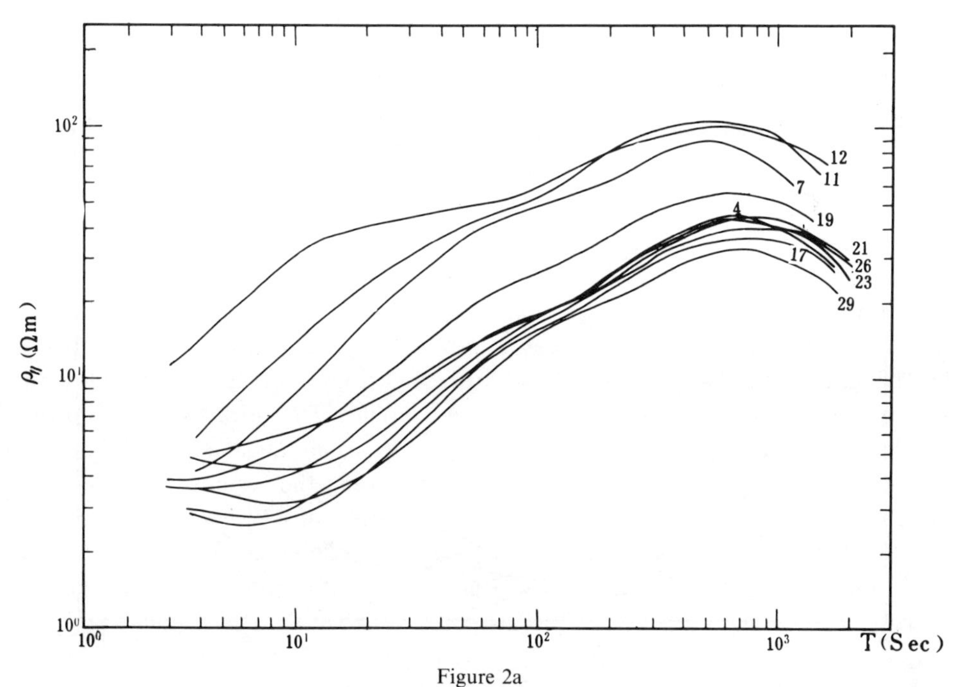

Figure 2a
Apparent resistivity Curve of E = polarization in the Lower Liaohe depression. Figure on curve shows a number of MT sites.

thickness cannot be determined because the recorded periods of MT signals are not long enough, but we can reasonably suppose the second conductive layer of the upper mantle has thickness enough. The first conductive layer of the upper mantle seems to disappear beneath the marginal area of this depression, where there exists the intracrustal conductive layer with a depth of 20 km, a thickness of 4–5 km, and resistivity of about 3–4 Ωm. The resistance layer between the two conductive layers of the upper mantle has a resistivity of 10^3 Ωm found by inversion of one-dimensional layered models. Figure 3 is the geoelectrical structure in the crust and upper mantle crossing the Lower Liaohe depression and its marginal area along NWW direction.

Nearly fifty MT sites were distributed within the middle and southern branches and surrounding areas of the North China Plain rift system (Figure 1). The typical apparent resistivity curves are shown in Figure 4. The result of the one-dimensional inversion for *E*-polarization curves indicate that the depth of the conductive layer with a resistivity of few Ωm in the upper mantle ranges from 50–60 km around the Bohai Gulf to more than 100 km in the surrounding mountain areas (Figure 5) (LIU GUODONG, 1983). According to the result of seismic data inversion, the low velocity zone with velocity value of 7.8 km/sec has been found at a depth of roughly 60–70 km in the North China Plain (SONG ZHONGHE, 1985). Thus we can say that both the conductive layer and the low velocity zone in the upper mantle may coincide in this

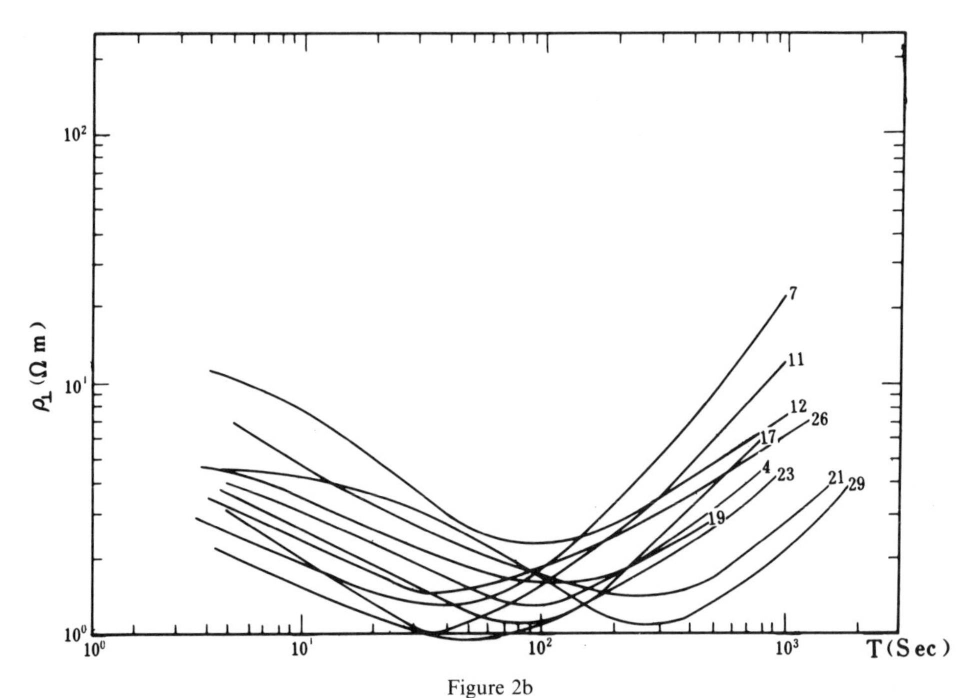

Figure 2b
Apparent resistivity curves of *H* = polarization in the Lower Laiohe depression. Figure on curve shows a number of MT sites.

area. A conductive thin layer of the upper mantle does not exist in this area, however, the intracrustal conductive layer has been measured at depths from 15–20 km in the plain to 25 km in the mountain areas (LIU GUODONG *et al.*, 1984). Figure 6 shows the distribution of resistivity from MT data and velocity from seismic data versus depth in different geological units of this area.

120 measurements of heat flow values in the North China Plain and the surrounding mountains (Figure 1) showed that the average heat flow exceeds 1.6 HFU inside the rift system. In particular, the heat flow values in the Bohai Gulf and its coastal areas range from 1.77 to 2.53 HFU, while the surrounding mountains average only 1.0 HFU (LIU GUODONG, 1985).

2. *Northeast region*

In the Northeast region the magnetotelluric soundings have been done at 50 sites, among them are 18 sites completed by the Department of Geophysics, Geological College of Changchun, distributed within the Song-Liao basin while approximately 30 sites are within the Hulin basin (see Figure 1).

The Song-Liao basin is a Mesozoic rift with a Jurasic-Cretaceous sediment of 5,000–6,000 m thickness and widely grown volcanic rocks. The rifting activity stopped at the beginning of the Cenozoic era, with Cenozoic deposits of only 200–400 m.

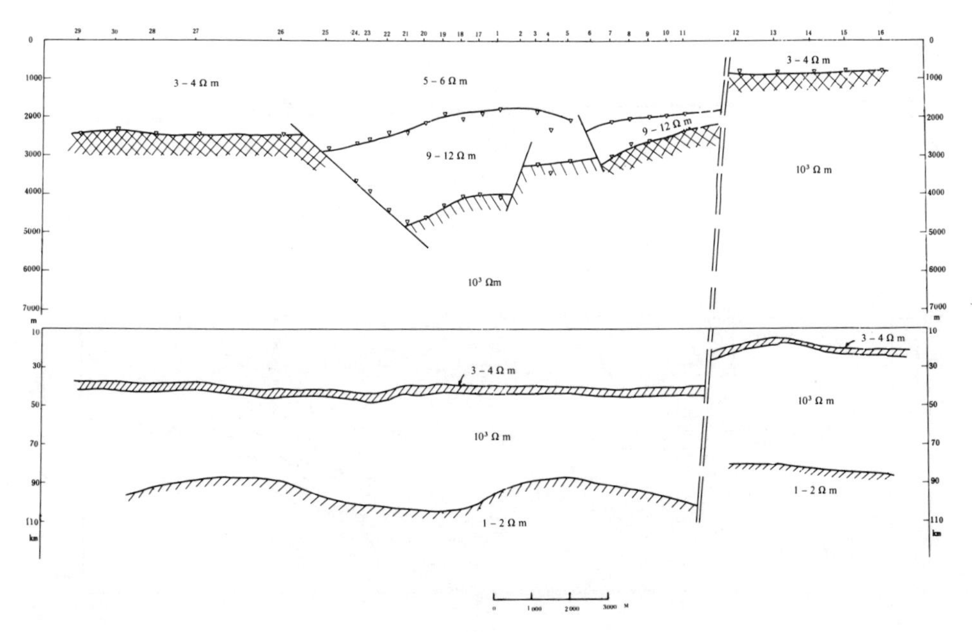

Figure 3
The geoelectrical structure in the crust and upper mantle crossing the Lower Liaohe depression and its marginal area along NWW direction. (Profile A–A′ on Figure 1)

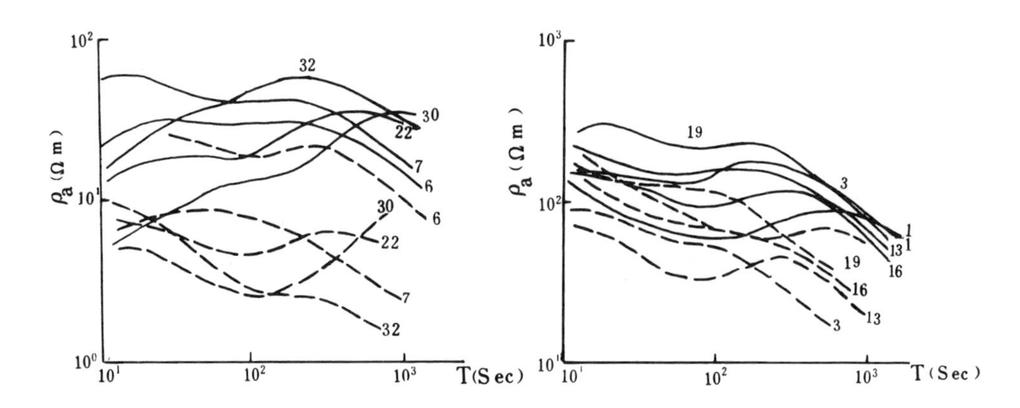

Figure 4
Some apparent resistivity curves of MTS in the North China Plain rift system (numbers 6, 7, 22, 30, and 32) and surrounding mountain areas (numbers 1, 3, 13, 15, and 19). Solid line is the curve of E = polarization Dashed line is the curve of H = polarization.

Magnetotelluric sounding results show that the conductive layer of the upper mantle within the Song-Liao basin appears at a depth 80–90 km while around it is at depths reaching 120–140 km. The values of heat flow measured at 10 sites located in this basin and in the surrounding area showed that the average value of heat flow within the basin is approxımately 1.90 HFU while that in the surrounding area is approximately 1.15 HFU. Obviously, there is a correspondent relationship between the depth of upper mantle conductive layer and the average value of heat flow.

Hulin basin is a small scale basin formed during the Cenozoic period and is located on the eastern border of the Northeast region, 250 km from the Sea of Japan. This basin has Cenozoic deposits with maximum depths more than 1000 m and a large amount of tertiary and quaternary basalt in the surrounding area. Figure

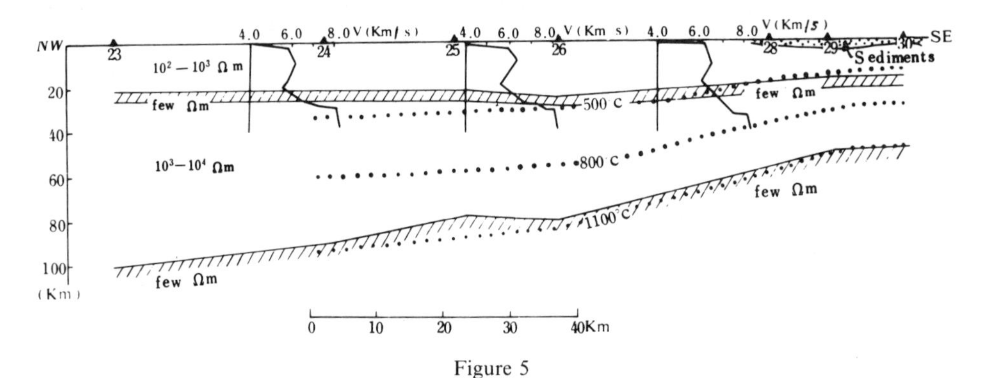

Figure 5
The geoelectric velocity, and geothermal structures in the crust and upper mantle from the northern coast of the Bohai Gulf to the Yenshan mountain area along NW direction. (Profile B–B′ on Figure 1)

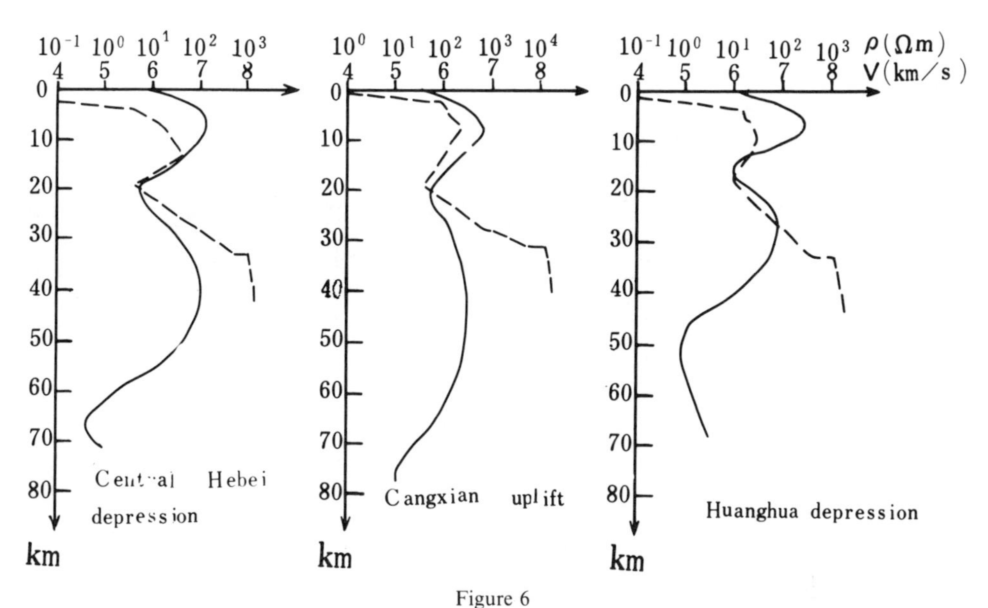

Figure 6
The distribution of resistivity from MTS data and of velocity from seismic data versus depth in different geological units of North China Plain rift system. MT data is fitted to one-dimensional continuous models.

7 shows the typical apparent resistivity curves from magnetotelluric sounding in the Hulin basin. The result of one-dimensional inversion on the curve of E = polarization shows that the depth of the first conductive layer of upper mantle is about 70 km with a thickness of 15 km and resistivity of 20 Ωm, while the depth of the second conductive layer is about 220 km with resistivity of few Ωm.

3. South China and Southwest region

South China and the Southwest region are the most stable areas of modern tectonic activity. There is no growth of large scale Cenozoic basins and very little Cenozoic magma activity, which is centered in the area of the southeast coast and Hainan island. The relevelling data show that the yearly uprising rate of this region is 1–2 mm/y and it has not experienced any strong earthquakes excepting in the southeast coast area.

There has been little data of upper mantle condutivity provided for this region, because MTS measurements have been completed at only 30 sites. These MTS sites are distributed along three profiles of NWW direction and one profile of NNE direction (see Figure 1). The regional geological tectonics trend for the most part in this region is NE-NNE. Figure 8a shows the geoelectrical structure of the crust and upper mantle along a NWW profile cutting laterally through the Mesozoic-Paleozoic Shichuan basin and the Pre-Cambrian Songpan block. Obviously, it can be seen

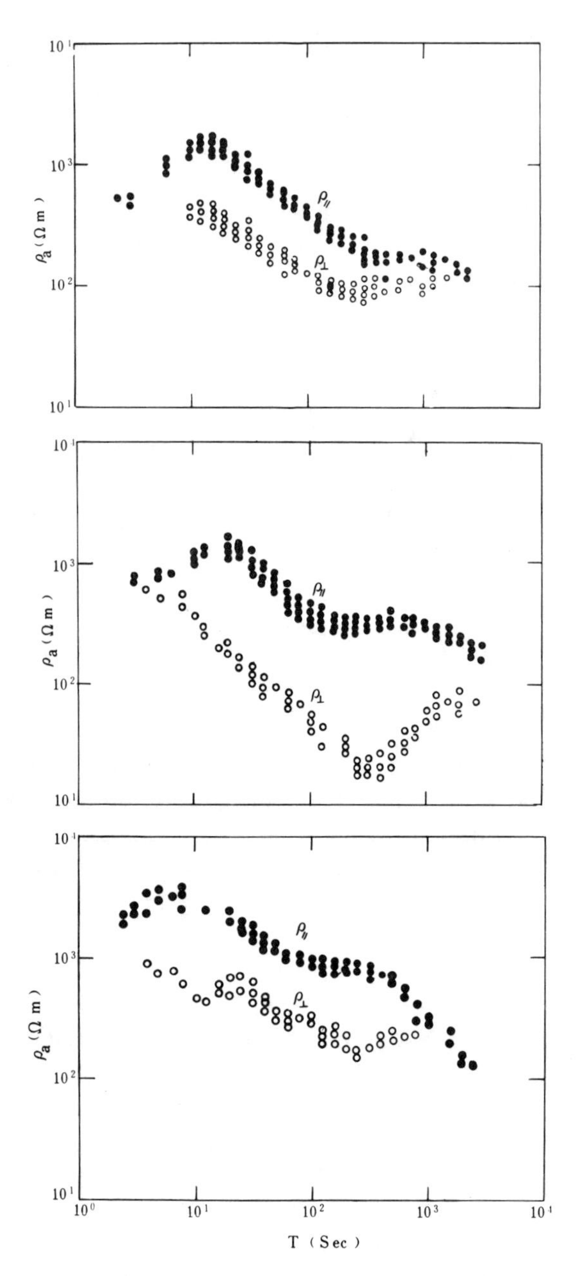

Figure 7
Typical apparent resistivity curves from MTS in the Hulin basin.

that a conductive thin layer exists under the Shichuan basin at a depth around 100 km with a thickness of about 10 km and resistivity of ten to few tens Ωm. After extending west through the north-south geological tectonic zone this layer deepens to 140 km under the Songpan block (LI LI and JIN GUOYOUN, 1985). Figure 8b shows the geoelectrical structure of the crust and upper mantle along a NWW profile cutting laterally through the Hunan Jiangnan old continent and the Caledonian folding zone. A relatively conductive thin layer exists under some MTS sites along this profile at a depth of 50–60 km with a thickness of few km and resistivity of few tens Ωm. Meanwhile, the conductive layer is usually found at depths of 190–220 km with a resistivity of few Ωm. Figure 8c shows a MTS profile perpendicular to the southeast coast with a length of only 120 km and with only four MTS sites. A conductive layer exists along this profile at a depth of 90–115 km, with its depth increasing from east to west, a thickness of about 25 km and a resistivity of few Ωm.

The distribution of heat flow values in South China and the Southwest region may be characterized by the following (see Figure 1): the heat flow values in the Shechunan basin and its surrounding area vary between 1.8 and 0.8 FHU with an average of 1.2 HFU; most of the heat flow values in the Hunan province vary within the range 0.8–1.2 HFU; the average heat flow value in the southeast coast area may possibly be larger than 1.5 HFU.

4. North-south geotectonic zone

The north-south geotectonic zone is the important boundary for the geological and geophysical field characteristics of the Chinese continent and is the zone of the strongest neotectonic activity as well. MTS measurements at more than 50 sites (see Figure 1) in the north segment of the north-south geotectonic zone have been

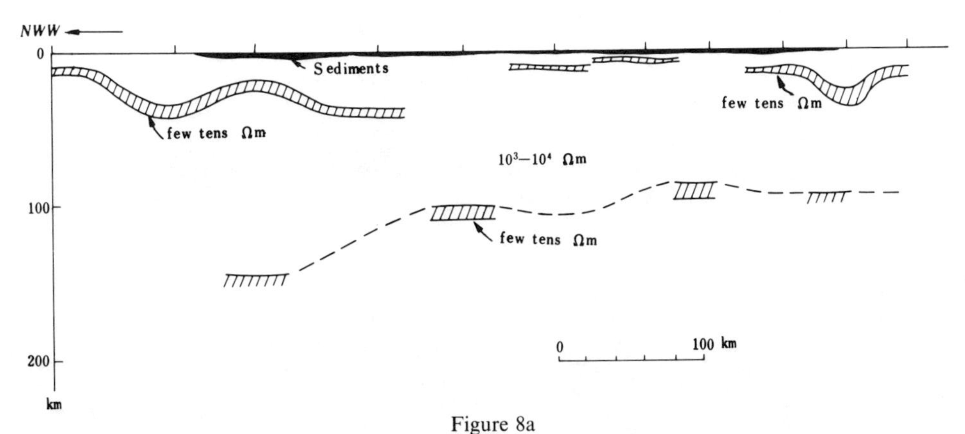

Figure 8a
Geoelectrical structure of the crust and upper mantle along NWW profile cutting laterally through the Mesozoic-paleozoic Shichuan basin and the pre-Cambrian Songpan block (LI LI, and YIN GUOYOUN, 1985). (Profile C–C′ on Figure 1)

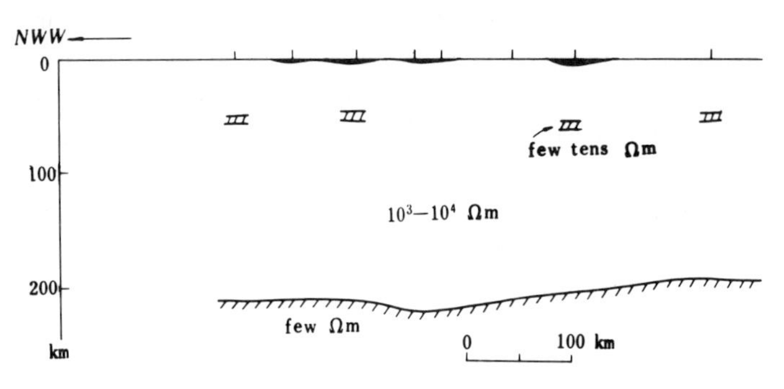

Figure 8b
Geoelectrical structure of the crust and upper mantle along NWW profile cutting laterally through the Jiangnan old continent and the Caledonian folding zone of the Hunan province. (Profile D–D′ on Figure 1)

completed, while in the south segment there has been only one MTS profile (Figure 8a). Since the beginning of historical records in 143, some 40 earthquakes of $M \geqslant 6$ have occurred in the north segment of the north-south geotectonic zone. The thickness of the Cenozoic deposits at some intervals in this segment reaches 7,000 m. Figure 9a shows the geoelectrical structure profile cutting laterally through the north segment of the north-south geotectonic zone while Figure 9b shows the geoelectrical structure profile along the direction of this zone. It can be seen from Figures 8a and b that the conductive layer of the upper mantle uplifts under the north-south geotectonic zone with a top depth of only 80 km, then deepens to 100–160 km to both the east and west, more rapidly to the west. The resistivity of the conductive layer of

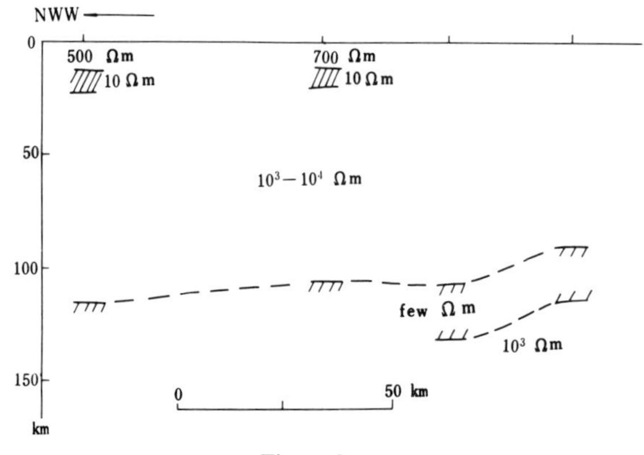

Figure 8c
Geoelectrical structure of the crust and upper mantle along NWW profile perpendicular to the southeast coast in Guangdong. (Profile E–E′ on Figure 1)

the upper mantle gradually increases from about 1 Ωm under the north-south geotectonic zone to more than ten Ωm at both sides of the east and west. The upper mantle conductive layer in the south segment of the north-south geotectonic zone shows a depth close to 80 km under the zone, which rapidly increases to 140 km on the west side (Figure 8a). Therefore, the north-south geotectonic zone also has rapid changes in the depth of the upper mantle conductive layer (GUO SHOUNIAN *et al.*, 1985; LI LI and JIN GUOYOUN, 1985).

5. Tibet plateau area

The Tibet plateau is known as the roof of the world. It has an average height of more than 4,500 m with mountain Xiamalaya as high as 8,848 m, which is the youngest mountain uplifting since the Tertiary era. Geologists and geophysicists

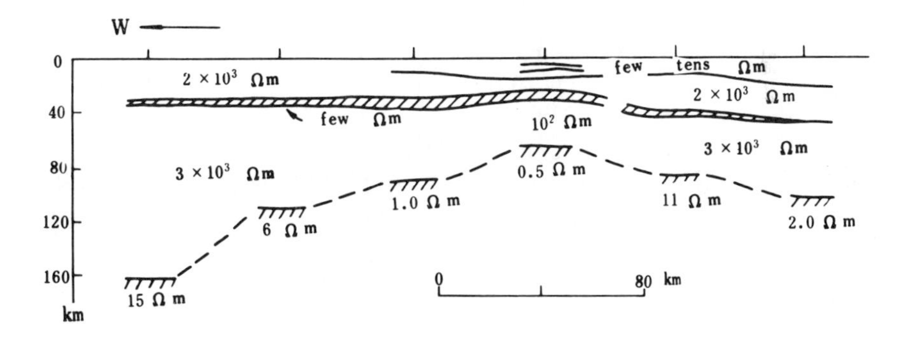

Figure 9a
Geoelectrical structure of the crust and upper mantle along W–E profile in the north segment of the north-south geotectonic zone (GUO SHOUNIAN, *et al.*, 1985). (Profile F–F′ on Figure 1)

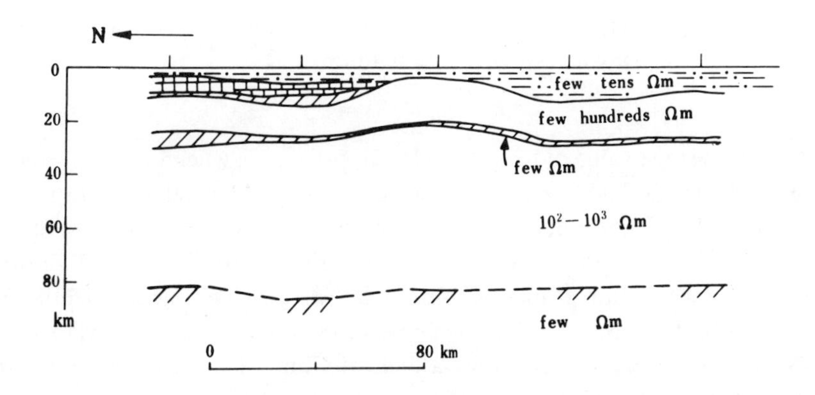

Figure 9b
Geoelectrical structure of the crust and upper mantle along N–S profile in the segment of the north-south geotectonic zone (GUO SHOUNIAN *et al.*, 1985). (Profile G–G′ on Figure 1)

have been widely interested for many years by the particular structure of the Tibet plateau. In recent years various research on deep geophysics of the Tibet plateau area has been conducted, the Chinese-France joint team has measured 8 MTS sites along a NS profile in 1980, and the Ministry of Geology and Mineral Resources of China has continuously measured 5 other MTS sites on the same profile. This MTS profile cuts laterally through the Tibet plateau, 400 km long, with 13 measuring sites (Figure 1). Figure 10 shows the apparent resistivity curves of the MTS sites while Figure 11 shows the profile of resistivity distribution with depth calculated by the Bostick inversion method on the *E*-polarization curves (LI LI and JIN GUOYOUN, 1985).

It can be seen from Figure 11 that the conductive layer commonly exists in the Tibet area crust. The upper mantle conductive layer near the south end of the profile is at the depth of about 120 km and uplifts to a depth of 70–80 km at the south of Yaluzangbu river. This layer significantly deepens to the north of Yaluzanbu river down to about 160 km under the C5 MTS sites. The results of deep seismic sounding show that there exists a low velocity layer in the crust at the Tibet plateau, with its top depth at 25–38 km and the crustal thickness about 70–75 km. Compare this to the crustal thickness under the Himalaya mountains of only some 53–55 km (TENG JIWEN, 1985). The measurements of heat flow at Pumoyon lake and Yanzhuoyon lake show that the average values of heat flow are 2.2 HFU and 3.5 HFU, respectively (SHEN XIANJIE *et al.*, 1985). The locations with these two heat flow values are correspondent to the south section of the MTS profile (see Figure 1).

4. The Relationship Between Upper Mantle Conductive Layer and Heat Flow Value and Upper Mantle Low Velocity Zone

From the data of the upper mantle conductive layer depth and the heat flow given in Figure 1, the relation between them can be found on a statistical basis, and show exponential dependence (Figure 12) as following

$$h = h_0 \, q^{-a} \text{ (km)} \tag{1}$$

$h_0 = 165$, q is average value of heat flow (HFU), $a = 1.35$; when $q \leqslant 1$ this relationship will not be valid. Such a result is basically consistent with the conclusion made by A ÁDÁM (1978). The result of inversion of seismic data shows that the depth of the upper mantle low velocity zone in China varies greatly. Figure 13 shows the distribution curve of upper mantle longitude wave velocity with depth (SONG ZHONGHE, 1985). It can be seen that the depth of the upper mantle low velocity zone in the North China Plain, the Yellow and East China Sea areas, the north-south geotectonic zone, and the Tibet plateau are 70 km, 60 km, 100 km, and 120 km, respectively; while no upper mantle low velocity zone has been found in the South China area. These results are basically consistent with those depths of upper mantle

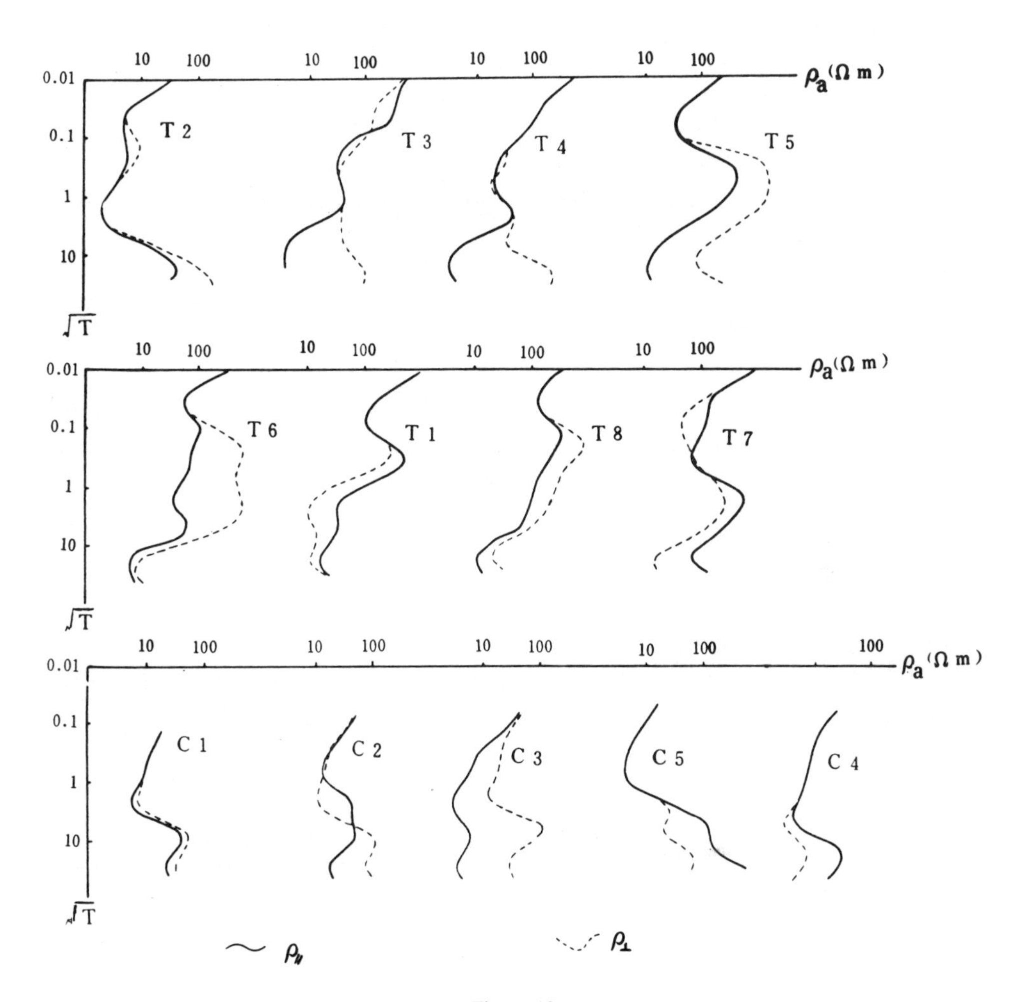

Figure 10
Apparent resistivity curves of MTS along N–S profile in the Tibet plateau (Li Li and Jin Guoyoun, 1985).

conductive layer given by MTS.

According to the upper mantle conductive layer depths determined by MTS, and the upper mantle conductive layer depths estimated by means of average heat flow values using formula (1); we have made a compilation of the maps of upper mantle conductive layer top depths in some parts of the Chinese continent (Figure 1), and recognize the following:

(1) Based on the above-mentioned results of the exponential correspondent relationship between the depth of upper mantle conductive layer and the average values of heat flow, the resistivity of the conductive layer being only 1-few Ωm,

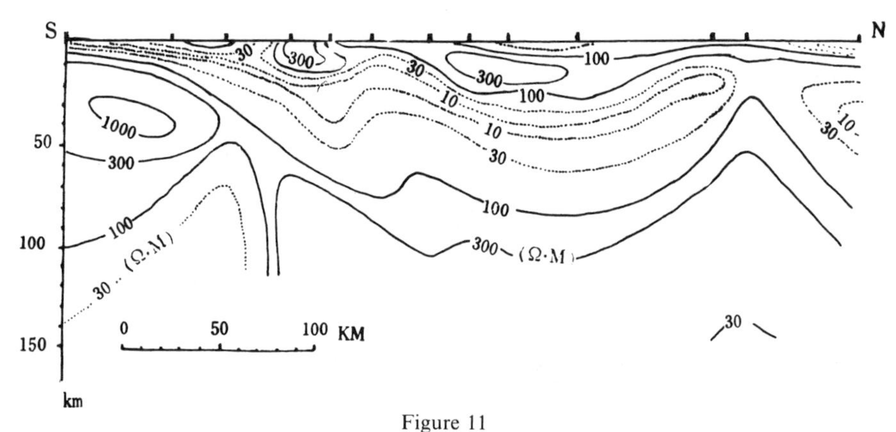

Figure 11
Profile of resistivity distribution with depth along N–S direction in the Tibet plateau (LI LI and JIN GUOYOUN, 1984). (Profile H–H′ on Figure 1)

and the depth of conductive layer being basically consistent with the depth of low velocity zone, the possible origin of the upper mantle conductive layer in most areas within the Chinese continent may be partial melting. For the area within South China with an upper mantle conductive layer depth and high velocity zone of more than 200 km and with an average value of heat flow less than 1.0 HFU, without the existence of low velocity zone, the possible origin of the upper mantle conductive layer is the phase change of mantle rocks (ÁDÁM, 1978).

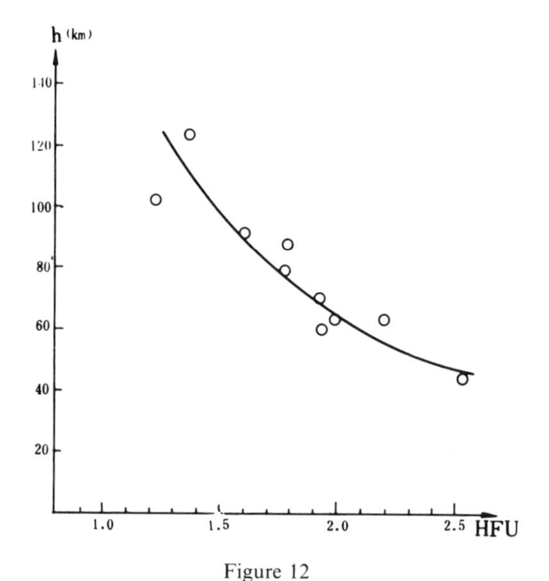

Figure 12
Relation between heat flow value and depth of upper mantle conductive layer in the Chinese continent.

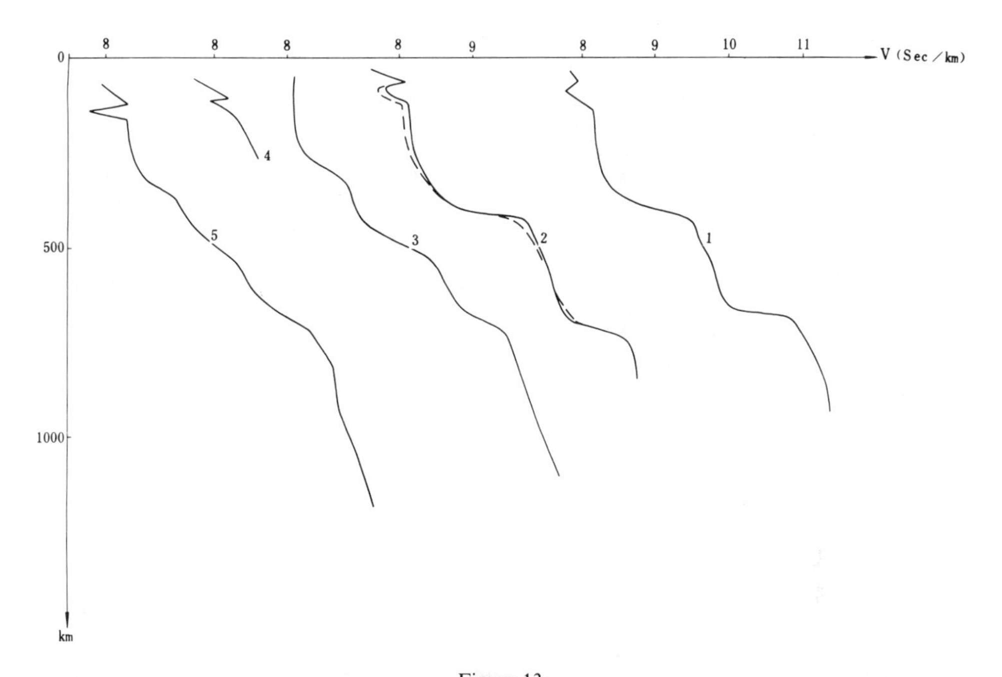

Figure 13
Distribution of upper mantle longitude wave with depth in China (SONG ZHONGHE, 1985). 1—North China Plain, 2—Yellow Sea and East China Sea, 3—South China, 4—South-north geotectonic zone, 5—Tibet plateau.

(2) The depth of the upper mantle conductive layer in the center of the Chinese continent varies greatly: from 50 km (at the Baohai Gulf) to more than 200 km (at the South China area). There has been a significant relationship between the top depth of this conductive layer and the tectonic activities since the Cenozoic period and the current seismicities. The conductive layer in the Cenozoic rift valley shows strong uplift. The strong seismicity zone either seems to be located at the zone where the depth of the conductive layer varies more or seems to be associated with the uplift of the conductive layer (except for the Tibet plateau).

(3) Two conductive layers have been found in the upper mantle of some areas, the first conductive layer is thin only a few km, with a resistivity of few to few tens Ωm. It is surmised that this thin layer of high conductivity might be the remains of partial melting from geological history; it also might be the layer where crystal defect or conductive minerals have been comparatively concentrated if this thin layer has a middle conductivity.

References

ADÁM, A. (1978), *Geothermal effects in the formation of electrically conducting zone and temperature distribution in the Earth*, Phys. Earth Planet. Inter. *17*, 21–28.

CHEN MOXIANG, HUANG GESHAN, and ZHANG WENYEN (1982), *Distribution characteristics of the geothermal resources in the northern area of the North China Plain*, Institute of Geology, Academia Sinica, Research on Geology *1*, 72–89.

GOUBAU, W. M., GAMBLE, T. D., and CLARKE, J. (1978), *Magnetotelluric data analysis: Removal of bias*, Geophysics *43*, 1157–1169.

GUO SHOUNIAN, ZHU ZUOQUAN, and LIN CHANGYOU (1985), *Relationship between the crustal and upper mantle conductive layers and seismicities*, (in press).

JIN GUANGWEN, WANG JIAYIN, and WANG TIANSHENG (1982), *A method for computing MT tensor impedance*, Acta Geophysica Sinica *25*, Supplement, 650–659.

LI LI and JIN GUOYOUN (1985), *MTS report on the crustal and upper mantle conductivity in Pan-Xi an Longmenshan mountain areas*, (in press).

LIU GUODONG (1985), *Cenozoic rift system of North China Plain and the deep internal processes*, Research on Recent Crustal Movement (1), Continental rifts and deep internal processes, Institute of Geology, State Seismological Bureau (Seismological Press, Beijing) pp. 17–25.

LIU GUODONG, GU QUN, SHI SULIN, SUN JIE, SHI ZHANGSONG, and LIU JINHAN (1983), *The electrical structure of the crust and upper mantle and its relationship with seismicity in the Beijing-Tianjin-Tangshan Region and adjacent area*, Acta Geophysica Sinica *26*, 149–157.

LIU GUODONG, and LIU CHANGQUEN (1983), *Structures of crust and upper mantle and their relation to Cenozoic tectonism in northern part of North China*, Scientia Sinica *26*, 550–560.

LIU GOUDONG, SHI SHULIN, and WANG BAOJUN (1984), *Conductive layer in the crust in North China and its relation to crustal tectonism*, Scientia Sinica *27*, 1093–1104.

LU ZAOXUN (1985), *Progress of geophysical prospecting study on the deep structure of Halcheng earthquake area*, Earthquake Research in China *1*, (1), 56–65.

SHEN XIANJIE, DENG XIOYUE, KAN WENHUA, (1985), *Corrections to the heat flow values measured in South-Xizang (Tibet) and the results*, Acta Geophysics Sinica *28*, Supplement (1), 80–92.

SIMS, W. E., BOSTICK, F. X., Jr., and SMITH, H. W. (1971), *The estimation of magnetotelluric impedance tensor elements from measured data*, Geophysics *36*, 938–942.

SONG ZHONGHE (1985), *P Wave velocity structure of the upper mantle in Chinese continent and its marginal areas*, (in press).

TENG JIWEN (1985), *An introduction to geophysical study on the Tibetan plateau area*, Acta Geophysica Sinica, *28*, Supplement (1), 11–15.

WANG TIYANG, WANG JIAN, (1985), *Geothermal measurements in Liaohe basin*, Kexue Tongbao *30*, 1008–1010.

WANG JUN, and LIU GUODONG (1985), *Terrestrial heat flow and deep geology in South China*, Acta Geophysica Sinica, (in press).

WU QIANFAN, XIE YIZHEN, ZU XINHUA, and WANG DOU (1985), *Terrestrial heat flow and seismicity in North China*, Research on Recent Crust Movement (1), Continental rifts and deep internal processes, Institute of Geology, State Seismological Bureau (Seismological Press, Beijing) pp. 133–141.

ZHANG RUHAI, XEI ZHENGWEN, WU JIXIN, and XEI YIZENG (1982), *The distribution of heat flow values in Tangshan and its surrounding*, Seismology and Geology *4*, (0), 57–68.

(Received 25th February, 1986, revised 19th March, 1986, accepted 29th April, 1986)

PAGEOPH, Vol. 125, Nos. 2/3 (1987) 0033-4553/87/030483-14$1.50+0.20/0

Application of Space Analysis of Electromagnetic Fields to Investigation of the Geoelectrical Structure of the Earth

M. S. ZHDANOV[1]

Abstract—Lateral composition inhomogeneities of the Earth's deep geoelectric structure require special consideration for any conductivity evaluation of a region. This paper presents a review of some theoretical techniques for determining both the vertical and horizontal conductivity profiles of a region using a spatial distribution of observed electromagnetic fields at the Earth's surface. Effects of shallow positioned anomalies upon a deep conductivity determination are also considered. An application of the procedure is illustrated by a conductivity study in the Soviet Carpathians.

Key words: Electromagnetic induction, Earth electric conductivity, electromagnetic anomalies, Soviet Carpathians.

Of all the geophysical methods used to study the terrestrial structure, those involving deep electromagnetic sounding are of particular concern. They provide data on the distribution of electrical conductivity in the interior of the Earth, which in turn convey exceedingly important information about the thermodynamic and phase states of deep-lying earth formations, hardly accessible to other geophysical techniques. That is why deep electromagnetic soundings are of great applied value.

It must be mentioned that interpretation of electromagnetic sounding data has long relied on simple one-dimensional models of geoelectrical sections wherein conductivity varies only with depth. Recent findings have revealed, however, that formal interpretation of electromagnetic sounding data, within one-dimensional models, may hardly yield faithful information about the electrical conductivity distribution in the Earth. Moreover, electromagnetic sounding curves plotted for isolated observation points are, as a rule, distorted markedly, due to the effect of horizontal geolectrical (both surface and deep) inhomogeneities and, hence, their formal one-dimensional interpretation may indicate false geologic structures. The only way to eliminate ambiguous interpretation of isolated point sounding data is to combine sounding and profiling. This means, eventually, application of a single method of deep electromagnetic investigations relying on simultaneous observation of components of the electric and magnetic fields (of natural or artificial origins)

[1] Institute of Terrestrial Magnetism, Ionosphere, and Radio-wave Propagation, Academy of Sciences of the USSR, 142092 Troitsk, Moscow Region, USSR.

along certain lines or over an area in a wide time or frequency range. The result of such observations is a space-time pattern of the electromagnetic field at the Earth's surface which makes it feasible, in principle, to reconstruct the behavior of electrical conductivity, both in the vertical and in the horizontal. Yet, this approach calls for quite different interpretation techniques, compared to the conventional procedures related to point electromagnetic soundings. The new interpretation methods are substantially similar to those employed in gravimetry and magnetometry and can be regarded as a kind of extension of geopotential field interpretation procedures to the electromagnetic case. These methods are based on space analysis of the fields, which relies on their integral and spectral transformations (BERDICHEVSKY and ZHDANOV, 1984).

Figure 1 is a general scheme of interpretation of electromagnetic data obtained above horizontally inhomogeneous geoelectrical sections. According to this scheme, interpretation of deep electromagnetic sounding data falls into two steps: (1) space analysis of the electromagnetic field, i.e., its separation into different parts, depending on the distribution of geoelectrical inhomogeneities; (2) solution of inverse problems, i.e., establishment of the parameters of an inhomogeneous geoelectrical section.

We will consider more closely the techniques employed in each step.

Step 1—Space Analysis of the Electromagnetic Field

This step of analysis includes separation of fields into external and internal, normal and anomalous, surface and deep parts (BERDICHEVSKY and ZHDANOV, 1984). Separation into external and internal parts is accomplished using the classical Gauss or Kertz-Siebert techniques. It merits remembering how the normal and anomalous components of the field are defined.

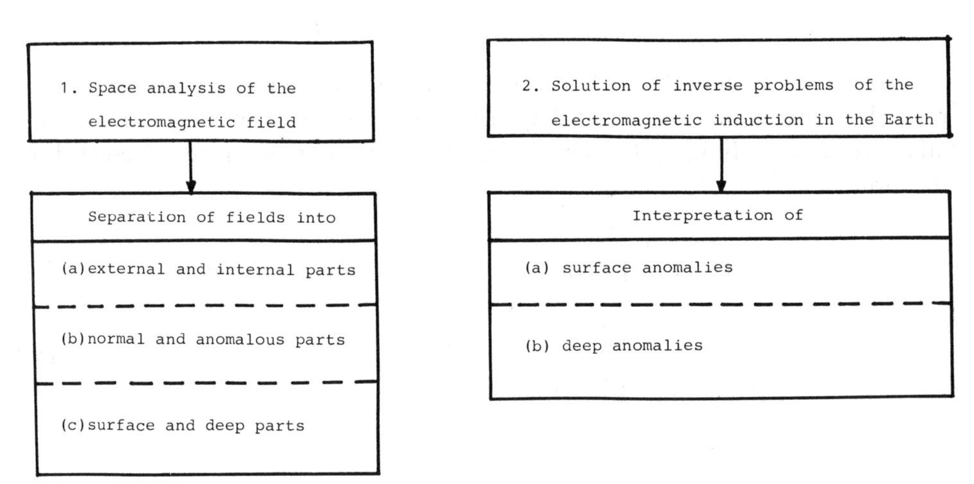

Figure 1
Scheme of interpretation of deep electromagnetic sounding data.

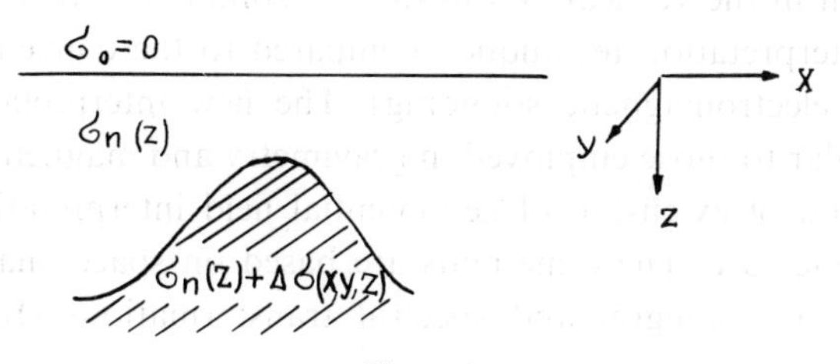

Figure 2
Geoelectrical model explaining the establishment of the normal and anomalous electromagnetic fields.

Take a model of Figure 2. Within this model, the horizontal plane $z = 0$ separates a nonconducting atmosphere, wherein extraneous electric currents j^Q are localized, from an inhomogeneous Earth of an arbitrary electrical conductivity distribution:

$$\sigma(x, y, z) = \sigma_n(z) + \Delta\sigma(x, y, z)$$

where σ_n and $\Delta\sigma$ are the normal and anomalous conductivities, respectively. The name of a normal electromagnetic field is ascribed to the field induced by an extraneous current j^Q in a horizontally homogeneous Earth of a conductivity $\sigma_n(z)$ varying only with depth. The normal field is distorted by horizontal geoelectrical inhomogeneities. These distortions are known as electromagnetic anomalies.

Thus, the electromagnetic field E,H can be represented as a sum of the normal E^n,H^n and anomalous E^a,H^a fields:

$$E = E^n + E^a, \qquad H = H^n + H^a. \tag{1}$$

The first problem in the analysis is to separate fields into normal and anomalous parts. This operation can be effected by means of various types of integral transformations of the field. In the general case, the integral transforms can be represented in the form:

$$\begin{aligned} H^{n,a}(x, y) &= \iint_{-\infty}^{+\infty} \hat{G}_H^{n,a}(x - x', y - y')\, H(x', y')\, dx'\, dy', \\ E^{n,a}(x, y) &= \iint_{-\infty}^{+\infty} \hat{G}_E^{n,a}(x - x', y - y')\, E(x', y')\, dx'\, dy' \end{aligned} \tag{2}$$

where $\hat{G}_{H,E}^{n,a}$ stands for the matrices of the kernels of the relevant integral transforms. These kernels imply space windows through which the observed fields are transmitted. We will show, as an example, calculation of the kernels of integral transforms for two-dimensional situations (in the most interesting case of E-polarization, where the field is uniform and the medium is homogeneous along the Y-axis) (BERDICHEVSKY and ZHDANOV, 1984).

$$\hat{G}^a_H(x) = \begin{bmatrix} G^a_{xx} & G^a_{xz} \\ G^a_{zx} & G^a_{zz} \end{bmatrix} = \frac{1}{2\pi} \int_{-\infty}^{+\infty} \mathscr{C} e^{-i\alpha x} \, d\alpha \tag{3}$$

where

$$\mathscr{C}^a = \begin{bmatrix} \dfrac{R^*|\alpha|}{R^*|\alpha| + n_1} & -\dfrac{i \operatorname{sign} \alpha \, n_1}{R^*|\alpha| + n_1} \\[2ex] \dfrac{i\, R^*\alpha}{R^*|\alpha| + n_1} & \dfrac{n_1}{R^*|\alpha| + n_1} \end{bmatrix} \tag{4}$$

Here α is the spatial frequency along the x-axis;

$$n_j = \sqrt{\alpha^2 + K_j^2}, \qquad \mathrm{j} = 1,3, \ldots, \mathscr{N}$$

is the wave number of the j-th layer of a normal geoelectrical section;

$$R^* = cotanh\left\{n_1 d_1 + arctanh\left[\frac{n_1}{n_2}\, cotanh\left(n_2 d_2 + \ldots + arctanh\,\frac{n_{N-1}}{n_N}\right)\ldots\right)\right]\right\}.$$

Expressions (3) and (4) show that the kernels of integral transforms (3) and (4) imply space windows through which the observed fields are transmitted. Take, for instance, the plots describing the shape of space windows within the model of a homogeneous Earth (Figure 3). It is evident from these plots and from formulas (3) and (4) that the size of the windows is determined by the parameters of a normal geoelectrical section as well as by the field variation period. The width of space windows is found (ZHDANOV and PLOTNIKOV, 1981) to range from 60 to 150 km. This dictates a desirable area of field observation, when it is to be separated into normal and anomalous parts.

Thus, the prime advantage, making the suggested methods of field separation into normal and anomalous parts superior to the conventional techniques of field separation into external and internal parts, is a narrower width of space windows of the corresponding integral transforms. This permits a practical solution of the problem of separating fields measured in limited areas.

Another problem tackled in the first step, i.e., in the course of space analysis of electromagnetic fields, is their separation into surface and deep parts.

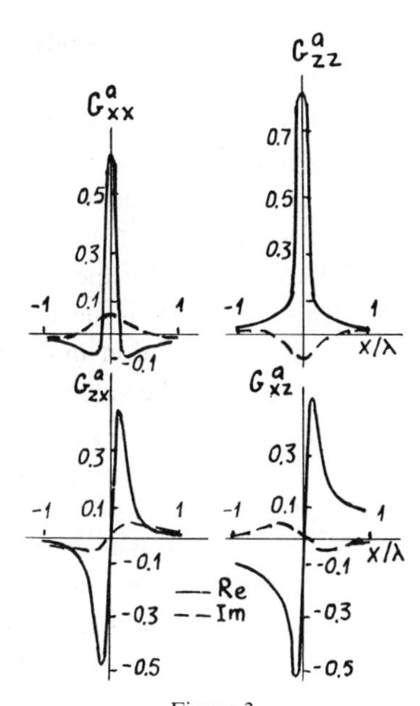

Figure 3
Plots of the kernels of integral transforms of geomagnetic fields used to separate the fields into normal and anomalous parts (electromagnetic wave length λ for σ_n = const).

By surface anomalies we mean the anomalies caused by the near-surface inhomogeneous geoelectrical layer formed by the coastal water of seas and oceans as well as by the inhomogeneity of the continental sediments.

Deep anomalies are attributed to conducting zones in the consolidated Earth' crust and upper mantle; whose origin is related to hydrothermal processes and partial melting of crustal formations at high temperatures and pressures.

As a rule, in practice we observe electromagnetic anomalies caused by the total effect of both surface and deep inhomogeneities. It is remarkable that when we study the deep structure of the Earth, surface anomalies act as an interfering factor, distorting the information about the deep section. Therefore, separation of anomalies of the surface H^s, E^s and deep H^d, E^d origins is one of the most important problems of analysis. This problem can be solved, provided the geoelectrical parameters of the near-surface inhomogeneous layer of the Earth are available. The most convenient model of this layer may be an inhomogeneous highly conducting thin Price layer. To separate fields within this model, one must know the integral conductivity $S(x, y)$ of the thin layer. Consequently, the procedure of field separation reduces, just as in the case of distinction between the normal and anaomalous components, to integral transformations of the field (BERDICHEVSKY and ZHDANOV, 1984):

$$\begin{Bmatrix} H^{s,d} \\ E^{s,d} \end{Bmatrix} = \iint_{-\infty}^{+\infty} \hat{G}^{s,d}(x - x^1, y - y^1) \begin{Bmatrix} H(x^1, y^1) \\ E(x^1, y^1) \end{Bmatrix} dx^1\, dy^1 \tag{5}$$

where $\hat{G}^{s,d}$ denotes the corresponding kernels dependent upon parameters of the normal geoelectrical section and $S(x, y)$.

It is noteworthy that the problem of field separation is solved with due reference to possible interaction of currents induced in the S-layer and in deep geoelectrical inhomogeneities. This is the principle difference between the methods of field separation which rely on the integral transforms of type (5) and the conventional techniques of considering near-surface inhomogeneities by means of numerical quasi three-dimensional (shell) modeling. This approach is developed, in particular, by M. Menvielle and Tarits (1986). But if the induction interaction of surface and deep currents is not taken into sufficient account, the value of separated anomalous fields may prove to be considerably distorted.

Separation of electromagnetic fields will be exemplified by the results of space analysis of deep electromagnetic sounding data carried out in the territory of the Voronezh crystalline strate by V. M. MAKSIMOV and V. N. GRUZDEV (MAKSIMOV *et al.*, 1976; MAKSIMOV and GRUZDEV, 1984). In this region, they had 62 magnetovariational (MV) profiling points and 30 MT-sounding points. The experimental data indicated two zones of higher electrical conductivity, which were clearly manifested in the behaviour of the induction vectors and characterized by an increased intensity of the horizontal components; as well as by the sign reversal of the vertical component of the field of bay-like perturbations.

The total longitudinal conductivity of the sedimentary section in the region of interest ranges from 10 to 3000 S, which argues in favor of allowance for the effect of near-surface inhomogeneities on the results of deep investigation. To this

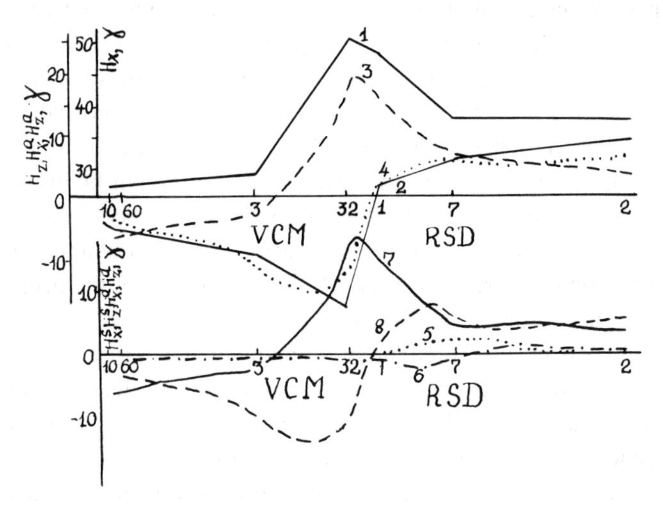

Figure 4
Geomagnetic anomaly in the Voronezh Crystalline Masses (VCM): 1,2—the H_x and H_z components of the observed field; 3,4—the H_x^a and H_z^a components of the anomalous field; 5,6—the H_x^s and H_z^s components of the surface anomalous field.

end, anomalies of the electromagnetic field were separated, using the above procedure, along one of the lines crossing the Voronezh Crystalline Masses.

At a glance Figure 4 shows the amplitude of the horizontal component of the magnetic field, H_x along this line increases markedly (curve 1) and the vertical component H_z of the observed field is inverted (curve 2). Meanwhile, the maximum values S related to the conducting sediments of the Ryazano-Saratovsky depression are not spatially consistent with the anomalous behaviour of the geomagnetic field. Separation of the total field into normal and anomalous parts made it possible to identify the normal background in the H_x and H_z components (H_x^n, H_z^n) and anomalous fields (H_x^a, H_z^a, curves 3 and 4 in Figure 4) characterized by a local extreme value of the horizontal component H_x^a and by the inversion of the vertical component H_z^a.

Figure 4 also depicts components of the surface anomalous field (H_x^s—curve 5 and H_z^s—curve 6), which have relatively small amplitudes (not exceeding $\frac{2}{3}$ nT). Thus, a conclusion can be reached that the deep anomaly H_x^d, H_z^d (curves 7 and 8) determines, first of all, the anomalous nature of the field observed which is seemingly due to the local conducting zone found in the consolidated Earth's crust.

Let us present brief characteristics of the methods employed in the second step-interpretation of data (Figure 1).

Step 2—Solution of Inverse Problems of Electromagnetic Induction in the Earth

In this step of data interpretation we deal with two groups of methods: interpretation of surface and deep anomalies. The methods used to study surface anomalies have been developed rather comprehensively by U. SCHMUCHER (1971). They rely largely on approximation of surface inhomogeneities by thin Price shells. It is generally assumed in surface anomaly interpretations that consideration is given to the range of periods wherein the field does not penetrate an area of developed deep geoelectrical inhomogeneities and, hence, their effect can be discarded.

Proceeding to a longer-period range of variations, the effect of surface inhomogeneities is taken into account by separating fields in the above manner. As a result, the main problem here is to establish the parameters of the deep geoelectrical section, which is just to be solved below.

Interpretation of deep electromagnetic anomalies involves three groups of methods (Figure 5): (1) approximate express methods of localizing a deep inhomogeneity; (2) methods of reconstructing the shape of an inhomogeneity; (3) automated methods of producing a geoelectrical section of a given inhomogeneous area.

The approximate express methods of localizing a deep inhomogeneity include, initially, analytical continuation of the field. The basis for the theory of these methods is closely treated by M. S. ZHDANOV and M. N. BERDICHEVSKY (ZHDANOV, 1984; BERDICHEVSKY and ZHDANOV, 1984). Now we will outline the essentials underlying

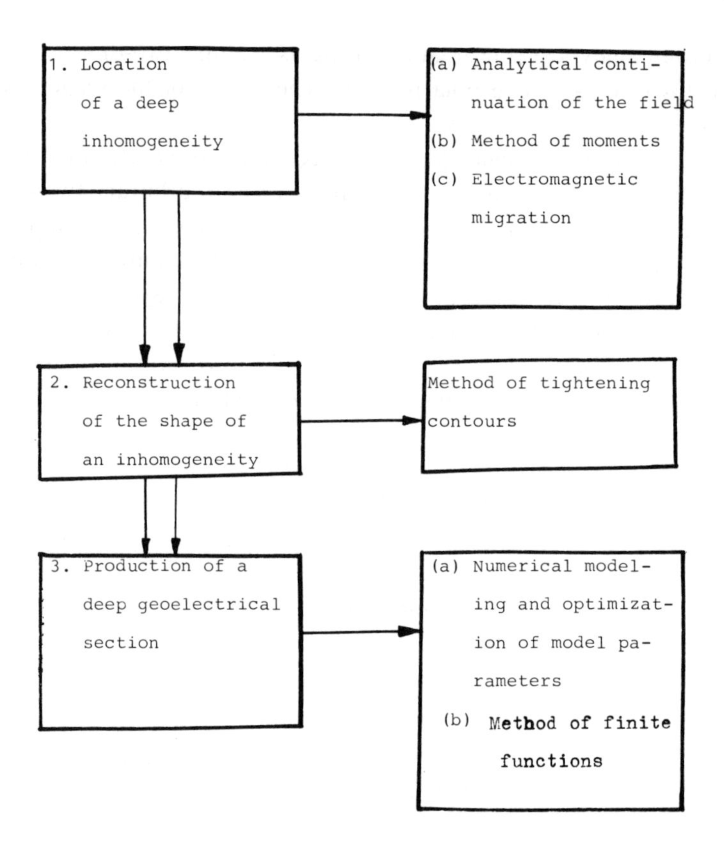

Figure 5
Methods of interpretation of deep electromagnetic anomalies.

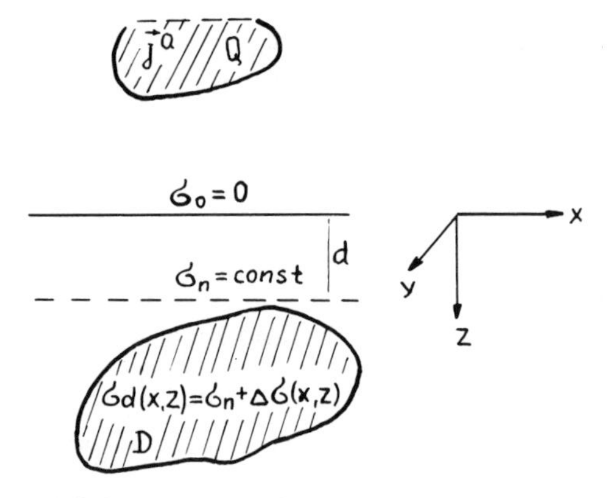

Figure 6
Analytical continuation of the electromagnetic field down to a deep inhomogeneity D.

the procedure of analytical continuation, using as example two-dimensional electromagnetic fields (the case of E-polarization). The model of the section is given in Figure 6.

A homogeneous well conducting Earth of a constant normal electrical conductivity σ_n contains an inhomogeneous deep domain D with an arbitrary two-dimensional distribution of conductivity $\sigma_d(x, z) = \sigma_n + \Delta\sigma(x, z)$. At the Earth's surface there are specified variations in the magnetic field H_x and H_z. It requires a continuation of this field analytically into the Earth right to the horizontal layer $0 \leqslant z \leqslant d$ (d being the distance from the Earth's surface to the domain D). To solve the problem, we pass to the spatial spectra h_x, h_z satisfying, as is known, within the homogeneous layer $0 \leqslant z \leqslant d$ the one-dimensional Helmholtz equations:

$$h_x'' = n^2 h_x, \qquad h_z'' = n^2 h_z$$

where

$$n = \sqrt{d^2 - i\omega\mu_0\sigma n}.$$

The last equation implies, in particular, the following representation for h_z:

$$h_z(z) = h_z^+ e^{nz} + h_z^- e^{-nz} \tag{6a}$$

whence

$$h_z'(z) = n(h_z^+ e^{nz} - h_z^- e^{-nz}) \tag{6b}$$

Taking $z = 0$ in equations (6a) and (6b) we derive two equations for two unknown coefficients h_z^+ and h_z^-:

$$h_z(0) = h_z^+ + h_z^-, \qquad h_z'(0) = n(h_z^+ - h_z^-). \tag{7}$$

Solving these equations we find

$$h_z^{\pm} = \frac{1}{2n}(nh_z(0) \pm h_z'(0)). \tag{8}$$

According to the third Maxwell equation div $H = 0$, we have for spectra

$$h_z' = i\alpha h_x. \tag{9}$$

Substituting (9) into (8) we write

$$h_z^{\pm} = \frac{1}{2\pi}(nh_z(0) \pm i\alpha h_x(0)). \tag{10}$$

Thus, we have found the unknown coefficient in formula (5). Upon substitution of (10) into (6a) and some transformation, we obtain in the final form

$$h_z(z) = cosh(nz)h_z(0) + i(\alpha/n)sinh(nz)h_x(0). \tag{11}$$

The corresponding expression for the spectrum of the horizontal component is derived from (11) by differentiating with respect to z and by allowing for (9)

$$h_z = (1/i\alpha)h_z' = cosh(nz)h_x(0) - i(n/K_x)sinh(nz)h_z(0). \tag{12}$$

The obtained formulas yield a solution to the problem of analytical continuation of the magnetovariational field into a horizontal layer. Indeed, taking the inverse Fourier transform of (11) and (12) we define the fields themselves inside the Earth:

$$
\begin{aligned}
H_x(x, z) &= \frac{1}{2\pi} \int_{-\infty}^{+\infty} \left[h_x(0) cosh(nz) - ih_z(0) \frac{n}{\alpha} sinh(nz) \right] e^{-i\alpha x} \, d\alpha \\
H_z(x, z) &= \frac{1}{2\pi} \int_{-\infty}^{+\infty} \left[h_z(0) cosh(nz) + ih_x(0) \frac{\alpha}{b} sinh(nz) \right] e^{-i\alpha x} \, d\alpha.
\end{aligned}
\qquad (13)
$$

Note that analytical continuation of the field into the lower half-plane is a typical example of an ill-posed problem. Indeed, with an increasing depth of continuation z, the exponents e^{nz} included in (13) point out high spatial frequencies in the field spectra, i.e., the high-frequency interference unavoidably present in practical observations is enhanced. To preclude this effect, it is desirable to resort to suitable regularizing algorithms which reduce, in a simple case, to low-frequency filtering of the observed field.

Analytically continued fields can be interpreted using two techniques: method of singular points and method of analysis of vector lines of the field in the vertical plane.

The method of singular points lies in finding singular points of the analytically continued field (i.e., points of focusing of the isolines of the continued field) and in locating therefrom anomaly-forming bodies.

The method of field vector lines involves construction of maps of vector lines of the real and imaginary vectors of the magnetic field, ReH and ImH, in the vertical plane. These maps represent the magnetic field in a clear-cut manner.

Assume that the domain D is highly conducting. Then, according to the boundary conditions on the boundary of the domain D, the normal component of the magnetic field is close to zero. Let us examine how this fact appears in the maps of vector lines of the real and imaginary parts of H.

In the whole space (except for the surface of a perfect conductor), the magnetic field is polarized elliptically. The vectors being conjugate radii of the polarization ellipse are not colinear and the corresponding vector lines intersect. At the surface of a perfect conductor, where $H_n = 0$, the magnetic field is linearly polarized. Hence, the vectors ReH and ImH are colinear, while the field vector lines ReH and ImH run together and coincide with the contour of a body. Thus, to delineate a body, it is sufficient to find the vector line ReH confluent with the vector line ImH. In practice, these lines are found visually. Therefore, the region occupied by a body is located somewhat roughly.

Consider, for example, a model within which the Earth, excited by an E-polarized plane wave, containing a horizontal conducting half-cylinder. Figure 7 is a vertical map of vector lines of the real and imaginary parts of the field. The vector lines ReH and those away from the conductor have different configurations and ImH intersect at angles close to $\pi/2$. As they approach the conductor, their forms become more and more similar and at the conductor surface they turn into horizontally extended ovals enclosing the body.

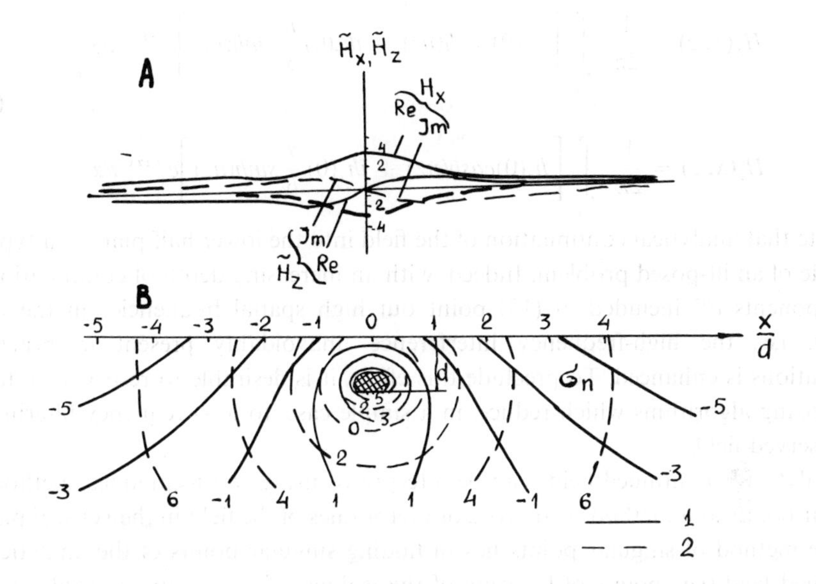

Figure 7

Plots (A) and vertical map (B) of vector lines of the real and imaginary parts of the magnetic field within the model of a homogeneous Earth containing a conducting half-cylinder.

Analytical continuation (in conjunction with the methods of moments (BEDRICHEVSKY and ZHDANOV, 1984) and of electromagnetic migration (ZHDANOV, 1984)) provide only a crude location and shape of a deep inhomogeneous domain D.

To define the shape and distribution of electrical conductivity inside D definitively, it is essential to employ the trial and error method. Here the direct problem is solved either by means of integral equations or by the lattice-point method. The sought parameters of the section are established by minimizing the function of the standard deviation $I(\Delta\sigma)$ of observed fields from the calculations in the frequency and space ranges. For instance, in the two-dimensional case of an E-polarized field, we have

$$I(\Delta\sigma) = \iint_{\Omega X} |H(\Delta\sigma) - H_0|^2 \, dx \, d\omega = \min \tag{14}$$

where $\Delta\sigma$ is the function of excess electrical conductivity (against the background of a certain specified normal section); X is the line of observation; Ω is the recorded frequency interval.

To regularize the solution of inverse problem, we introduce a stabilization functional

$$S(\Delta\sigma) = \iint_D (\Delta\sigma - \Delta\sigma_0)^2 \, dS \tag{15}$$

where $\Delta\sigma_0$ is a certain initial distribution of excess electrical conductivity. Then the solution of the inverse problem reduces to minimization of the Tikhonov parametric function

$$M_\alpha(\Delta\sigma) = I(\Delta\sigma) + \alpha S(\Delta\sigma) \tag{16}$$

where α is the regularization parameter.

The above scheme is implemented in a set of computer programs intended basically for solution of the direct problem.

Application of the above procedure will be exemplified by the deep electromagnetic sounding data obtained in the Soviet Carpathians. The Carpathian region has long been an object of particular interest to geophysicists. The anomaly of the natural electromagnetic field found here is one of the most highly developed in the territory of the Soviet Union. It is related to concentration of currents in a region of higher electrical conductivity. The magnetotelluric and magnetovariation investigations performed by a representative group of Soviet scientists (A. P. BONDARENKO, M. N. BERDICHEVSKY, A. I. BILINSKY, M. S. ZHDANOV, S. N. KULIKOV, I. I. ROKITYANSKY, L. M. ABRAMOVA, V. S. SHNEER *et al.*) made it possible to trace the anomaly over most of the length of the Folded Carpathians. The Soviet Carpathians, Transcarpathians, and the adjacent part of the Eastern-European platform have been studied in the most detailed fashion: over 60 magnetovariational profiling points and 20 magnetotelluric sounding points have been interrogated on several lines (ZHDANOV *et al.*, 1986). Figure 8 presents a typical profile of observed values of variations in the geomagnetic field for a 1-h period.

The above method was employed to examine the electromagnetic field along a line, i.e., to identify the anomalous part of the field and to separate the latter into a surface and deep components (Figure 9). The deep component was found to be almost twice as large as the surface component. To locate the sources of the deep anomaly, the latter was continued analytically into the lower half-plane (Figure 9). The isolines of the continued field clearly indicate the location of a crustal zone of higher electrical conductivity lying at a depth of nearly 10–12 km in the joining zone of the Folded Carpathians and Transcarpathian depression. This information was allowed for while developing a model of the whole geoelectrical section. The structure of the crustal well-conducting zone and deep geoelectrical strata was optimized within the scope of the automated trial and error method using lattice-point algorithms to solve the direct problem (ZHDANOV *et al.*, 1986). The final model is depicted in Figure 10 and the corresponding calculation curves are plotted in Figure 8. The

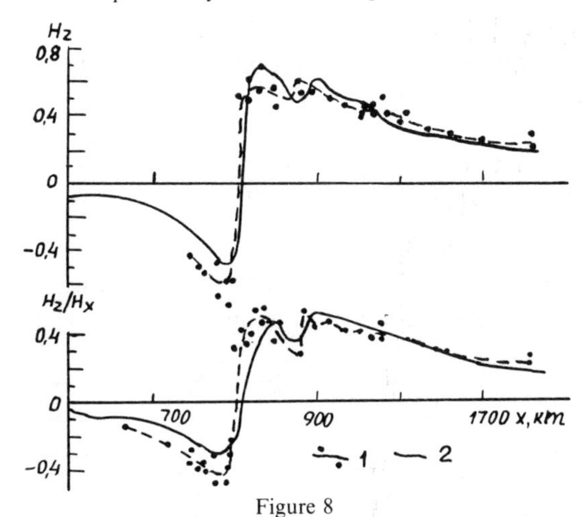

Figure 8
Geomagnetic anomaly in the Soviet Carpathians: The amplitudes of H_z and H_z/H_x for a variation period $T = 1$ hr: 1—observation results; 2—finite-difference modeling results for the section of Figure 10.

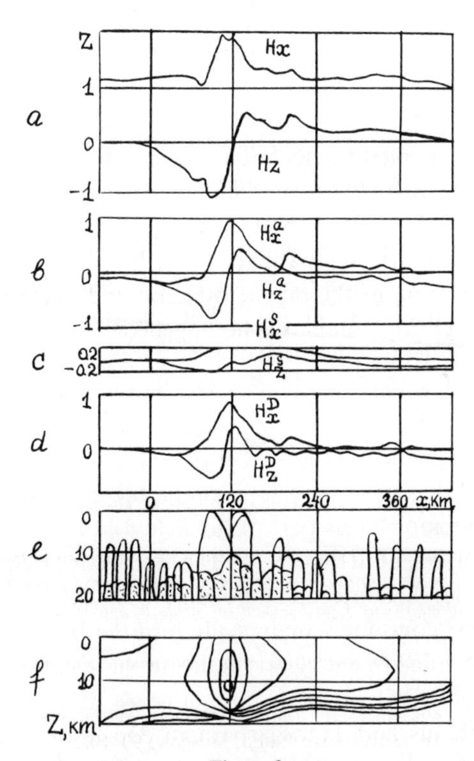

Figure 9
Space analysis of the geomagnetic field in the Soviet Carpathians: a: H_z,H_x—components of the total field; b: H_z^a,H_x^a—anomalous part of the field; c: H_z^s,H_x^s—surface anomaly; d: H_x^d,H_z^d—deep anomaly; e: analytical continuation of the vertical component H_z^d into the conducting Earth (map of isolines of the field H_z^d in the vertical plane); f: map of the flux function of the analytically continued field in the vertical plane.

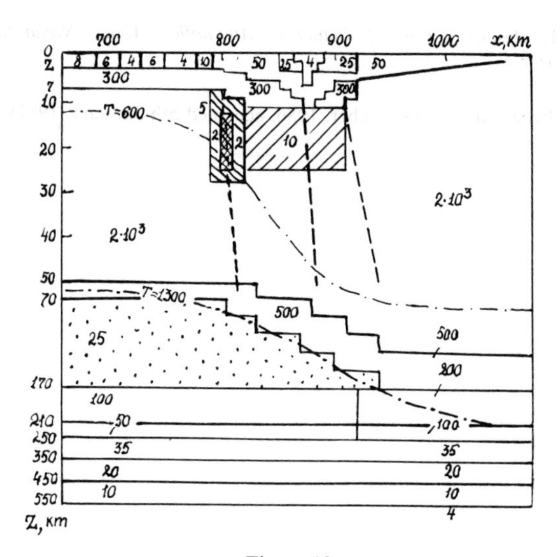

Figure 10
Geoelectrical model of the deep structure of the Carpathian region: 1—geoelectrical boundaries and resistivities of blocks (0hm.m); 2—deep faults; 3—deep isotherms (°C).

trial and error method enabled the location and structure of the crustal anomaly of electrical conductivity to be defined more closely. It also served to recognize a deep conducting stratum (asthenosphere) highly developed in the Pannonian basin and pinching out under the Folded Carpathians.

Thus, a joint application of the magnetotelluric and magnetovariation methods permits investigation of fairly complex inhomogeneous geoelectrical sections.

References

Berdichevsky, M. N. and Zhdanov, M. S. (1984), *Advanced Theory of Deep Geomagnetic Sounding.* Amsterdam; Elsevier, pp. 408.

Zhdanov, M. S. and Plotnikov, S. V. (1981), *Separation of the Variations in the Geomagnetic Field into a Normal and Anomalous Parts in a Limited Territory.* Gheomagnetism i aeronomiya, *XXI* (6), 1064–1069.

Zhdanov, M. S. (1984), *Analogs of the Cauchy-type Integrals in Theory of Geophysical Fields.* Moscow; Nauka, pp. 326.

Zhdanov, M. S. *et al.* (1986), *2-D Model Fitting of a Geomagnetic Anomaly in the Soviet Carpathians.* Annales Geophysicae, *4*, 335–342.

Maksimov, V. M. *et al.* (1986), *Magnetovariation anomalies in the Voronezh crystalline masses.* In *Geomagnetic Investigations. 16*, (Sovetskoye Radio, Moscow) 90–102.

Maksimov, V. M. and Gruzdev, V. N. (1984), *Oboyansk anomaly of geomagnetic variations and its relationship to the main geological and geophysical features of the structure of the Kursk magnetic anomaly.* In *Geophysical Investigations into the Kursk Magnetic Anomaly* (Publ. Voronezh State University) 21–29.

Menvielle, M. and Tarits, P. (1986), *The Andean Conductivity Anomaly Reexamined.* Annales Geophysicae *4*, 63–70.

SCHMUCKER, U. (1971), *Interpretation of Induction Anomalies Above Nonuniform Surface Layers.* Geophysics **36**, 156–165.

(Received 29th December, 1986, accepted 5th January, 1987)